Restoration and Recovery of Shallow Eutrophic Lake Ecosystems in The Netherlands

Developments in Hydrobiology 74

Series editor

H. J. Dumont

Restoration and Recovery of Shallow Eutrophic Lake Ecosystems in The Netherlands

Proceedings of a conference held in Amsterdam, The Netherlands, 18–19 April 1991

Edited by

Louis Van Liere & Ramesh D. Gulati

Reprinted from Hydrobiologia, vol. 233 (1992)

Kluwer Academic Publishers
Dordrecht / Boston / London

Library of Congress Cataloging-in-Publication Data

Restoration and recovery of shallow eutrophic lake ecosystems in the Netherlands : proceeding[s] of a conference held in Amsterdam, The Netherlands, 18-19 April 1991 / edited by Louis van Liere & Ramesh D. Gulati.
p. cm. -- (Developments in hydrobiology ; v. 74)
"Reprinted from hydrobiologia, vol. 233 (1992)."
ISBN 0-7923-1697-5 (hb)
1. Lake ecology--Netherlands--Loosdrecht--Congresses. 2. Water quality--Netherlands--Loosdrecht--Congresses. 3. Lake renewal--Netherlands--Congresses. 4. Eutrophication--Control--Netherlands--Congresses. I. Liere, Louis van. II. Gulati, R. D. III. Hydrobiologia. IV. Series: Developments in hydrobiology ; 74.
QH159.R47 1992
333.91'63153'09492--dc20 92-12305

ISBN 0-7923-1697-5

Published by Kluwer Academic Publishers,
P.O. Box 17, 3300 AA Dordrecht, The Netherlands.

Kluwer Academic Publishers incorporates
the publishing programmes of
D. Reidel, Martinus Nijhoff, Dr W. Junk and MTP Press.

Sold and distributed in the U.S.A. and Canada
by Kluwer Academic Publishers,
101 Philip Drive, Norwell, MA 02061, U.S.A.

In all other countries, sold and distributed
by Kluwer Academic Publishers Group,
P.O. Box 322, 3300 AH Dordrecht, The Netherlands.

The image on the front cover is a LANDSAT Thematic Mapper water quality concentration map on the basis of the 5 July 1987 band 2 [green] image (left) and the red band 3 image (right). For further explanation see the article of Dekker *et al.* on page 137.

Printed on acid-free paper

Printed in Belgium

Dedicated to Sikko Parma
on the occasion of his retirement
April 1, 1992

Contents

a critical look at eutrophication research, more from a practical point of view. Most of these studies were directed to additional measures, since they were carried out at institutes dealing with regional water problems, and at restoration measures such as flushing, dredging, chemomanipulation and biomanipulation, either in isolation or in combination.

The Symposium, which ended with a vigourous but stimulating discussion on the contributions, was chaired by Dr H. M. De Boois (the Chairwoman of the Water Pollution Control Authority of the Amstel and Gooiland) with as members of the panel: Ir. M. A. De Ruiter; Dr H. W. Kroes; Dr S. Parma (Director of the Limnological Institute and Secretary of the WQL Board); Dr L. Van Liere (the WQL's Research-Coordinator, and one of the undersigned) and Drs P. J. T. Verstraelen (the Water Pollution Control Authority of the Amstel and Gooiland). The deliberations reaffirmed the complex nature of the shallow lake ecosystem, especially those driven to a state of hypertrophy by decades of uninterrupted eutrophication. That the eutrophication should be reduced primarily by decreasing the inputs of external nutrients, and that secondary measures can hasten the recovery process and reduce the ecosystem resilience, was a general consensus.

The Symposium Organization Committee consisted of Jeannine Ebert, Cecilia C. C. Janssen-Kroon, Herman J. Gons and Henk De Haan, all from the Limnological Institute and was chaired by Louis Van Liere from the National Institute of Public Health and Environmental Protection. The Symposium was financially supported by the Dutch Ministry of Housing, Physical Planning and Environment, the Limnological Institute and the National Institute of Public Health and Environmental Protection.

We, as editors, are highly obliged to the referees for their critical reviews and suggestions. Each manuscript was sent to at least two referees, one of them a native of an English-speaking country for linguistic improvements; the second was generally a Dutch colleague, acquainted with the geographical situation and regional problems. They all responded very promptly, sparing us the need to send extra reminders, or look for alternative referees. The authors too were generally cooperative, with a few exceptions.

We are very grateful to Cecilia C. C. Janssen-Kroon. Not only was she one of the organizers of the Symposium, she also skilfully handled the editorial correspondence.

On the front cover of this volume there is an image of the LANDSAT Thematic Mapper. We thank Arnold Dekker, who prepared the image for the Loosdrecht lakes area; The International Journal of Remote Sensing is acknowledged for the permission to use the image.

From the inception of the WQL Dr S. Parma (Director of the Limnological Institute) has been one of the driving forces; as a Secretary of the WQL Board he exhibited great interest in WQL and contributed significantly to its successful outcome. We dedicate this volume to Sikko Parma on the occasion of his retirement on April 1, 1992.

January 6, 1992

LOUIS VAN LIERE, *Bilthoven*
RAMESH D. GULATI, *Nieuwersluis*

List of referees

Auer, M. T., Michigan Technol. University, Dept. of Civil Engineering, Michigan, U.S.A.

Bailey-Watts, A. E., Institute of Terrestrial Ecology, Edinburgh Research Station, Scotland, U.K.

Blom, G., Agricultural University at Wageningen, Dept. Of Nature Conservation, The Netherlands.

Boers, P. C. M., National Institute of Integral Freshwater Management, The Netherlands.

Buyse, J. J., IWACO Consultants, The Netherlands.

Cooke, G. D., Kent State University, Dept. of Biological Sciences, Ohio, U.S.A.

Curran, P. J., University of Sheffield, Dept. of Geography, U.K.

De Haan, H., Limnological Institute, The Netherlands.

DePinto, J. V., SUNY at Buffalo, New York, U.S.A.

Donze, M., Technical University at Delft, Limnological Laboratory, The Netherlands.

Gibson, C. E., Fresh Water Biological Unit. Dept. of Agriculture, Northern Ireland, U.K.

Gons, H. J., Limnological Institute, The Netherlands.

Gulati, R. D., Limnological Institute, The Netherlands.

Higler, L. W. G., National Institute of Nature Conservation, The Netherlands.

Hosper, S. H., National Institute for Integral Freshwater Management, The Netherlands.

Jewson, D., The University of Ulster, Limnology Laboratory, Northern Ireland, U.K.

Keizer, P., Limnological Institute, The Netherlands.

Klapwijk, S. P., Water Authority of Rijnland, The Netherlands.

Latour, J., National Institute of Public Health and Environmental Protection, The Netherlands.

Loogman, J. G., European Parliament, Luxembourg.

Malthus, T. J., Univ. of Nottingham, Dept. of Physiol. and Environmental Sciences, U.K.

Moss, B., Univ. of Liverpool, Dept. of Environmental and Evolutionary Biology, U.K.

Nawalany, N., University of Warsaw, Poland.

Rhee, G.-Y., School of Public Health, SUNY at Albany, NY State Dept. of Health, Albany, U.S.A.

Parma, S., Limnological Institute, The Netherlands.

Roijackers, R. M. M., Agricultural University at Wageningen, Dept. of Nature Conservation, The Netherlands.

Schot, P. P., University of Utrecht, Dept. of Environmental Sciences, The Netherlands.

Ter Braak, K., National Institute of nature Conservation, The Netherlands.

Van Donk, E., Agricultural University at Wageningen, Dept. of Nature Conservation, The Netherlands.

Van Huet, H., University of Nijmegen, The Netherlands.

Van Liere, L., National Institute of Public Health and Environmental Protection, The Netherlands.

Van Straaten, G., Agricultural University at Wageningen, Dept. of Agricultural Techniques and Agricultural Physics, The Netherlands.

Van Tongeren, O. F. R., Limnological Institute, The Netherlands.

Van Vierssen, W., Institute for Hydraulic and Environmental Engineering, The Netherlands.

Verstraelen, P. J. T., Water Authority of Amstel and Gooiland, The Netherlands.

Vijverberg, J., Limnological Institute, The Netherlands.

Walsby, A. E., University of Bristol, Dept. of Botany, U.K.

Winfield, I., Institute of Freshwater Ecology, The Windermere Laboratory, U.K.

Young, T. C., Clarkson University, Dept. of Civil and Environmental Engineering, New York, U.S.A.

Hydrobiologia **233**: xiii–xviii, 1992.
L. Van Liere & R.D. Gulati (eds), Restoration and Recovery of Shallow Eutrophic Lake Ecosystems in The Netherlands.

Sikko Parma: limnologist, environmentalist and scientific manager

Louis Van Liere[1,2], Ramesh D. Gulati[1], Bert Z. Salomé[1], Joop Ringelberg[3], Cees Davids[4] & Kees Bakker[5]
[1]*Limnological Institute, Rijksstraatweg 6, 3631 AC Nieuwersluis, The Netherlands;* [2]*National Institute of Public Health and Environmental Protection, P.O. Box 1, 3720 BA Bilthoven, The Netherlands (address for correspondence);* [3]*Dept. of Pure and Applied Ecology, University of Amsterdam, Kruislaan 320, 1098 SM Amsterdam, The Netherlands;* [4]*Dept. of Aquatic Ecotoxicology, University of Amsterdam, Kruislaan 320, 1098 SM Amsterdam, The Netherlands;* [5]*Delta Institute for Hydrobiological Research, Vierstraat 28,4401 EA Yerseke, The Netherlands*

It's work, the most important thing is work,
Work, the most important thing is work.
Lou Reed: 'Songs for Drella'

We dedicate the Proceedings of the final Water Quality research Loosdrecht lakes Symposium 'Restoration and recovery of shallow eutrophic lake ecosystems in The Netherlands' to Dr. Sikko Parma who retired on April 1, 1992 as director of the Limnological Institute of the Royal Netherlands Academy of Arts and Sciences. Here we attempt to give a thumbnail sketch of the person of Sikko Parma, including his early career, as well as his accomplishments as a limnologist, as a director of the Limnological Institute, and as an environmentalist: in all Sikko is 'mission-oriented'.Sikko was an active member of the WQL right from its inception and secretary of the WQL-board.His job as a director of the Limnological Institute from 1978 to March 31, 1992 was none-too-easy, in view of the financial, scientific and the reorganisation problems, the government-run research laboratories were faced with. To lead a group of limnologists, one must be a generalist. Sikko Parma, though specialist by his research training, is a generalist with main interests, besides, in general limnology, in problems of water pollution and environmental deterioration.Being a dedicated environmentalist, Sikko Parma was member of various advisory committees of the national government, but also of action groups to arouse public awareness and warn the national government on environmental problems.

Sikko Parma, student of biology

Sikko Parma was born in November of 1931 at IJmuiden in the Province of North Holland. He passed the State High School at the 'Paardenmarkt', Alkmaar in 1950. He was a very enthusiastic scout but also member of a Youth Club for Nature Study, so much that he failed to pass his third class at school on his first attempt. After finishing high school, Sikko joined the undergraduate course in biology (then with only 20 other students) at the University of Amsterdam. He successfully completed the requirements for a 'candidate in biology' (1950-1955, 'cum laude'). He majored in animal physiology (Prof. Dr. A. Punt), with a minor in hydrobiology (Dr. A. G. Vorstman). Besides hydrobiology, his special interests were bats, midges and Prof. Punt's technician Netteke. Netteke and Sikko were married in 1956. During his study period Sikko was an assistant during several field courses and excursions abroad (Ambleteuse, Pont du Gard, Bangor, Dubrovnik) and in The Netherlands (Limburg and the Zoological Station at Den Helder). He passed his doctoral examination in 1959 'cum laude', and was appointed, without having the time for a holiday, as a scientist at the Laboratory of Animal Physiology (Prof. Dr. A. Punt). During that time Sikko was active in the scientific 'camps' of the Netherlands Hydrobiological Society in the Biesbosch, and ecologically unique tidal area in the Rhine and Meuse estuaries.

Sikko Parma, Limnologist

In 1961 he was appointed as a zoologist at the Hydrobiological Institute of the Royal Netherlands Academy of Arts and Sciences at Nieuwersluis. The choice of research object fell on *Chaoborus flavicans*, the midge larvae, which were abundant in the Lake Vechten, the main object of research at the institute at that time. On December 10, 1971, Sikko Parma successfully defended his doctorate thesis entitled '*Chaoborus flavicans* (Meigen) (Diptera, Chaoboridae): an autecological study' at the State University of Groningen.

From 1971 to 1978 he worked at the Delta Institute for Hydrobiological Research of the Royal Netherlands Academy of Arts and Sciences at Yerseke. Later, he became leader of the working group 'Structure and dynamics of instable aquatic ecosystems'. Because of his continuing interests in the life cycle of dipteran larvae he studied the distribution of chironomid larvae in brackish waters. In addition he got involved in actions against the complete damming (damning?) of the Eastern Scheldt estuary, which would have led to its transformation into a freshwater lake with loss of estuarine flora and fauna. Due to the campaign of 'mission-oriented scientists' like Sikko Parma and other environmentalists the choice fell on a unique tidal surge barrier-dam allowing exchange with the North Sea, so that essential features of the former estuary are preserved.

Fig. 1. Sikko Parma at one of the Eastern Scheldt excursions.

'Sikko is a biologist in his heart of hearts, and no water is safe to him. On holidays he invariably carries at least one net and a series of vials with him. In the course of time he also has become an enthusiastic bird watcher, and now he never travels without his binoculars' (Netteke Parma, pers. comm.).

Sikko Parma, a scientific manager

Sikko Parma has the inherent qualities of a leader, a manager. He is a staunch devotee of fundamental sciences, more so if it is socially relevant and morally acceptable. He is a good listener; and could often bring order to situations in which working-group leaders and co-workers experienced confusion and chaos. He took time to listen to staff members and their problems, irrespective of their age or experience. The door of his room was 'always open' and people could generally walk in without a prior appointment.

When Sikko Parma rejoined the Limnological Institute in 1978 the research organization was somewhat in disarray, lacking in a clear integral approach to studies. Soon on his arrival Sikko streamlined the institute's organization by setting up different working-groups. He undertook many actions to set up new projects, which were often financed by sources outside the institute. This paid rich dividends so far as attracting young scientists working for their Ph.D. degrees. The Limnological Institute has now become a 'Centre for Limnology', both nationally and internationally. Sikko has the ability not only to inspire the scientists but also to exploit their talent for the best interest of science. More than ten years ago Sikko initiated a monthly bulletin for the institute's personnel to provide in a nutshell first hand information on the different developments and activities, scientific and others, including memoranda of working group meetings, summaries of scientific lectures, etc.

As a director of the Limnological Institute Sikko Parma fulfilled multifarious functions outside the institute. He is actively involved in sociopolitical aspects of environmental problems.

The Limnological Institute had two establishments; the Vijverhof Laboratory at Nieuwersluis and the Tjeukemeer Laboratory at Oosterzee. With his office at Nieuwersluis it was not easy for Sikko to pay equal or optimal attention to both the laboratories. He, however, succeeded in preventing the two from drifting apart in their research programmes. He developed plans to merge the two departments, and expand and reorganize the research facilities at Nieuwersluis. Thanks to Sikko Parma, the new facilities at Nieuwersluis have now been realized and personnel from 'Tjeukemeer' institute has shifted since 1 January 1992 to Nieuwersluis. Also, he was involved in the founding of a new national institute, the Netherlands Institute of Ecology (NIE) in January 1992 (incorporating the three former ecological institutes of the Royal Netherlands Academy of Arts and Sciences: the Institute of Terrestrial Ecology, the Limnological Institute and the Delta Institute for Hydrobiological Research). Sikko showed a lot of patience, perseverance and steadfast attitude in defending the interests of limnologists in the world of ecology at large.

The Limnological Institute with its new name the Centre for Limnology, has since January 1, 1992, become an integral part of the NIE.

Fig. 2. Sikko Parma in his office at the Limnological Institute.

International contacts

Sikko Parma initiated many international contacts. To name just a few examples, in the research project 'Water Quality research Loosdrecht lakes' (WQL), he encouraged foreign colleagues to come to Nieuwersluis, providing them research facilities to participate in the WQL-research programme, and guest facilities at the institute. During his directorship scientists from several countries (Soviet Union, Poland, Czechoslovakia, Hungary, Spain, U.K., New Zealand, U.S.A., Canada, China, India, Sri Lanka, Sweden, Japan, Israel, Libya, Germany, Belgium and Finland) participated in the Institute's research programme. Besides, he encouraged contacts with limnologists from the Peoples Republic of China, India and Sri Lanka through exchange of visits by the scientist from the Institute to these countries and *vice versa*. He also is a stimulating international congress organizer. In 1990 he put up the facilities of the Limnological Institute at the disposal of the GAP (Group on Aquatic Productivity) workshop of the SIL (Societas Internationalis Limnologiae). A group of about 20 international scientists performed research on photosynthetic activity and metabolism of phytoplankton, especially in relation to the diurnal light-dark cycle. In 1990 he chaired the organizing committee of the international conference: 'Biomanipulation: Tool for Water Management' (Hydrobiologia Vol. 200/201). Sikko Parma has been also a national representative at SIL, and an active one at that.

Sikko played an important role in initiating and stimulating scientific cooperation in the European Community, starting with the national-based research programme 'Water Quality research Loosdrecht lakes; studying and modelling the impact of water management measures on the internal nutrient cycle'. This led the Limnological Institute to play a coordinating role in the EC group 'Ecosystem Research in Freshwater Environment Recovery' with the Italian Institute of Hydrobiology, the Institute of Agricultural Entomology of the University of Milano, the School of Environmental Sciences of the University of East Anglia, the University of Liverpool, the Institute of Inland Water Management and Waste Water Treatment, and the Limnological Institute. Sikko was also involved in setting up scientific cooperation with the Peoples Republic of China, and in formally organizing cooperation between the Royal Netherlands Academy of Arts and Sciences and with its counter parts in Poland and the former Soviet Union. Also for the other main projects of the Limnological Institute (Lake Vechten, Tjeukemeer, etc.) he initiated and stimulated international cooperation.

All and sundries

The foregoing memoir on Sikko, rather than his biography, does not have the contention of completeness, nor of perfection. It contains some reminiscences of a part of Sikko's life devoted to science and professional pursuits.

From the 'coffee break gossip' at the Institute it was clear that Sikko has a great sense of humour, and capacity to mimic and imitate, hum and sing and crack jokes. He has a tremendous memory especially in remembering names and keeping an updated diary of addresses of professional colleagues and friends. He is apparently a fanatic television fan with special interest in the current affairs, satirical programmes and football.

He has a large collection of Dutch stamps, which he collects and catalogues with the same enthusiasm as the snaps taken at the SIL congresses and other scientific meetings.

He reads a great deal, his interest varying from detectives and thrillers to literature and poetry. He recites poems of Adriaan Roland Holst, but also those of John O'Mill during coffee-breaks. Despite all these very social and sociable characteristics, Sikko is modest, compromising if the occasion demands, and resisting to be involved into controversies.

Before returning to the Limnological Institute as a director he played badminton and volleyball. The latter was quite surprising since he never joined the lunchtime volleyball-game for which the Limnological Institute was locally well known. While lunching he often strolled in the garden. He took great interest if it came to preserving the flora in the garden of the old mansion 'Vijverhof', the institute's main building.

Membership of committees and advisory boards

Sikko Parma is an active member of the councils engaged in scientific and socio-political aspects of environmental issues, and intends to continue in some of these after his retirement. Because of his broad knowledge, both scientific as well as sociopolitical, he was invited to join many such groups. He considered some of these as an important platform for interaction with the policy makers and planners. His subtle humour served as a great help when scientific arguments did not quite succeed, and there was a stalemate. Thus, he was also invited many a time to chair discussion groups at conferences and symposia. A selection of commissions and committees in which Sikko took active part are listed alphabetically.

- Advisory Council for the environment (section on Water)
- Biological Research in The Netherlands (BION) (Member of the Board and Chairman subgroup Aquatic Ecology)

- Biological Council
- Central Committee of Fisheries Research (NRLO)
- Commission for Ecological Standards in Water Management (Health Council: Chairman)
- Commission Biological Problems Cooling Water (Chairman)
- Commission on Hydrological Research (CHO/TNO)
- Committee of Nature- and Landscape Conservation
- Committee on the Inventarization of the River Vecht and the Eastern and Western Areas of the River Vecht
- Coordination Committee of Nature-, Landscape- and Environmental Conservation in Zealand (Chairman)
- Coordination Committee for the Execution of the Surface water Pollution Law (CUWVO)
- Advisory council for Nature- and Environmental Research (RMNO)
- Foundation for Nature and Environment (chairman)
- Health Council
- Hydrobiological Society
- International Water Tribunal, Rotterdam, The Netherlands
- Naples-Roscoff Committee (KNAW)
- Netherlands Zoological Society (Secretary)
- Planning Committee for Soil Research
- Royal Zealand Academy of Sciences (Secretary)
- Society of the Conservation of Nature Monuments in the Netherlands
- Study Group Environmental Management (Ecological Society)
- Water Quality research Loosdrecht lakes (Secretary)

Publications

As an active scientist, Sikko Parma, besides publishing fundamental studies on *Chaoborus*, also published several papers relating to environmental issues, which revealed in an early stage his 'mission orientation'. As a director of the Limnological Institute he was time-limited both to carry out active research or to publish his earlier research studies. He gave his consent to be a co-author of a joint paper only if he had actively participated in the preparation of the manuscript, and not *per se* because he was director. He greatly stimulated his staff and students to publish, and kept an updated inventory of the papers 'in press'.

Scientific papers of Sikko Parma, and those in which he was involved as a co-author are listed in the references below in a chronological order. Since we did not check with Sikko some omissions and errors are not ruled out.

Acknowledgements

The most important thing is work, but it is not only work. However, the citation at the beginning refers to the creativity of Andy Warhol, and his ability to inspire the artists at The Factory to reach for a higher level. And that needs work. Sikko Parma can work, and he likes it. Sikko gave his consent to chair a national symposium on water management on April 8, i.e. just one week after his retirement.

We acknowledge on behalf of very many colleagues all the work that Sikko Parma did: for science, for limnology, for the Limnological Institute, and for the environment and its 'Sustainable Future'.

Sikko, we wish you a very happy retired but active life, which undoubtedly will be devoted partly to science and the environment. As a good start we offer you this book.

References

Punt, A. & S. Parma, 1964. On the hibernation of bats in a marl cave. Publ. Natuurhist. Genootsch. Limburg. Reeks 13: 45–59.

Parma, S., 1967. Oorzaken van de verspreiding van *Chaoborus*-larven in de bodem van meren. Meded. Hydrobiol. Ver. 1: 28–30.

Parma, S., 1967. Het onderscheiden van watertypen. Vakbl. Biol. 47: 141–150.

Parma, S. & K. F. Vaas, 1968. Verslag van het hydrobiologisch onderzoek in de Biesbosch, voorwoord. Meded. Hydrobiol. Ver. 2: 87–91

Parma, S., 1968. Hydrografie van de Biesbosch. Meded. Hydrobiol. Ver. 2: 92–145.

Parma, S., 1969. Notes on the larval taxonomy, ecology, and distribution of the Dutch *Chaoborus* species (Diptera, Chaoboridae). Beaufortia 17: 21–50.

Parma, S., 1969. The life cycle of *Chaoborus crystallinus* (De Geer) in a Dutch pond. Verh. int. Ver. Limnol. 17: 888–894.

Roth, J. C. & S. Parma, 1970. A *Chaoborus* bibliography. Bull. Entomol. Soc. Am. 16: 100–110.

Parma, S., 1971. The morphology of the larval instars of *Chaoborus flavicans* (Meigen, 1818) (Diptera, Chaoboridae). Beaufortia 18: 173–182.

Parma, S., 1971. *Chaoborus flavicans* (Meigen) (Diptera, Chaoboridae): an autecological study. Thesis, University of Groningen, 128 pp.

Parma, S., 1972. Muggenplagen op en om het Zeeuwse Meer. Vakbl. Biologen 52: 134–140.

Parma, S. & A. J. Cavé, 1972. De consequenties van de mening van den Hartog en de Jonge (VB 11, 1972): de afsluiting van de Oosterschelde. Vakbl. Biol. 52: 344–346.

Swüste, H. F. J., R. Cremer & S. Parma, 1973. Selective predation by larvae of *Chaoborus flavicans* (Diptera, Chaoboridae). Verh. int. Ver. Limnol. 18: 1559–1563.

Parma, S., 1973. De afsluiting van de Oosterschelde opnieuw in discussie. Natuur en Landschap 27: 29–48.

Parma, S., 1974. De Oosterschelde, open of dicht. Maritiem Journaal 1974: 189–194.

Parma, S., 1975. De Oosterschelde, open of dicht. Maritiem Journaal 1975: 153–159.

Parma, S. & B. P. M. Krebs, 1977. The distribution of chironomid larvae in relation to chloride concentration in a brackish water region of The Netherlands. Hydrobiologia 52: 117–126.

Boissevain, E. C. & S. Parma, 1977. Hoe kan de Oosterschelde gered worden? Nat. Milieu 1: 12–15.

Parma, S., 1978. Political aspects of the closure of the Eastern Scheldt estuary. Hydrobiol. Bull. 12: 163–175.

Parma, S., 1978. Kreken en wielen in Zeeland. Zeeuws Nieuws over Natuur, Landschap en Milieu 3: 5–10.

Parma, S., 1980. The history of the eutrophication concept and the eutrophication in The Netherlands. Hydrobiol Bull. 14: 5–11.

Parma, S., 1982. The twenty-fifth anniversary of the Limnological Institute, The Netherlands (1957–1982). Hydrobiologia 95: 1–9.

Gulati, R. D. & S. Parma (eds), 1982. Studies on Lake Vechten and Tjeukemeer, The Netherlands. Hydrobiologia 95: 1–383. (Also published as: Developments in Hydrobiology 11: 383 pp. 25th Anniversary of the Limnological Institute of the Royal Netherlands Academy of Arts and Sciences).

Parma, S., H. M. Van Embden & J. Castelein (eds), 1983. Oecologie van meren en plassen. Biologische Raad Reeks. PUDOC, Wageningen. pp. 157.

Parma, S., 1983. Aquatisch-oecosyteemonderzoek in wetenschappelijk en maatschappelijk kader. In: S. Parma, H. M. Van Embden & J. Castelein (eds) Oecologie van meren en plassen. Biologische Raad Reeks pp. 7–18. PUDOC, Wageningen.

Parma, S., 1983. Wensen voor Natuur- en Milieuonderzoek. Nat. Milieu 7: 12–15

Van Liere, L., S. Parma, L. R. Mur, P. Leentvaar & G. B. Engelen, 1984. Loosdrecht lakes restoration project: an introduction. Verh. int. Ver. Limnol. 22: 829–834.

Weekenstroo, J. E. & S. Parma, 1984. Aspects of the role of *Chaoborus flavicans* (Meigen) in Lake Vechten (The Netherlands). Verh. int. Ver. Limnol. 22: 904–908.

Parma, S., 1987. Ontwikkelingsmogelijkheden voor aquatische ecosystemen. In: Produktiebeperkingen in de landbouw; nieuwe kansen voor natuur en landschap! Verslag van het WLO-lustrumcongres, Wageningen. Werkgemeenschap Landschapsoecologisch Onderzoek. pp. 29–43.

Parma, S., 1988. Management of water quality in surface waters of The Netherlands. Verh. int. Ver. Limnol. 23: 1465–1470.

Parma, S., 1989. Strategie van onderzoek voor de milieubeweging. In: J. G. Waardenburg, W. Dijkman & N. G. Van Vennedaal-Koevoets (eds), Samenleving en onderzoek. Verslag van een studiedag op 31 October 1986, Zoetermeer, Commissie van Overleg Sectorraden, p. 37–43.

Parma, S. & A. J. Murk, 1990. Inleiding: het advies van de gezondheidsraad over Ecologische Normen Waterbeheer. pp. 9–17. In: A. J. Murk, A. A. A. Van der Schaaf, H. A. M. De Kruijf *et al.* (eds), Strategieën voor ecologische normstelling waterbeheer: het spel en de knikkers. SDU, 's Gravenhage.

Van Liere, L., S. Parma & R. D. Gulati, 1992. Working group Water Quality Research Loosdrecht Lakes: its history, structure, research programme, and some results. Hydrobiologia 233: 1–9.

Wolff, W. J., S. Parma, 1974. Oosterschelde pleidooi voor veilig voorbestaan Stichting Zeeuws Coordinatieorgaan voor Natuurlandschap en milieubescherming Middelburg, 95 p.

Editor of Progress Reports of the Limnological Institute since 1978.

Hydrobiologia **233**: 1–9, 1992.
L. Van Liere & R.D. Gulati (eds), Restoration and Recovery of Shallow Eutrophic Lake Ecosystems in The Netherlands.

Working group Water Quality Research Loosdrecht Lakes: its history, structure, research programme, and some results

Louis Van Liere,[1,2]* Sikko Parma[1] & Ramesh D. Gulati[1]
[1]*Limnological Institute, Rijksstraatweg 6, 3631 AC Nieuwersluis, The Netherlands;* [2]*National Institute of Public Health and Environmental Protection, P.O. Box 1, 3720 BA Bilthoven, The Netherlands; (* author for correspondence)*

Key words: interdisciplinary research, lake restoration

Abstract

In 1984 the external phosphorus load to the Loosdrecht lakes ecosystem was decreased substantially. A working group, Water Quality research Loosdrecht lakes (WQL), was formed to study the consequences for recovery of the lakes, and to evaluate the water management measures taken. Its history from the start in 1979 is described. The working group had an interdisciplinary character; its organization structure (project management) is depicted. From March 1983 a coordinate multidisciplinary research programme was started, carried out uninterrupted up to December 1990. The programme paid special attention to the flow of phosphorus through the lake ecosystems. The results were presented at several international conferences. The WQL was also involved in the international working group ERiFER (Ecosystem Research in Freshwater Environment Recovery). Important financial support was given by Dutch ministries, the Commission of European Communities, regional water authorities, and private funds.

The measures taken to counteract eutrophication were not successful. It has been the task of the WQL to explain the ecosystem resilience.

Introduction

Loosdrecht lakes have been subjected to severe eutrophication since the beginning of this century (Gulati *et al.*, 1991; Hofstra & Van Liere, 1992). The water quality lakes had already deteriorated in the early sixties. One of the lakes of the area served as source of raw material for the drinking water production of the city of Amsterdam. Around 1970 the demand for drinking water of the city had increased and increasingly more water from the lake area had to be used. This meant a higher demand for extraneous (phosphorus-rich water) to maintain the water level. This led to fears about rapid rise in the deterioration of the water quality in Loosdrecht lakes. The Stichtse Milieufederatie (The Sticht Environmental Federation) warned about the consequences, upon which the Province of Utrecht formed a working group to assess the water-quality problems of Loosdrecht lakes. This working group consisted of representatives of the Province of Utrecht, Amsterdam Municipal Waterworks, the Research Institute of Nature Conservation, the National Institute of Drinking Water Supply and the Limnological Institute. Soon thereafter the group reported its assessment about the state of the environment (Province of Utrecht, 1973). A second working group was instituted to explore the various alternatives for external water supply in order to safeguard the water quality of the lakes in the future (Province of Utrecht, 1980). The model that was chosen envisaged on its application on a drastic reduction in external phosphorus loading. Some members of the working groups opted to carry out detailed ecological studies of the lakes, and founded the working group WQL (Water Quality research Loosdrecht lakes). The history, structure, research programme of the working group, and some results are subject of this paper.

WQL, founding and first years (1979–1983)

The first meetings to found WQL were called by Ing. Wolter Kats and Dr. Ir. Anton Graveland (Amsterdam Municipal Waterworks) on March 7, 1979 followed by a second meeting on May 3, 1979. No notes were made. The first 'official' meeting of the working group was held on June 15, 1979. The goals of the working group were 'to qualify and quantify the effects of a considerable preduction on the functional and structural aspects of a shallow lake ecosystem, to translate these effects in terms of water quality and to contribute, if necessary, to water management measures with a view to attaining and maintaining the desired water quality'. The objective was to perform a baseline study prior to measures relating the phosphorus reduction came into effect, and to perform a follow-up study thereafter to evaluate the effects of the measures taken.

Already in July 1979 a fourweekly monitoring programme was started by the Amsterdam Municipal Waterworks. About twenty physical and chemical parameters were monitored at ten sampling stations spread over the whole Loosdrecht lakes system. This programme was carried out uninterrupted up to December 1990. From 1981 phytoplankton was counted routinely, in samples from a selected number of stations. Zooplankton research was started in summer 1981. In 1982 preliminary projects started to the possibilities of hydrological research and exchange at the sediment water interface.

The structure of the working group to start with was rather loose, which is not surprising, since the group owed its foundation to several personal initiatives, lacking finances to have a steering function. A board was formed which convened two to three times a year meetings of the scientists involved to discuss the results and future plans of work. The participants of the working group came from various disciplines. University departments and institutes belonging to the Royal Netherlands Academy of Arts and Sciences, which were assigned to perform fundamental research. Laboratories of water authorities and water boards, and consultant firms which are confined to applied aspects, i.e. to problem solving research.

Since the scientist is generally 'curiosity-oriented', the funds to perform such research becomes a limiting factor in terms of economics. The Dutch policy makers demand rightly, that research institutes pay attention to social relevant problems. This does not necessarily mean that the level of scientific research is lowered. On the contrary, such 'mission-oriented' research may be equally valuable as 'curiosity-oriented' research. In the Dutch scientific jargon it is often called 'strategic fundamental research'. The primary motivation, at least for the Limnological Institute, was to perform fundamental research, to understand the structure and functioning of the lake ecosystems.

Such studies are by definition multidisciplinary in character, with their roots in hydrology, physics, chemistry, biology and mathematics. However, from experience we learned that it is rather difficult for scientists to cooperate, especially if the involved disciplines are remote. The integral level of ecosystem research is such that it is difficult to put forth a joint question, and this is even true for the practical research.

Multidisciplinary ecosystem research has an excellent chance if the objective has the carrying capacity of the environmental concern. The eutrophication problem was the basis of executing fundamental research in WQL, although some participants were inclined to carry out purely problem-solving research. However, the computed drastic reduction in external phosphorus load to the Loosdrecht lakes was an experiment conducted at the ecosystem level. It has been a chance to study effects, but also to get a better understanding of the causal processes. And, if water management measures were ineffective WQL would not only explain this, but suggest additional measures.

A main advantage of 'mission oriented research' is that it brings together scientists, policy makers and water quality managers. It would not be honest to say that also financial motives did not play a role. Eutrophication research, which is socially relevant, attracts financial support.

A few years after the start of the studies it was felt that separate projects, as they were carried out, would not meet the objectives. Ecosystem research is only possible if logistics and scientific demands are coordinated, let alone of integration of the various disciplines. The Ministry of Education and Science funded the appointment of a research-coordinator (first author of this paper) to WQL from 1983 to 1987, in order to streamline the functioning of WQL. The Ministry of Housing, Physical Planning and Environment enabled the coordinator's work to be extended.

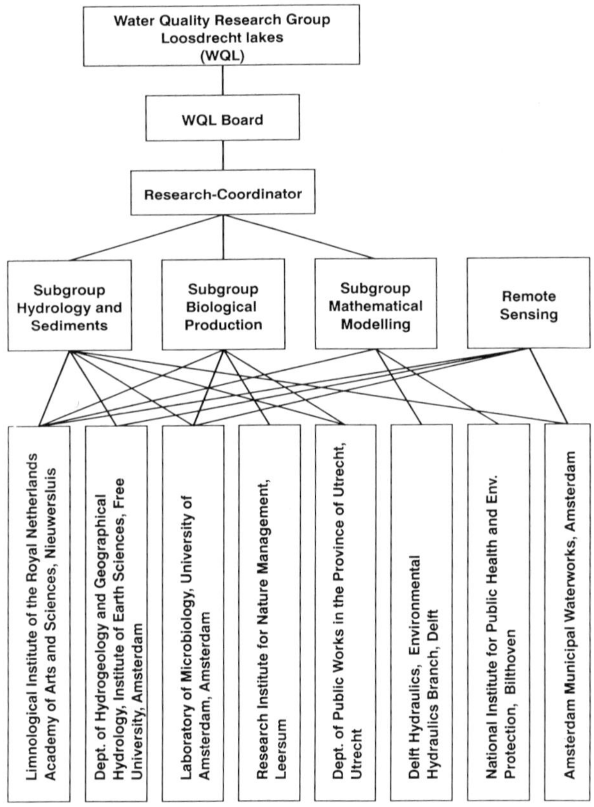

Fig. 1. Organization scheme of Water Quality research Loosdrecht lakes.

WQL, organization structure

For optimizing exchange of knowledge and setting up new programmes, an organization structure was developed to enable the participating scientists to meet and discuss results. Representatives of the eight participating institutes were arranged in working groups according to four themes (Fig. 1). In these group study plans, progress of research and evaluations were discussed. Several members participated in more than one group in order to optimize exchange of knowledge. All WQL-participants were kept informed of the progress of the four groups through notes and internal reports. The research-coordinator chaired the working group meetings, each group meeting four to five times per year. Only the Remote Sensing group met *ad hoc*. The research-coordinator was added to the Board, which consisted of the directors/working-group leaders of the main WQL institutes. Twice a year all participants of WQL, as well as policy makers and water managers, met in a plenary session, in which papers of various WQL-subjects were presented and discussed. WQL consisted of about 60 members. Depending on the number of projects in operation about 30 scientists were actually working in the field or in the laboratories, or both.

WQL, integrated research (1983–1990)

In 1983 the coordinated research programme was set up. It consisted of fourweekly measurements of physical, chemical and biological parameters and processes. Because of the available manpower and apparatus it was possible to run the routine programme only at four sampling stations, two in the Lake Loosdrecht, the other two were years the Loosdrecht lakes have changed from oligotrophic ecosystems, with a clearly visible bottom, into highly eutrophic ones, with low transparency (Secchi-disc depth <0.4 m). The course of eutrophication of the Loosdrecht lakes is described below as a series of phases. A similar approach has been used by Moss (1989) to describe the eutrophication of the Norfolk Broads (UK). Although the exact distinction of the phases is arbitrary, Moss' concept (1989) is supported by other studies (Van Liere *et al.*, 1991; Janse & Aldenberg, 1990; Scheffer, 1989 and Gulati *et al.*, 1991).

Phase 1
Time span: 1800–1920
Characterization: strongly nutrient limited (oligotrophic)

Clear water (Heymann, 1922); a virtual absence of submerged waterplants (Best *et al.*, 1984); a estimated fish population biomass of about 50 kg ha^{-1} (Lammens, pers. comm.);

Phase 2
Time span: 1930–1955
Characterization: nutrient limited; domination by submerged macrophytes (mesotrophic)

Clear water (Van Heusden, 1942); well-developed submerged vegetation of *Characeae* (open water) and *Potamogetanaceae* (periphery) (Van Heusden, 1942); a relatively large amount of zooplankton compared to phytoplankton (Geelen, 1955); a well-developed population of zebra mussels (*Dreissena polymorpha*) (Leentvaar & Mörzer Bruijns, 1962); a large population of hibernating waterbirds, especially pochards (*Anthya ferina*), tufted ducks (*Anthya fuligula*) and coots (*Fulica atra*) (Leentvaar & Mörzer Bruijns, 1962); a resident population of otters (*Lutra lutra*) (Creutzberg *et al.*, 1969); an estimated fish biomass of about 250 kg ha^{-1}, the dominant species being pike (*Esox lucius*, 45 kg ha^{-1}) and bream (*Abramis brama*, 50 kg ha^{-1}) (Lammens, pers. comm.); annual average concentrations of chlorophyll *a* about 25 $\mu g\,l^{-1}$ (Geelen, 1955; CUWVO, 1988; Hofstra, 1990) and total phosphorus about 5.5 $\mu g\,P\,l^{-1}$ (Van Liere *et al.*, 1991); a mineral nitrogen concentration of about 17 $\mu mol\,l^{-1}$ (summer average) (Van Liere *et al.*, 1991);

In Lake Breukeleveen Phase 2 was followed by a period of abundant growth of tall, submerged, floating and emergent macrophytes, especially *Potamogetonaceae* (Prud'homme Van Reine & Van Der Meulen, 1961). This period lasted from 1955 to 1965.

Phase 3
Time span: 1960–1980
Characterization: nutrient limited; cyanobacteria domination (eutrophic)

No waterplants submerged (Best *et al.*, 1984); dense cyanobacterial growth, limited in its growth by light-energy, the dominant species being *Oscillatoria redekei*, *Oscillatoria limnetica* and possibly the recently discovered prochlorophyte *Prochlorothrix hollandica* (Van Liere *et al.*, 1991); a large population of small zooplankters (Gulati, 1984); no fresh water mussels; no otters, few hibernating waterbirds (Morel, 1989; Hofstra, 1990); a fish biomass of about 300 kg ha^{-1}, dominated by bream (180 kg ha^{-1}) and pike-perch (*Stizostedion lucioperca*, 40 kg ha^{-1}) (Lammens, pers. comm.); annual average concentrations of chlorophyll *a* about 100 $\mu g\,l^{-1}$ and total phosphorus about 100 $\mu g\,P\,l^{-1}$; a mineral nitrogen concentration of about 60 $\mu mol\,l^{-1}$ (summer average) (Van Liere *et al.*, 1991);

Phase 4
Time span: 1980–present
Characterization: high internal phosphorus loading rate; detritus domination (eutrophic)

Like Phase 3 but seston ($<150\ \mu m$) is now dominated (50–75 per cent) by detritus (Van Liere *et al.*, 1991).

Phase 4 is at present a stable state of the lakes. Measures that have been taken to reduce the external phosphorus loading rate to the Loosdrecht lakes have resulted in only a small decline of total phosphorus concentration since 1984. The ecosystem is apparently well-buffered against changes in external nutrient supply, due to high internal nutrient loading rates (Van Liere & Janse, 1992).

Accumulation of heavy metals in the Loosdrecht Lakes

The main sources of heavy metals to the Loosdrecht lakes are the inlet of allochthonous water (carrying polluted sediment), atmospheric deposition, agricultural effluents and local emissions (Kühnel, 1985). The inlet of water from the R. Vecht caused peak concentrations of heavy metals in the sediments in the Kievits Area and in the southern part of Lake Loosdrecht (Table 1). The average concentration of heavy metals in the sediment of the Loosdrecht lakes as a whole is lower. Between 1979 and 1984 the cadmium concentration in the sediment of the Loosdrecht lakes has increased sharply (Kühnel, 1985).

Destruction of banks

Topographical maps show that most residual banks, the strips of land that remained after the industrial peatmining, have disappeared. The destruction of the banks as well as eutrophication has contributed to the present amount of detritus in the Loosdrecht lakes. On windy days significant amounts of detritus are resuspended lowering the transparency of the water. The finer fraction of detritus, which consists merely of remnants of algae, may remain suspended for days or even weeks (Gons & Van Keulen, 1989).

Table 1. Concentrations (mg kg^{-1}) of cadmium, zinc and lead in the sediments of the Loosdrecht lakes (spatial average of six stations) and Kievits Area (spatial average of six stations) (Kühnel, 1985), estimated background concentrations (Hofstra, 1990), and General Environmental Quality (GEQ) target values for fresh water systems from the Dutch Water Management Plan (Ministry of Traffic and Public Works, 1989à.

	Cadmium	Zinc	Lead
Loosdrecht lakes	6.3	17	90
Kievits area	17.5	928	168
Background (reference value)	0.3	100	30
Objective (GEQ)	2	480	530

Disturbance of the fauna by recreation and habitation

Since World War II the Loosdrecht lakes area has become increasingly populated. Recreation in the area is still on the increase. In recent decades yachting harbours, camping sites and holiday cottages have been constructed.

A decline in the breeding populations of Sedge Warbler (*Acrocephalus schoenobaenus*), Bittern (*Botaurus stellaris*), Little Bittern (*Ixobrychus minutus*), Night Heron (*Nycticorax nycticorax*) and Purple Heron (*Ardea purpurea*) and the extinction of the otter (*Lutra lutra*) may be partially attributed to disturbance (Morel, 1989; Creutzberg *et al.*, 1969). However, a study by Reijnen (1988) on the environmental aspects of water recreation demonstrated that quantified dose-effect relationships are not yet available.

Policy aims and ecological target values for fresh waters in The Netherlands

A 'General Environmental Quality' (GEQ; 'Algemene Milieukwaliteit') is the policy objective for all ecosystems in The Netherlands, both terrestrial and aquatic. GEQ guarantees 'health and maintenance of plants, animals, goods and forms of use' (Ministry of Housing, Physical Planning & Environmental Protection, 1985). In some areas a 'Specific Environmental Quality' (SEQ; 'Bijzondere Milieukwaliteit') is required, corresponding to specific functions, such as nature, drinking water supply, recreation or fisheries The Water Authorities of the Province of Utrecht have assigned the following functions to the Loosdrecht lakes: nature (Loosdrecht lakes), drinking water supply (Polder Bethune and Lake Loenderveen) and recreation (Loosdrecht lakes). Therefore a SEQ seems appropriate for the area.

Ecological target values for the Loosdrecht lakes

Ecological target values for various freshwater types, including shallow lakes, matching to SEQ target values, have been drafted by a workgroup 'CUWVO' (1988). The CUWVO target values, however, should merely be regarded as general guidelines and not as operational tools for environmental quality assessment (Kroes, 1987). So far, concrete and verifiable ecological target values, for both biotic and abiotic parameters, have been set only for a few Dutch aquatic ecosystems. The Loosdrecht lakes, unfortunately, do not belong to them. To assess the state of the environment of the Loosdrecht lakes operational ecological target values have first to be set.

Lakes originating as a result of peatmining and receiving seepage water could have clear, nutrient-poor water with corresponding plant and animal communities (CUWVO, 1988). Therefore Phase 2 above, corresponding to the situation around 1940 and being a well-documented period in the ecological history of the Loosdrecht lakes, has been chosen as a 'frame of reference'. Target values for 32 ecological parameters that prevailed during this period have been chosen (Hofstra, 1990; Hofstra *et al.*, 1991) and are summarized in Table 2 in which also the present, Phase 4, values of the different parameters are summarized.

Criteria for the selection of parameters

A first choice of the ecological parameters was made according to expert judgement. A final selection of the parameters was made according to the method of Klijn *et al.* (1990). The criteria of Klijn *et al.* (1990) were: firstly, parameters should correspond to specific environmental functions of the area in study, especially nature, drinking water, recreation, agriculture, forestry and fishery. Secondly, parameters should reflect the environmental problems of the area. Thirdly, data on the parameters should be available and fourthly, parameters should appeal to the interest of policy makers.

A graphical presentation of the state of the lake environment

In the Dutch Water Management Plan (Ministry of Traffic and Public Works, 1989) a graphical presentation technique, called the 'AMOEBE approach' was introduced. AMOEBE is an abbreviation for 'Algemene Methode voor OEcologische BEoordeling', which means 'General Method for Ecological Assessment'. The method was designed for policy makers and environmental authorities and has been applied to the North Sea, the River Rhine and some terrestrial ecosystems (Ten Brink & Hosper, 1989; Klijn *et al.*, 1990; Latour *et al.*, 1990). In Fig. 2 an AMOEBE, containing the parameters from Table 2, is presented. The sequence of the parameters corresponds to the method of Klijn *et al.* (1990).

Evaluation of the state of the environment

The AMOEBE (Fig. 2) clearly demonstrates that the state of the environment of the Loosdrecht lakes does not match with the frame of reference. Some parameters, total phosphorus, cadmium, cyanobacteria, diatoms and seston have increased since 1940. Others, transparency, green algae, zebra mussel larvae, *Characeae*, pike, sedge warbler and otter, are below their target value. Only a few parameters, oxygen, calcium-ions and salinity, approach their aims.

The AMOEBE-approach forces the choice of concrete target values for ecological parameters. A complete and verifiable set of target values can only be obtained if the study area is well-documented. The Loosdrecht lakes are a good example of such an area. But even if historical data are lacking or incomplete the setting of ecological aims is an indispensable step in environmental quality assessment (Public Health Council, 1989). In fact, it should be the first step in

Table 2. A survey of the present concentrations/values (Phase 4) and target values (Phase 2) for 32 ecological parameters (after Hofstra, 1990).

Parameter	Present concentration/ value		Desired concentration/ value	
Abiotic parameters				
Transparency (yearly average, Secchi-disc depth, m)	0.4	*a	1.9	*a
Sestion < 150 μm (yearly average, mg C l^{-1})	8.0	*b	0.2	*cd
Total phosphorus (yearly average, mg P l^{-1})	0.1	*a	0.0054	*ef
Soluble Reactive Phosphorus (yearly avg., mg P l^{-1})	0.002	*b	0.0015	*g
Mineral nitrogen (*1, summer average, mg N l^{-1})	0.840	*a	0.238	*a
Salinity (*2, summer average, mg l^{-1})	305	*a	231	*a
Calcium-ions (summer average, mg l^{-1})	49.0	*a	40.8	*a
Sulphate (summer average, mg l^{-1})	19.2	*a	33.6	*a
Bicarbonate (summer average, mg l^{-1})	153	*a	92	*a
pH (summer average, pH-units)	8.9	*a	7.8	*a
Oxygen (summer average, mg O_2 l^{-1})	10.4	*a	11.0	*a
Cadmium (in sediment, mg kg^{-1})	6.3	*h	0.3	*f
Plankton (yearly averages)				
Zebra mussel larvae (fresh weight, mg l^{-1})	0	*c	0.0038	*e
Total zooplankton (carbon , mg C l^{-1})	0.575	*d	0.075	*de
Total cyanobacteria (fresh weight, mg l^{-1})	30.6	*b	1.25	*e
Total diatoms (fresh weight, mg l^{-1})	0.81	*b	0.06	*e
Total green algae (fresh weight, mg l^{-1})	0	*b	0.48	*e
Chlorophyll *a* (mg l^{-1})	0.1	*a	0.025	*fi
Vegetation-groups unit: Total covering according to Tansley (1946) (sum of Tansley abundance terms)				
Characeae (4 species)	0	*fjk	15	*fjk
Potamogeton-group (10 species)	1	*fjk	11	*fjk
Menyanthus-group (7 species)	2	*fjk	14	*fjk
Thelypteris-group (15 species)	6	*fjk	25	*fjk
Nymphaea-group (7 species)	14	*fjk	18	*fjk
Butomus-group (6 species)	11	*fjk	15	*fjk
Caltha-group (10 species)	14	*fjk	24	*fjk
Fauna				
Pike (fresh weight, kg ha^{-1})	1	*l	45	*l
Bream (fresh weight, kg ha)	180	*l	50	*l
Otter (*3, nr. individuals)	0	*f	15	*m
Pochard (*4, nr. individuals)	750	*f	7500	*o
Sedge Warbler (*5, breeding pairs)	5	*n	50	*f
Little Bittern (*5, breeding pairs)	1	*n	5	*f
Great Crested Grebe (*5, breeding pairs)	70	*n	20	*f

Remarks
*1 Sum of the ions NO_2^-, NO_3^- and NH_4^+
*2 Sum of ions SO_4^{2-}, HCO_3^-, Cl^-, Mg^{2+}, Na^+ and K^+
*3 Number of individuals in Southern Vecht lakes area (Crutzberg *et al.*, 1969)
*4 Number of individuals hibernating in the Loosdrecht lakes ares (Leentvaar & Mörzer Bruijns, 1962)
*5 Number of breeding pairs in Soutern Kievits Area (Morel, 1989)

References supporting this table
*a Van Liere *etal.*, 1991; *b Breebaart *et al.*, 1989; *c Gulati 1990, pers. comm.; *d Gulati, 1984; *e Geelen, 1955; *f Hofstra, 1990; *g Amsterdam Municipal Waterworks, 1940; *h Kühnel, 1985; *i CUWVO, 1988; *j Best *et al.*, 1984; *k Runhaar *et al.*, 1987; *l Lammens, pers. comm.; *m Creutzberg *et al.*, 1969; *n Morel, 1989; *o Leentvaar & Mörzer-Bruijns, 1962.

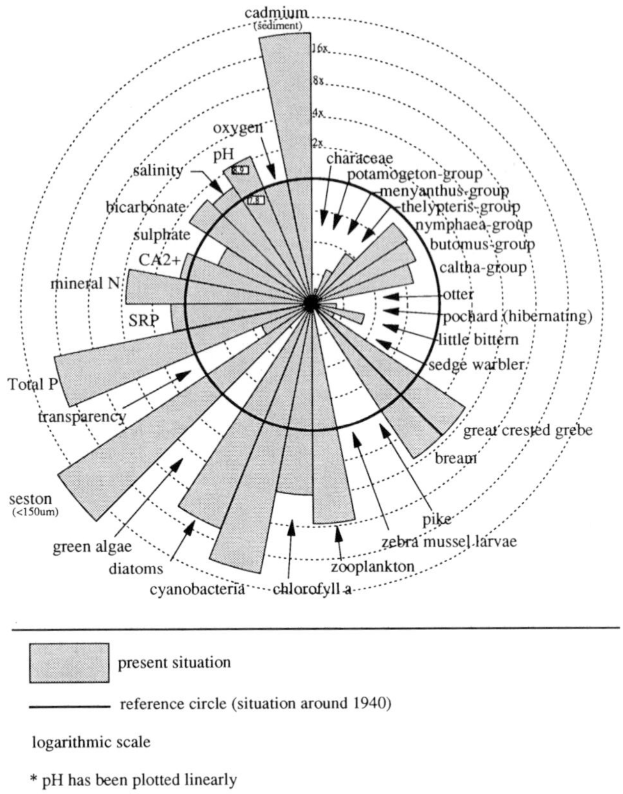

Fig. 2. 'AMOEBE' presentation of the state of the environment of the Loosdrecht lakes. The circle (thicker line) corresponds to a state of the ecosystem (around 1940; see 'Phase 2' under the text), used as an ecological frame of reference. The shaded parts represent the present state of the environment ('Phase 4'). The magnitude of deviations from the circle indicates the extent of unfavourable environmental conditions, which are due to a number of courses, described in the text, such as eutrophication, disturbance of the fauna and destruction of banks. Note that the graph has been plotted on a logarithmic scale except for pH.

translating general policy aims to directives for management.

One must bear in mind, however, that the target values that have been set in this study are based on a variety of sources, including historical reports and data from Lake Western Loenderveen (Hofstra, 1990). No attempts have been made to describe seasonal fluctuations or spatial variation in each parameter. Therefore, the target values should be considered as a rough draft of Specific Environmental Quality for the Loosdrecht lakes. Exact upper and lower limits for each parameter have not been established, nor could have been at this stage. In the future a more precise definition of ecological aims for the area might be desirable, for instance if, by proper management, the lakes approach their Phase 2 situation again.

Although we are unable presently to formulate exact ecological aims, a general conclusion can be generated from this study: the state of the environment of the Loosdrecht lakes is poor with respect to the Specific Environmental Quality that is required, especially with respect to its main functions, nature and recreation. Measures that combat eutrophication are strongly recommended.

Conclusions

1. The Loosdrecht lakes have changed in the past century from a oligotrophic ecosystem with very clear water into a highly eutrophic one with very low transparency.
2. The state of the environment of the Loosdrecht lakes is poor with respect to the Specific Environmental Quality target values that are required for the area.
3. The situation around 1940, described in the text as 'Phase 2', is recommended as a concrete and verifiable ecological frame of reference or target, corresponding to the policy aim of Specific Environmental Quality.
4. The AMOEBE presentation technique, devised for policy makers, is a valuable tool in environmental quality assessment as it visualizes in one glimpse the state of the environment of an area as well as a frame of reference, i.e. level desired.

References

Amsterdam Municipal Waterworks, 1940. Rapport 1940. Inzake de watervoorziening van Amsterdam – Report 1940. The watersupply of the city Amsterdam. Amsterdam. 575 pp.

Best, E. P. H., D. De Vries & A. Reins, 1984. The macrophytes in the Loosdrecht lakes: a story of their decline in

the course of eutrophication. Verh. Int. Ver. Limnol. 22: 829–834.
Breebaart, L., J. Ebert & L. Van Liere, 1989. Waterkwaliteitsonderzoek Loosdrechtse Plassen. Basisgegevens 1983–1988 – Water Quality research Loosdrecht lakes. Data from the period 1983–1988. WQL-report 1989–5. Limnological Institute, Nieuwersluis. 153 pp.
Creutzberg, F., P. Leentvaar, R. Rense, C. T. B. Rikkert de Koe, H. A. de Vries & K. W. R. Zwart, 1969. De Zuidelijke Vechtplassen. Flora en fauna – The Southern Vecht lakes. Flora and fauna. P. Leentvaar (ed) RIVON-verhandeling 7. 205 pp.
CUWVO (Coördinatiecommissie Uitvoering Wet Verontreiniging Oppervlaktewateren. Werkgroep V-1, 1988. Ecologische normdoelstellingen voor Nederlandse oppervlaktewateren – Ecological objectives for Dutch surface waters. 212 pp.
Engelen, G. B., 1986. Interaction of hydrological systems and eutrophication of the Loosdrecht lakes. Hydrobiol. Bull. 20: 17–25.
Geelen, J. F. M., 1955. Het plankton in de plassen in 1954 – The plankton of the Loosdrecht lakes in 1954. Amsterdam Municipal Waterworks. Report 38. 74 pp.
Gons, H. J. & R. van Keulen, 1989. De relatie tussen doorzicht en slib in de Loosdrechtse Plassen – The relation between transparency and sludge in the Loosdrecht lakes. WQL-Report 1989–4. Limnological Institute, Nieuwersluis. 35 pp.
Gulati, R. D., 1984. The zooplankton and its grazing as measures of trophy in the Loosdrecht lakes. Verh. int. Ver. Limnol. 22: 863–867.
Gulati, R. D., L. Van Liere & K. Siewertsen, 1991. The Loosdrecht lake system: Man's role in its creation, perturbation and rehabilitation. In: O. Ravera (ed.), Terrestrial and aquatic ecosystems. Perturbation and recovery. Ellis Horwood, New York, London. pp: 593–606.
Heymann, J. A., 1922. Onderzoek van de Loosdrechtsche Plassen op maandag 12 juni 1922 – Research in the Loosdrecht lakes on monday, June 12th, 1922. Amsterdam Municipal Waterworks. 7 pp.
Hofstra, J. J., 1990. De milieutoestand van de Loosdrechtse Plassen weergegeven door middel van een 'AMOEBE' – State of the environment of the Loosdrecht lakes reflected by means of an 'AMOEBE'. RIVM-report No. 751901004. 72 pp.
Hofstra, J. J., J. H. Janse & L. Van Liere, 1991. De milieutoestand van de Loosdrechtse Plassen. Toepassing van de presentatietechniek AMOEBE en het eutrofiëringsmodel PCLOOS – The state of the environment of the Loosdrecht lakes. Application of the presentation technique 'AMOEBE' and the eutrophication model 'PCLOOS'. H_2O 24: 260–265.
Janse, J. H. & T. Aldenberg, 1990. PCLOOS: a eutrophication model of the Loosdrecht lakes. WQL-report No. 1990–1. RIVM-report No. 714502001. 91 pp.
Klijn, F., J. B. Latour, M. I. Nip, H. A. Udo De Haes, M. M. H. E. Van Den Berg & J. J. Hofstra, 1990. Milieukwaliteit van ecodistricten. Deel 2: Methode en aanzet tot uitwerking – Environmental quality of 'ecodistricts'. Part 2. Method and initial application results. RIVM-report No. 751901003/CML-report No. 63. 110 pp.
Kroes, H. W., 1987. Van biologische waterbeoordeling naar ecologische normdoelstelling: de verbinding tussen wetenschap en beleid. In: Biologische waterbeoordeling. Instrument voor waterbeheer? – From biological water quality assessment toward ecological target values: the link between science and policy. P. F. M. Verdonschot en L. W. G. Higler (red.). RIN, Leersum. pp: 51–62.
Kühnel, V., 1985. Zware metalen in het Loosdrechtse Plassen-sediment – Heavy metals in sediment of the Loosdrecht lakes. WQL-report No. 1985–5. 36 pp.
Latour, J. B., J. J. Hofstra & M. I. Nip, 1990. Toepasbaarheid van de AMOEBE-benadering op terrestrische ecosystemen – Application of the 'AMOEBE-approach' on terrestic ecosystems. RIVM-report No. 751901001. 32 pp.
Leentvaar, P. & M. F. Mörzer Bruijns, 1962. De verontreiniging van de Loosdrechtse Plassen en haar gevolgen – Pollution of the Loosdrecht lakes and its consequences. De Levende Natuur 65: 42–67.
Loogman, J. G. & L. Van Liere (eds.), 1986. Proceedings of the WQL-symposium: Restoration of shallow lake ecosystems, with emphasis on Loosdrecht lakes. Hydrobiol. Bull. 20: 1–259.
Ministry of Housing, Physical Planning & Environmental Protection, 1985. Indicatief Meerjarenprogramma Milieubeheer 1986–1990 – Indicative Programme of Environmental Management 1986–1990. The Hague. 141 pp.
Ministry of Traffic & Public Works, 1989. Water voor nu en later. Derde Nota Waterhuishouding – Dutch Water Management Plan. The Haque. 297 pp.
Morel, G. A., 1989. De Zuidelijke Kievitsbuurt: Moerasvogels in een natuurgebied – Southern Kievits Area: marshbirds in a nature area. De Kruisbek 5: 166–173.
Moss, 1989. Water pollution and the management of ecosystems: a case study of science and scientist. Reprinted from: Toward a more exact ecology. The 30th symposium of the British Ecological Society. pp: 401–422. Blackwell Scientific Publications. London.
Province of Utrecht, 1975. Kwaliteit oppervlaktewater Utrecht. April 1964 tot en met maart 1975 – Quality of surface water in the Province of Utrecht. April 1964 until March 1975. 105 pp.
Province of Utrecht, 1982. Kwaliteit oppervlaktewater provincie Utrecht 1975–1980. Quality of surface water in the Province of Utrecht 1975–1980. 154 pp.
Province of Utrecht, 1989. Waterkwaliteitsverslag 1980 t/m 1986. Het waterkwaliteitsonderzoek in het oppervlaktewater van de Provincie Utrecht. Deel 1: tekst, deel 2: tabellen. Quality of surface water in the Province of Utrecht 1980–1986. Volume 1: text, volume 2: tables. 66 pp. 106 pp.
Prud'homme Van Reine, W. F. & P. E. Van Der Meulen, 1961. Verslag van het onderzoekje naar het voorkomen van

Chara spec. in het plassengebied van Loosdrecht – *Characeae* in the Loosdrecht lakes. RIN-report. 2 pp.

Public Health Council, 1989. Ecologische normen waterbeheer. Deeladvies III: beschrijving van de parameters – Ecological target values for water management. Part 3. Description of the parameters. Report 89/21. The Hague. 408 pp.

Reijnen, 1989. Invloed van watersport op de natuur – Influence of waterrecreation on nature. RMNO-report No. 37. 107 pp.

Runhaar, J., C. L. G. Groen, R. Van Der Meijden & R. A. M. Stevers, 1987. Een nieuwe indeling in ecologische groepen binnen de Nederlandse flora – A new way of dividing the Dutch flora into ecological plant communities. Gorteria 13: 277–359.

Scheffer, M., 1989. Alternative stable states in eutrophic shallow freshwater systems: a minimal model. Hydrobiol. Bull. 23: 73–83.

Tansley, A. G., 1946. Introduction to plant ecology. Allen & Unwin, London.

Ten Brink, B. J. E. & S. H. Hosper, 1989. Naar toetsbare ecologische doelstellingen voor het waterbeheer: de AMOEBE benadering – Towards verifiable ecological target values for water management: the AMOEBE-approach. H_2O 22: 612–617.

Van Heusden, G. P. H., 1942. Planktongroei in en fosfaat- en nitraatgehalte van de Loosdrechtse Plassen – Growth of plankton and concentration of phosphate and nitrate in de Loosdrecht lakes. Amsterdam Municipal Waterworks. Report 11. 11 pp.

Van Liere, L., 1992. Publications of the workgroup WQL (Water Quality Research Loosdrecht Lakes) 1983 - present. Hydrobiologia 233: 179–186.

Van Liere, L., J. Ebert, W. Kats & J. J. Buyse, 1991. The water quality of Loosdrecht lakes, a review. Mem. Ist. ital. Idrobiol. 48: (in press).

Van Liere, L. & J. H. Janse, 1992. Restoration and resilience to recovery of the Lake Loosdrecht ecosystem in relation to its phosphorus flow. Hydrobiologia 233: 95–104.

Hydrobiologia **233**: 21–38, 1992.
L. Van Liere & R.D. Gulati (eds), Restoration and Recovery of Shallow Eutrophic Lake Ecosystems in The Netherlands.

The hydrology of the Loosdrecht lakes area

G. B. Engelen, B. F. M. Kal, J. J. Buyse & F. G. M. Van Pruissen
Free University, Fac. of Earth Sciences, Dept. of Hydrology, De Boelelaan 1085, 1081 HV Amsterdam, The Netherlands

Key words: lake restoration, hydrological systems, nutrient balance, groundwater modelling, water management

Abstract

The hydrological research of the Loosdrecht Lakes and their surroundings over the period 1982–1990 was based on the integrated systems approach to surface- and ground water, with respect to their quantity and quality.

The description of the water systems includes an evaluation of the structure and interactions of water systems in and around the lakes since the beginning of the Holocene until recent times, focussing progressively from the regional to the local systems and from the lake complex to the individual lakes.

Quantitative research concerned: monthly and annual water balances for the complex of lakes and for individual lakes; numerical two- and semi-three-dimensional groundwater flow models; field experiments on turbulence and diffusion; and experimental laboratory studies on wind-driven currents.

Qualitative research concerned: behaviour of environmental isotopes in groundwater; heavy metals in lake bottom sediments; phosphorus balances of the complex of lakes and of individual lakes for the existing and the simulated conditions; and remote sensing studies for the water quality pattern recognition.

The interaction between the various water systems in the area has been influenced strongly in historic and recent times by human interference on both the local and regional scales of the water resources, *e.g.* through the reclamation of lakes, the extraction of water for public water supply, the regional transfer of water, the agricultural and urban activities etc. Two major remedial measures have been taken thus far: construction of sewer systems, and the supply with dephosphorized water from the Amsterdam-Rhine canal. These measures resulted in a decrease of the external P-load but did not result in a significant decrease in eutrophication as yet.

Scenario's to improve this situation either pertain to tackling the fundamental causes of the eutrophication or to redressing their effects. The first category includes: drastic reductions in groundwater abstraction for public supply in the Gooi catchment area of the lakes and (partial?) flooding of adjacent deep polder areas to reduce the leakage losses from the lakes (both measures aim at restoration of the water balance). The second category involves less drastic water management alternatives for lake restoration: dredging P-enriched bottom sediments; reinforced flushing during summer with dephosphorized Amsterdam-Rhine Canal water; interception and rerouting of polluted inflows from the watershed; or additional P-removal from yet untreated inflows.

Introduction

The hydrological studies in the interdisciplinary workgroup WQL (Water Quality research Loosdrecht lakes) on restoration of the Loosdrecht Lakes have been carried from 1982–1988 (Kal *et al.*, 1984; Engelen & Kal, 1985; Kal, 1986). The aim of the study was to provide an integrated description of the abiotic part of the water systems and its contribution to the functioning of the aquatic ecosystems in the Loosdrecht Lakes (Engelen, 1986).

The results had to serve the following purposes:

General framework. The integrated hydrological studies had to provide a general frame of reference for the other project participants representing other disciplines as well as for the national, regional and provincial authorities which are in charge of planning and management of the lakes area.

Input data for biological and management models. The data on water quantity and quality were to be used as: part of the input of a eutrophication simulation model for the Loosdrecht lakes (PCLOOS, Janse & Aldenberg, 1990); input for the simulation models for scenario's of surface water management (Buyse, 1989); input for the simulation models for scenario's of groundwater management (Van Pruissen, unpublished); input for the development of plans for engineering works in the surface water network, related to different management scenario's.

The Loosdrecht Lakes are connected to their surroundings by natural, semi-natural and artificial flows of surface- and groundwaters, nutrients, *etc.*, which act as external boundary conditions for the development of the ecosystems in the lakes (Engelen & Schot, 1990).

The complexity of the interactions between groundwater and surface water in the area – amplified by intensive human interferences through the water management activities in the area – required a methodological approach in which the interactions of surface waters and groundwater are considered quantitatively and qualitatively and within a wide range of spatial and time scales.

The present hydrological situation in the area is the result of a complex sequence of events of natural and man-made origin. The present distribution of nutrients in the water systems of the area can be understood only when the hydrological history is traced back. One has to start at the beginning of the Holocene period and to follow up closely the increasing interference of man. Over the last centuries and decades man has been changing the flow systems around the lakes by digging peat, draining lakes, extracting groundwater, maintaining fixed surface water levels in polders and transferring water from adjacent rivers and canals to supply the lakes to maintain their water levels.

This paper presents a description of the Loosdrecht lakes hydrology, its history and interactions with other systems. A water- and matter-balance is set up for the whole lake area, as well for its compartments. Since the measures taken did not led to the result expected, additional measures, from a hydrological point of view, will be presented.

The study revealed that the main forcing functions for the response of the ecosystems and the size and distribution of the eutrophication are external factors like hydrological structure of the area and its modifications by water management practices and superposed pollution. The hydrological studies had both purely scientific and applied aspects. Moreover a large number of public agencies at various levels of administration have been involved: national agencies dealing with the environment, the provincial government of Utrecht, the water management board of the lakes, the public water supply company of the city of Amsterdam, *etc.* The water systems approach provided the logical background for the integration of all different aspects.

Methodology

The complicated interactions between natural and human, biotic and abiotic factors in the area require an interdisciplinary, holistic systems approach. The hydrological integration has been achieved by applying the regional hydrological systems concept (Engelen & Jones, 1986). In this concept the hierarchy of flow systems of the surface waters and groundwater and their coupling are emphasized. Moreover the quantity- and quality aspects of both are intimately related and studied as such. The integration of hydrology and ecology takes advantage of the close correlation between hydrological and ecological zoning.

The large number of sub-themes of this hydrological research can be summarised under the following themes.

Genesis, types and distribution of the water systems. Descriptive and semi-quantitative studies of the water systems in the area concerning the development of the major surface water and groundwater flow systems in the area with time after the end of the last glaciation in the Pleistocene period (Engelen & Schot, 1989).

Research on the flow quantities and the flow processes in individual water systems. The surface water flows into, between and within the lakes have been studied by the study of records of pumping stations, the recording of water levels, areal surveys of electrical conductivity and water chemistry of the water bodies, the monitoring of changes in electrical conductivity and chemistry in connecting canals and passages, the application of dyes and other tracers (Teunissen, 1985; Fisser, 1986; Van Der Spek, 1986). The effects of different water management practices as well as of wind on currents and turbulent mixing in the lakes were investigated *in situ* (Buyse, 1986).

Physical scale models of the lakes were used in the laboratory to get an idea about the currents generated by wind and by water management practices and to compare their outcome with field data (Wasser, 1988). The remote sensing study contributed as well to this type of information (Dekker *et al.*, 1991). Water balances for monthly, seasonal and annual periods have been calculated for the lake area as a whole and for individual lakes within the complex of lakes. The reader is referred to Engelen *et al.* (1988) and Buyse (1989) for full details on the calculation methods for the balances of water, chloride and phosphorus and for their results. Chloride has been used as a conservative tracer to check and correct the water balance calculations. On the basis of those water balances and the concentration data from field measurement the phosphorus balances have been calculated. Balances have been calculated for the system of lakes as a whole and for individual lakes within it for the following periods (Engelen *et al.*, 1988; Buyse, 1989):

- 1978–1983: a period with supply of strongly polluted water from the River Vecht;
- 1983–1987: a period with supply of less polluted water from the Amsterdam-Rhine Canal in combination with phosphate removal.

Hettling (1985a) surveyed the isotope situation in the Loosdrecht lakes area to study the interaction between the surface water- and groundwater systems and to determine groundwater ages. After the qualitative identification of the hierarchy of groundwater flow systems in the project zone the systems have been modelled. This was done first roughly with the two-dimensional, finite difference model FLOWNET (Engelen & Jones, 1986; Hettling, 1985b; Kühnel, 1985a) and subsequently with the multi-layer, finite element model MICROFEM (Van Pruissen, 1989). The models were calibrated by means of surface-water-level- and hydraulic head data, water balances of polder units, isotopes and hydrochemical information. The following situations have been modelled: a wet (winter-) situation; a dry (summer-) situation; a situation after dredging the top of the lake bottom sediments; and a situation with a flooded Polder Bethune.

Research on remedial water management measures to combat the eutrophication of the lakes. The effects have been quantified and evaluated of: (1) the installation of sewer systems in the area; (2) the substitution in 1986 of water supply from the polluted River Vecht by water supply from the Amsterdam-Rhine canal and of phosphate removal on the nutrient balances of the lakes; (3) a future drastic reduction in groundwater abstraction in the catchment area of the lakes to restore the original groundwater flow patterns and areas of upward seepage as existed at the beginning of the century; (4) (partial) flooding of adjacent deep polder areas, restoring thus partly the hydrological situation with the Lake Horstermeer and especially the Bethune area, before their reclamation into deep polders; (5) the phosphate removal from the lake system by dredging of nutrient-rich lake sediments. This may result in an increased leakage from the lakes and a shift in the water- and nutrient balances of the lakes. The possible effect of such a measure has been evaluated (Higler *et al.*, 1987); (6) the reinforced flushing of the lakes during summer with dephosphorized water from the Amsterdam-Rhine Canal to progressively deplete the stock of P in the lake complex; (7) the diversion of the nutrient-rich surface water flows into the lake complex from the foothills of the Gooi recharge area directly into the River Vecht. It resulted in recommendations for changes in the water management and adaptation of the surface water network (see below); and (8) the additional P-removal from yet untreated inflows would decrease the P-load on the lakes further.

Apart from these main themes the hydrological studies formed a framework for other research, such as investigations to patterns of water quality in the surface and groundwater systems. These were established for different seasons and water management conditions (Engelen *et al.*, 1988; Buyse, 1989). The quality aspects were focussed on the major inorganic components and on the nutrients phosphorus and nitrogen. Some work has been done on heavy metals in the lake sediments (Kühnel, 1985b). The distribution of the heavy metals Fe, Zn, Cd, Pb, Cu and Zn was measured in 1984 in 17 core-samples of river- and lake sediments. The mineralogy of three samples was studied by differential thermal analysis and X-ray diffraction methods. In that period the artificial summer supply of water to the lakes was still from the polluted River Vecht via three intake sluices. Many different diffuse sources of heavy metals occur in the lake area, the main point sources at that time were the sluices with Vecht water intake.

Description, functioning and interactions of the water systems

Development phases of the water systems in the Loosdrecht lakes area

The four major phases in the development of the landscape and of the water systems in and along the western border of the ice-pushed hills of 't Gooi/Utrecht since the Pleistocene period are given in Fig. 1 (Engelen & Schot, 1989). In the first phase, at the end of the Pleistocene epoch

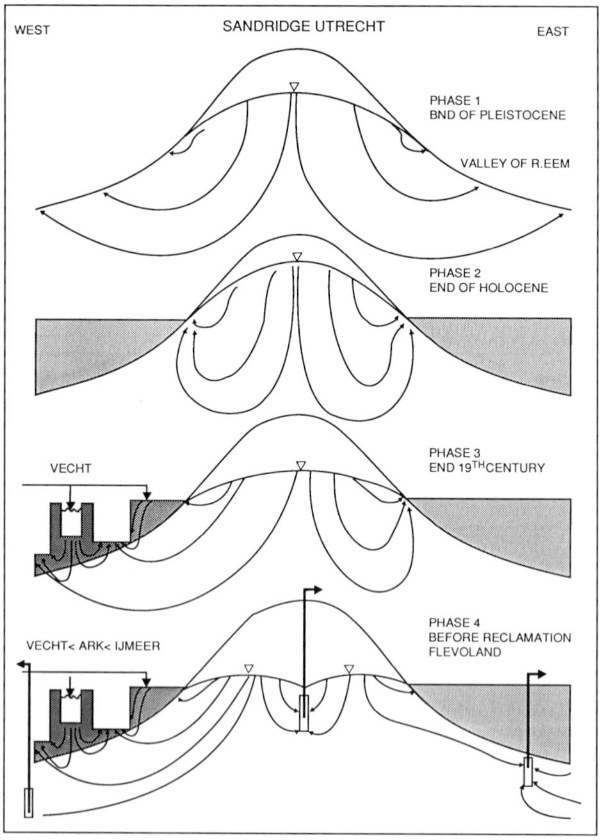

Fig. 1. Four phases in the development of the groundwater flow systems in the Gooi/Utrecht ice-pushed hills and the Loosdrecht Lakes area: 1. End Pleistocene 2. End Holocene 3. End 19th century 4. Recent, before the construction of the Flevo polders to the north. (After Engelen & Schot, 1989).

with low sea level conditions, the infiltration in the pervious hills of ice-pushed older fluviatile sands and gravels generated a large groundwater flow system with flow branches to the adjacent valleys. In the second phase, at the end of the Holocene period and before the historic Middle Ages, a semi-permeable cover of Holocene sands, clays and peat had filled in the low areas on both sides of the ice-pushed hills. The outlet zones of the Gooi/Utrecht Hills groundwater flow system had retracted to the base of the slope at the transition between the Pleistocene and Holocene landscapes. Groundwater emerged in upward seepage zones and was discharged by small rivers across the surrounding peat belt to larger meandering distributary branches of the River Rhine such as the River Vecht. In the third phase (Fig. 2), at the end of the 19th century, lakes and marshes had been formed by peat extraction for fuel since the Middle Ages and some (Horstermeer and Bethune area) had been reclaimed as deep polders. The Gooi/Utrecht Hills groundwater flow system extended again with additional flow branches towards the deep polders. Moreover the lakes had become the recharge areas for new groundwater flow systems by leakage towards the deep polders.

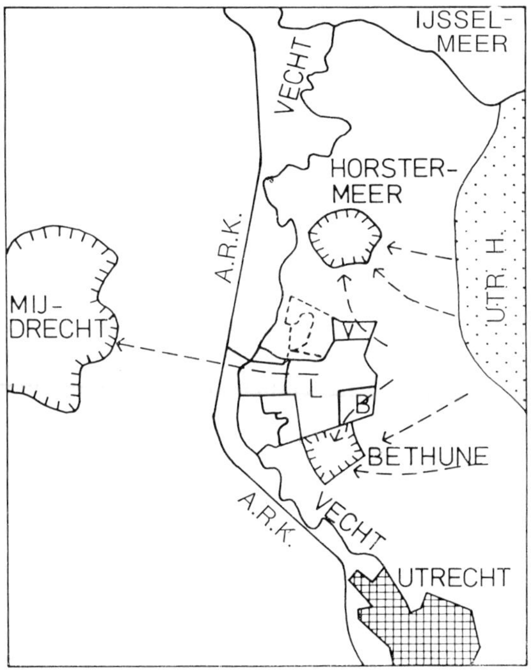

Fig. 2. Situation of the Loosdrecht Lakes with respect to the most important hydrological sybsystems in the region (After Engelen, 1986).

In the fourth phase, especially since the middle of the 20th century, the natural outflow of the Gooi/Utrecht Hills groundwater flow system has decreased considerably because of groundwater extraction and has been polluted by urbanisation and agriculture (Witmer, 1989; Schot, 1991). In the summer the intake of strongly polluted water from the river Vecht was necessary to maintain the water level of the lakes. Thus the losses due to evapotranspiration decreased groundwater inflow and the heavy leakage losses to the nearby deep polders were compensated. The eutrophication of the lakes was the result. Since 1984 the polluted supply from the Vecht has been replaced by supply from the Amsterdam-Rhine canal with P-removal before entering the Loosdrecht Lakes complex. Figs. 3–6 show in flow charts the structure of the interaction of the various surface- and groundwater flow systems of the Loosdrecht Lakes area from the 15th century to the present.

Surface water flow systems

The relative location, form, depth and size of the individual lakes in the lake complex and their hydraulic connections play a very important role in the physico-chemical and biological behaviour of the lakes (Fig. 7). All lakes have rather steady leakage losses via a 'cascade' of coupled surface water and groundwater systems to the adjacent lower polders throughout the year (Fig. 8). Roughly during the winter half year, however, there is a water surplus due to decreased evapotranspiration and increased seepage from the Gooi/Utrecht Hills catchment. The excess water is pumped out to the River Vecht. The simulated flow pattern during this period is given in Fig. 9.

In the summer half year the lake complex has a more or less negative water balance depending upon the meteorological conditions, requiring supply from outside to maintain the lake levels.

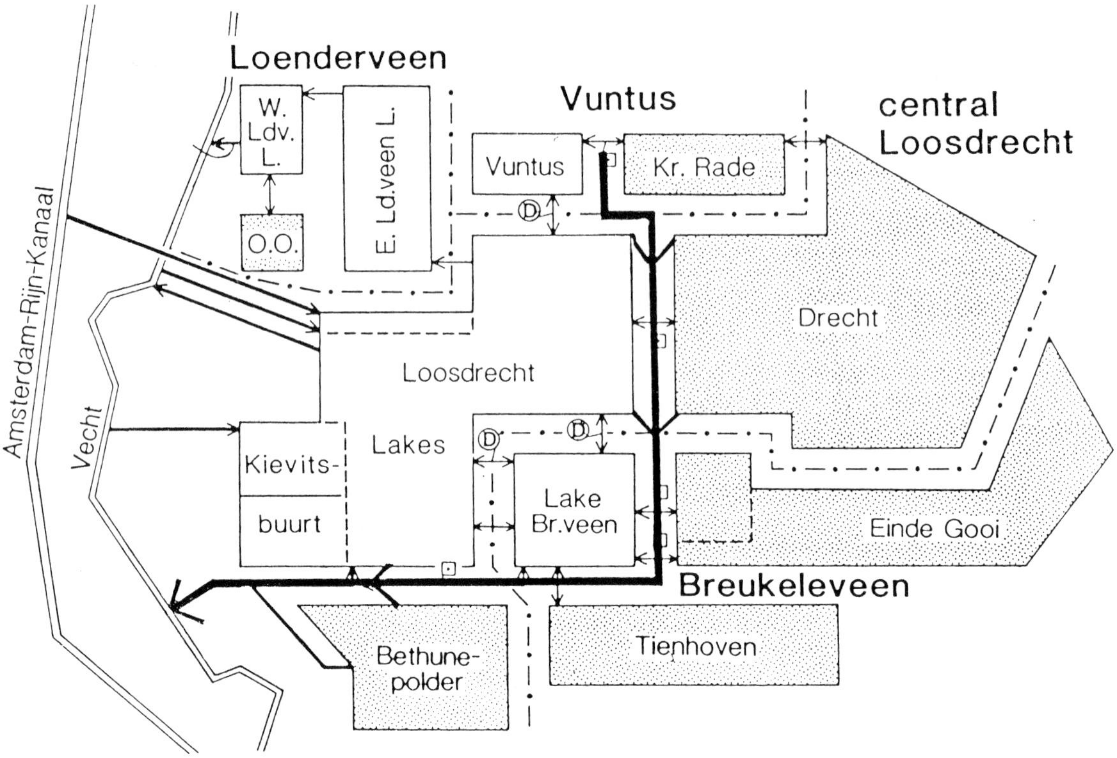

Fig. 7. Layout of the Loosdrecht Lakes complex (After Engelen *et al.*, 1988).

ent items of the water balance in the period 1978–1987. The external supply of water from the Amsterdam-Rhine Canal varies on average between 20% of the total input on an annual basis to 90% during the summer season. The annual input to the Loosdrecht Lakes from the catchment area in the adjacent hills amounts to 40%. The output of the system of lakes consists of 40% through forced discharge by pumping in winter, 25% through infiltration losses to the groundwater flow systems and 20% is return flow to the upstream part of the catchment during the dry summer periods.

Phosphorus balances

The results from the balance calculations for total phosphorus indicate that the external P-load on the whole lake system amounts to 0.8 mg m^{-2} d^{-1} under the present (1984–1987) conditions. The contributions from various sources calculated for 1984–1987 are: 30% from the catchment in the hills, 30% from the intake of water from Polder Bethune, 10% supplied by Amsterdam-Rhine Canal water; 30% from leakage inflows through sluices, precipitation, water-based tourism and unsewered housing.

The annual phosphorus balance, though inaccurate, indicates that a netto flux of P of about 50% of the total P-input disappears through downward seepage, or is adsorbed by the lake sediments. However, monthly balances indicate that there is phosphorus release from this stock into the supernatant water during the summer months in the order of 0.2 to 0.4 mg P m^{-2} d^{-1}. Fig. 13 shows the contribution to the P-load by the supply water before and after the start of P-removal. In 1983 the summer input of P from the supply of Vechtwater is still dominant in the P-balance. In 1986 the supply component has been reduced greatly by the change to Amsterdam-Rhine-Canal water. The inputs to the Loosdrecht lakes are notably high in the winter half year due

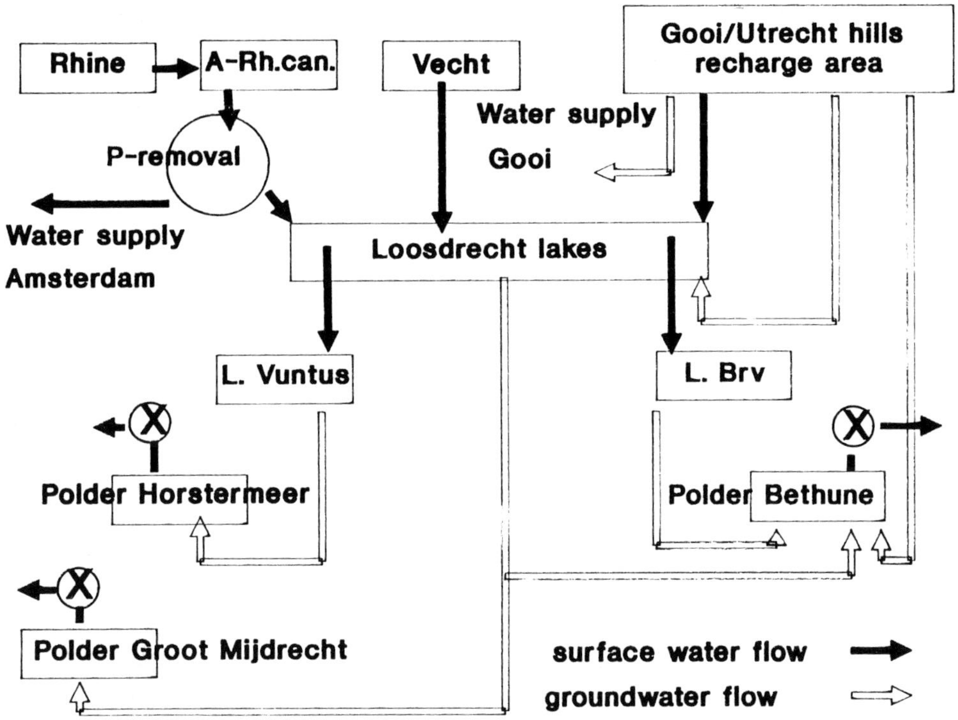

Fig. 8. Cascades of surface water- and groundwater flow systems in the Loosdrecht Lakes area due to water supply to compensate for leakage losses.

to polluted surplus flow from the Gooi/Utrecht Hills catchment. During the summer half-year the upwelling P-enriched groundwater in the Bethune Polder is largely used for public water supply by Amsterdam Municipal Waterworks. In the winter half-year part is discharged into the Lake Loosdrecht.

Heavy metals in the lake sediments

The occurrence of heavy metals in the lake sediments is of two types: (a) comparatively high but localized concentrations of heavy metals bound to rapid settlement of suspended sediments of the River Vecht, dragged into the lakes by the previous water intake from the River Vecht during dry spells in summer. The concentrations show fairly steep decreases in the channels downstream of the intake sluices (the point sources) over distances of 1 to 2 kilometers; (b) more evenly distributed lower concentrations over the entire lake area due to diffuse sources and from the Vecht-intake points.

The Vecht-water heavy metals are absorbed chiefly on ferri-hydroxydes and expandable clays. Therefore the heavy metals entering the lakes through intake points did not spread farther than 2 km downstream of the intake points due to rapid settling of the suspended river sediments and an increase in pH due to the mixing of Vecht water and lake water. The latter factor was limiting the mobility of most metals.

Dispersed, suspended Fe reached the lakes in considerable quantities but accumulated rather rapidly in the sediment due to adsorption of phosphates depending upon the turbulence conditions. In non-turbulent environments like the Kievitsbuurt area of the lake the highest Fe- and P-concentrations were measured in the sediments.

The behaviour of Cd differed slightly from that of the other heavy metals. A significant fraction of the Cd in the Vechtwater remained in solution at first as uncomplexed bivalent Cd and precipi-

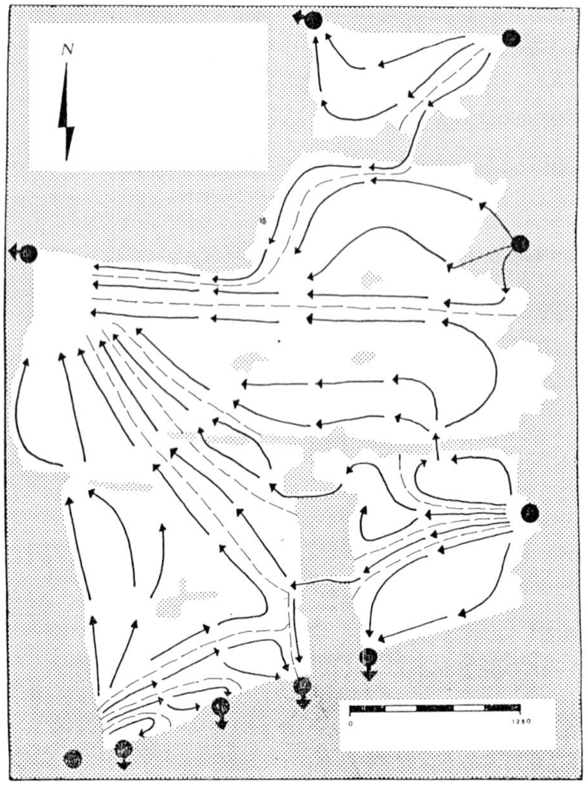

Fig. 9. Simulated flow pattern in the Loosdrecht Lakes complex with inflow from the eastern catchment area, leakage losses and pumped discharge for the winter situation without wind, in a physical model at the scale 1:10 000. (After Wasser, 1988).

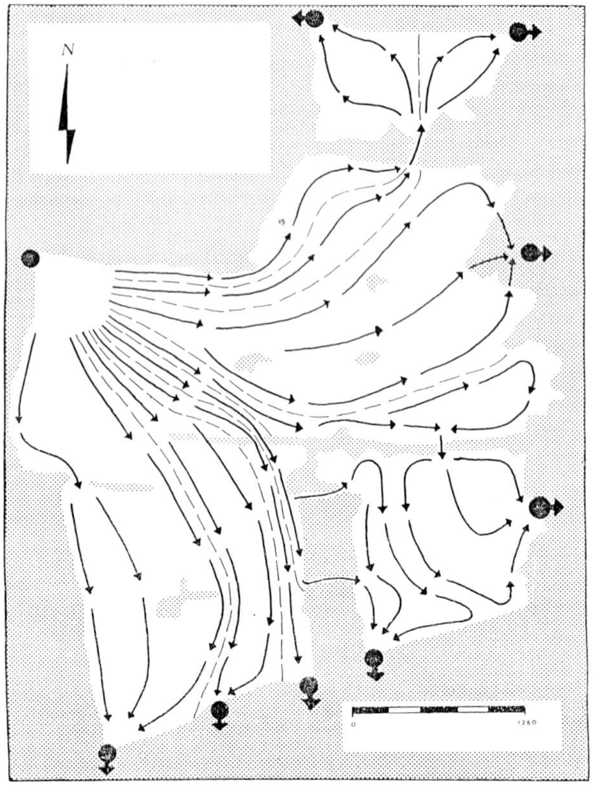

Fig. 10. Simulated flow pattern in the Loosdrecht Lakes complex with water supply from the Amsterdam-Rhine canal, leakage losses and reversed flow to the eastern catchment area for the summer situation without wind, in a physical model at the scale 1:10 000 . (After Wasser, 1988).

tated farther on in the lakes, mostly as carbonate. The maximum value of about 70 ppm Cd (in dry weight minus loss-on-ignition sample) in the Kievits Area at the SW intake point is probably due to the cumulative effects of intake at that point of both Vecht water and of water from the agriculturally-used Polder Bethune. In the period 1979–1984 Cd has been the metal whose concentration in the sediment of the 'open' compartments of the Lake Loosdrecht has most strongly increased, viz. from 1.6 to 6.3 ppm. Fe follows in second place in this respect. The concentrations of the other heavy metals did not change significantly over that period.

No data are available on the situation of heavy metals in the sediment after the shift from Vecht-water supply to dephosphatized Amsterdam-Rhine Canal water. The point sources of the Vecht sluices have strongly decreased in strength since only leakage of the sluices and water exchange due to ships passing the locks remain here as pollutant fluxes.

It may be assumed that the bulk of heavy metal load in the water supplied from the Amsterdam-Rhine canal will be trapped nowadays in the process of phosphate removal by flocculation, coprecipitation and settling upon addition of iron chloride. Consequently the total input of heavy metals to the lakes has probably decreased strongly.

Scenario's for water quality restoration

Remedial measures until 1984

The construction of sewer systems in the catchment area of the lakes and discharge of the efflu-

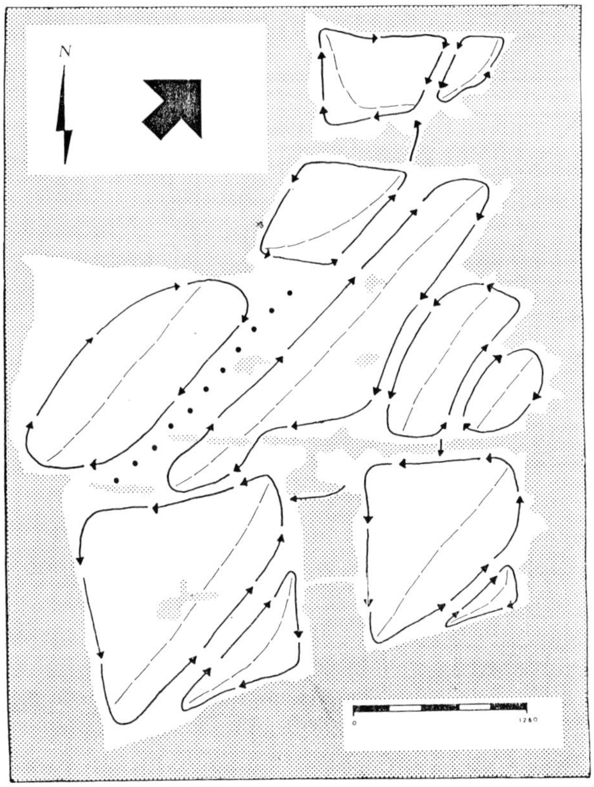

Fig. 11. Simulated flow pattern in the Loosdrecht Lakes complex, induced by a southwestern wind in a physical model at the scale 1:10 000 in a wind tunnel. (After Wasser, 1988).

ent outside the catchment. The sewage load from the permanent population in the villages around the lakes has been removed largely by the installation of sewer systems, sewage treatment and discharge of the effluent outside the catchment of the lakes in the period 1970–1986. The sewage load from heavy recreational use of the lakes during summer has been reduced by the installation of sanitary facilities. As a result of these measures the P-input from sewage decreased from 2360 kg P y^{-1} in 1980 to 613 kg P y^{-1} in 1987. Currently the P-loading due to remaining unsewered discharges is about 13% of the total external P-load in summer and 6% in winter.

Since 1984 additional water supply to the lakes has changed from the intake through sluices of highly polluted Vecht water to supply by pipeline from the Amsterdam-Rhine canal via a P-removal facility of Municipal Waterworks of the city of Amsterdam. The annual external P-loading of the lakes has decreased due to these two measures considerably: from 1.2 to 0.4 g P m^{-2} y^{-1}. However, those measures have not resulted as yet in a visible improvement of the water quality of the lakes.

Optional restoration measures, removing the original cause of eutrophication

The results of the balance studies indicated which additional measures might be taken for a further reduction of the P-load. The first category of radical solutions aims at the removal of the original cause of the eutrophication problem: the negative water balance of the lakes in the summer half year which requires an additional supply to maintain the lake levels. This could be achieved by: (1) restoring the original groundwater influx; and (2) decreasing/terminating the leakage losses from the lakes by flooding anew (parts of) the empoldered former Lakes Horstermeer and especially the nearby Polder Bethune.

Drastic reductions in groundwater abstraction in the Gooi catchment area of the lakes (Engelen *et al.*, 1989; Engelen & Schot, 1989; Witmer, 1989; Wassen, 1990; Schot, 1991). Reductions will lead to increased seepage input of good quality and thus less input of water from Amsterdam-Rhine Canal. The provincial government of North Holland has initiated recently a policy to reduce the groundwater abstraction in the northern part of the Gooi/Utrecht Hills area. Such a policy is still lacking in the province of Utrecht.

(Partial) flooding of deep polders (former lakes or marshy areas) The possible hydrological effects of restoring the original hydrological situation by flooding Polder Bethune were considered as an option in the groundwater modelling. The upward seeping flow branch of the regional groundwater flow system along the eastern border and within the central part of the Lake Loosdrecht would be restored to a large degree. The need for water supply in the summer half year to compensate for the leakage losses would be reduced completely, because under wet conditions

Construction of additional P-removal facilities

Instead of directing the polluted influx from the hills of the catchment in winter directly to the River Vecht, this polluted water could be collected and processed for P-removal and then discharged into the Loosdrecht Lakes. In addition, the input of 30% of the external P-load from the intake of water from the Polder Bethune could also be largely eliminated if an additional treatment plant for phosphorus removal were built or if these waters were routed through the existing treatment plant before discharge into the lake complex. It would require additional technical infrastructure works and adaptation of the water extraction schemes of the Amsterdam Municipal Waterworks, which uses part of the excess seepage waters of the Polder Bethune.

The last major balance item of 30% of the external P-load consists of a combination of loads from various sources of a distributed nature. These are very difficult to tackle by systematic collection- or treatment schemes. The 10% load from the supply water could be reduced further by technical improvements in removing phosphorus from the supply water in the treatment plant. However, the very large quantity of needed supply water produces – even at the current, already fairly high level of phosphate removal in the treatment plant – still about 10% of the total present load.

Discussion

The scenario's have different social and political feasibilities, financial and technical boundary conditions, time spans to become effective, and net effects on the solution of the eutrophication problem. The proposed rating is rather approximate because systematic and complete feasibility studies of the scenario's which take into account net effect on P load, technical/financial constraints and social/political feasibility have not been made as yet: (1) additional P-removal scenario; (2) deviation scenario; (3) flushing scenario; (4) dredging scenario; (5) Bethune Lake scenario; (6) Combination scenario (Fig. 14).

The full additional P-removal scenario would remove a considerable portion (60%) of the remaining external P-load and constitute no additional P-burden on the receiving waters as the deviation scenario would do. The flushing scenario requires major adaptations in the water management infra-structure (pumping capacity and relocation of inlets and outlets). It would tackle the internal P-load of the lakes by depleting the stock of P in the bottom sediments and flushing out biomass. During a transitional phase it might be the necessary trigger to overcome a for the biological systems critical boundary value of *e.g.* light penetration, necessary for the restoration of meso- to oligotrophic conditions and submerged vegetations. It would loose its significance once the lake water quality should be restored. The partial dredging scenario would affect the internal P-load by removal of part of the P-enriched bottom sediments and improve visibility by a decrease in suspended sediment concentrations. Costs and disposal of the dredged sediments form major obstacles. Dredging of the lakes would be expensive and technically difficult by stirring the mud due to the very low cohesion of the upper layer of the bottom sediments. Proper disposal or use of the huge quantities of dredged material would constitute another practical and environmental problem to be solved. However from a hydrological point of view no major objections against dredging should be raised. The Bethune Lake scenario would be a very fundamental and thorough solution, but difficult to realize. It would require substantial physical planning efforts, evacuation and relocation of the sparse agricultural population of the polder and the cooperation of the Amsterdam Municipal Waterworks to relinquish its rights to the excess seepage water of the present Polder Bethune. The advantages could be great however in terms of a fundamental solution to the eutrophication problem, improved water supply to the nature protection areas along the Gooi/Utrecht Hills and the extension of part of the crowded recreational facilities to the new lake.

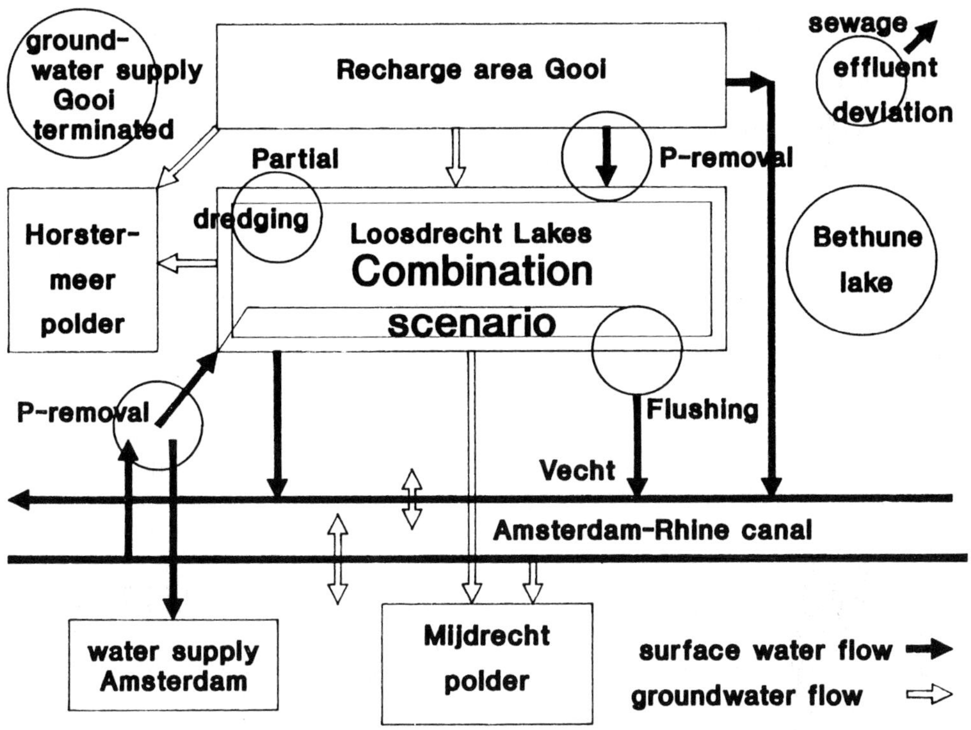

Fig. 14. Water flow and remedial measures in the Loosdrecht Lakes area in a comprehensive combination scenario.

Finally the combination scenario contains all scenario's as an overview and could be the end result of successive steps in the process of restoration of the Loosdrecht lakes area by measures affecting the P-load. P-removal as a single approach still remains a far cry from a return to a complete restoration of the original ecological situation because of the considerable remaining input of other components to the lake complex in the supply water, *e.g.* the continuing overloading of the lake complex with other nutrients like nitrogen, dissolved organic compounds, etc.

References

Buyse, J. J., 1986. Turbulente diffusie in de Loosdrechtse Plassen (Turbulent diffusion in Loosdrecht lakes, in Dutch). WQL-report 1986-12. Free University, Institute of Earth Sciences, Amsterdam, 42 pp.

Buyse, J. J., 1989. Water- en stofbalansmodellen voor de Loosdrechtse Plassen (Water- and matter balances in Loosdrecht lakes, in Dutch) WQL-report 1988-7. Free University, Institute of Earth Sciences, Amsterdam, 71 pp.

Dekker, A. G., T. J. Malthus, S. W. M. Peters & E. Seyhan, 1991. The remote sensing Loosdrecht Lakes project. BCRS Report no. 90-29. Free University, Institute of Earth Sciences, Amsterdam, 105 pp.

Engelen, G. B., 1986. Interaction of hydrological systems and eutrophication of the Loosdrecht Lakes. Hydrobiol. Bull. 20: 17–25.

Engelen, G. B. & G. P. Jones (eds), 1986. Developments in the analysis of groundwater flow systems. IAHS Publ. no. 163, 356 pp.

Engelen, G. B. & B. F. M. Kal, 1985. Hydrologisch onderzoek in WOL kader van het Loosdrechtse Plassengebied in verband met beleid en beheer voor waterkwaliteitsherstel (Hydrological research in the Loosdrecht lake area in relation to policy and watermanagement for lake restoration, in Dutch). WQL-report 1985-1. Free University, Institute of Earth Sciences, Amsterdam, 37 pp.

Engelen, G. B., B. F. M. Kal, J. J. Buyse, A. Wakjira & S. O. Los, 1988. Waterhuishoudkundige beleids- en beheersmogelijkheden in verband met waterkwaliteitsherstel van het Loosdrechtse Plassen-system (Water management and policy in relation to Loosdrecht lakes' restoration, in Dutch). WQL-report 1988-1. Free University, Institute of Earth Sciences, Amsterdam, 43 pp.

Engelen, G. B. & P. P. Schot, 1989. Analyse van de watersystemen in het Goois/Utrechts stuwwallengebied, de Vechtstreek en het Eemdal: probleemsignalering (Analysis of watersystems in the Gooi/Utrecht sandridge area; the Vecht area and the Valley of the Eem, defining the problem, in Dutch). CHO/TNO Rapporten en Nota's 22: 21–31.

Fisser, C. H. B., 1986. Hydrogeografisch vooronderzoek van de Vuntus (Preliminary research of Lake Vuntus. WQL-report 1986-5. Free University, Institute of Earth Sciences, Amsterdam, 89 pp.

Gons, H. J. & R. Van Keulen, 1989. De relatie tussen doorzicht en slib in the Loosdrechtse Plassen in verband met de zwemwaternorm: afsluitend onderzoek (Relation between water clarity and suspended matter in relation to swimmingwater standards: final report, in Dutch) WQL-report 1989-4. Limnological Institute, Nieuwersluis. 35 pp.

Hettling, H. K., 1985a. Isotopic groundwater study of the Loosdrecht Lakes area (the Netherlands). WQL-report 1985-4. Free University, Institute of Earth Sciences, Amsterdam, 55 pp.

Hettling, H. K., 1985b. Een hydrogeologische inventarisatie van het Loosdrechtse Plassengebied (A hydrogeological inventarisation of the Loosdrecht lakes area, in Dutch). WQL-report 1985-2, Free University, Institute of Earth Sciences, Amsterdam, 116 pp.

Higler, L. W. G., S. Parma & L. Van Liere (eds), 1987. Baggeren in de Loosdrechtse Plassen (Dredging in Loosdrecht lakes, in Dutch). WQL-report 1987-5, National Inst. of Nature Protection, Leersum. 18 pp.

Janse, J. H. & T. Aldenberg, 1990. Modelling phosphorus fluxes in the hypertrophic Loosdrecht lakes. Hydrobiol. Bull. 24: 69–89.

Kal, B. F. M., G. B. Engelen & Th. E. Cappenberg, 1984. The Loosdrecht lakes restoration project: hydrology and physico-chemical characteristics of the lakes. Verh. int. Ver. Limnol. 22: 835–841.

Kal, B. F. M., 1986. Monthly mass balances for compartments of the Loosdrecht Lakes system: approach and preliminary results. Hydrobiol. Bull. 20: 27–39.

Kühnel, V., 1985a. Toepassing van het grondwaterstromingsmodel FLOSA binnen het Loosdrechtse Plassen restauratie-project (The application of the groundwaterflowmodel FLOSA in the Loosdrecht lakes restoration project, in Dutch). WQL-report 1985-3. Free University, Institute of Earth Sciences, Amsterdam, 66 pp.

Kühnel, V., 1985b. Zware metalen in het Loosdrechtse Plassen-sediment (Heavy metals in Loosdrecht lakes' sediments, in Dutch). WQL-report 1985-6. Free University, Institute of Earth Sciences, Amsterdam, 36 pp.

Schot, P., 1991. Solute transport by groundwater flow to wetland ecosystems. Ph.D. thesis. State University at Utrecht, 134 p.

Van Pruissen, F. G. M., 1989. Modelleren met 'MICRO-FEM'. H_2O 22: 618–623.

Van Der Spek, T. R., 1986. Onderzoek naar de mogelijkheid van het simuleren van oppervlaktewater-stromingen in het Loosdrechtse Plassen-gebied met behulp van een Hele-Shaw model (Research to the possibilities of the simulation of surface-currents in the Loosdrecht lakes area by means of a Hele-Shaw model, in Dutch). WQL-report 1986-4. Free University, Institute of Earth Sciences, Amsterdam, 61 pp.

Teunissen, A. C., 1985. De Breukeleveense Plas. Een hydrogeografisch vooronderzoek (Lake Breukeleveen. A hydrogeographic study, in Dutch). WQL-report 1985-6. Free University, Institute of Earth Sciences, Amsterdam, 103 pp.

Wassen, M. J., 1990. Water flow as a major landscape ecological factor in fen development. Ph.D. thesis, State University at Utrecht, 199 p.

Wasser, H. J., 1988. Fysisch-model onderzoek naar door wind aangedreven stroming in het Loosdrechtse Plassen gebied (Physical model research to wind-driven currents in Loosdrecht lakes, in Dutch). WQL-report 1986-6. Free University, Institute of Earth Sciences, Amsterdam, 27 pp.

Witmer, M. C. H., 1989. Integral water management at regional level. An environmental study of the Gooi and the Vechtstreek. Ph.D. thesis, State University at Utrecht, 172 pp.

Hydrobiologia **233**: 39–50, 1992.
L. Van Liere & R.D. Gulati (eds), *Restoration and Recovery of Shallow Eutrophic Lake Ecosystems in The Netherlands.*

Phosphorus in the sediment of the Loosdrecht lakes and its implications for lake restoration perspectives

Peer Keizer & Anja J. C. Sinke
Limnological Institute, Rijksstraatweg 6, 3631 AC Nieuwersluis, The Netherlands

Key words: phosphate, iron hydroxide, dredging, downward seepage, mineralization, eutrophication

Abstract

In 1984 the external phosphorus load of the shallow eutrophic Loosdrecht lakes was reduced from 3.3 to 1.0 mg m^{-2} d^{-1}. The effect of phosphorus release from the sediment on lake restoration was investigated. Diffusive release under aerobic conditions (20 °C) decreased from 1 mg m^{-2} d^{-1} in 1984 to 0.3 mg m^{-2} d^{-1} in 1990. The generation of inorganic phosphorus due to mineralization during summer equals 3 mg m^{-2} d^{-1}, which is much higher than the measured rate of diffusive release. Despite that, the phosphorus release is hardly stimulated by anaerobic conditions, which indicates that only a small amount of phosphorus is adsorbed by ferric iron in the top sediment layer. This apparent discrepancy is probably caused by the uptake of inorganic phosphorus uptake during resuspension and the loss of inorganic phosphorus with downward seepage.

The estimated removal of phosphorus due to downward seepage of 0.8 mg m^{-2} d^{-1} agrees well with the average phosphorus retention in the lake. This indicates that sediment burial and diagenesis are unimportant mechanisms for withdrawing phosphorus from the nutrient cycle.

Between 1982 and 1991 the total phosphorus content of the upper 2 cm of the sediment decreased from 0.94 to 0.60 g kg^{-1} DW. At present, about 20% of total phosphorus in this layer is potentially bioavailable, but largely incorporated in easily degradable organic matter. This pool is much smaller in deeper layers. Based on the estimated and measured rates and pool sizes, the annual average phosphorus cycle in the lakes was modelled to evaluate the effects of various restoration measures. The main predictions of the model are: 1) further reduction of the external load may cause a gradual decrease of the total phosphorus concentration in the lake water; 2) dredging and iron addition, without reduction of the external load, may give a rapid improvement followed by a slow return to the present situation; and 3) reduction of the external load, combined with a cut off of downward seepage will not improve the water quality.

Introduction

Most surface waters in the Netherlands are eutrophic or even hypereutrophic, having total phosphorus concentrations exceeding 0.15 g m^{-3} (Lijklema *et al.*, 1988). During the last decade, management measures to improve water quality have been focussed mainly on a reduction of the external phosphorus load. The response of a lake to a reduction of the phosphorus load depends on the recycling of phosphorus from the sediment (Marsden, 1989; Sas, 1989). The phosphorus release from the sediment, i.e. internal loading, is identified as a major drawback for lake recovery (Ryding, 1985; Cullen & Forsberg, 1988).

The diffusive release of phosphorus from the sediment, measured in columns under standardized conditions, has been thoroughly investigated (Holdren & Armstrong, 1980; Boers & van Hese, 1988). Release rates are stimulated by high temperature (Kamp-Nielsen, 1974; Sinke & Cappenberg, 1988) and a low redox potential (Mortimer, 1941; Tessenow, 1972; Redshaw *et al.*, 1990). In shallow lakes resuspension of the upper layer of the sediment can be especially important for internal loading (Ryding & Forsberg, 1977; Søndergaard *et al.*, in press). To predict the release rates of phosphorus, several workers used the total phosphorus content of the sediment (Nürnberg, 1988; Sas, 1989). However, it is likely that the remobilization of phosphorus from the sediment is determined by its speciation. Especially the loosely adsorbed fraction (Søndergaard, 1989) and the iron bound fraction (Hosomi *et al.*, 1982; Nürnberg, 1988; Shaw & Prepas, 1990) are subject to mobilization. Boström *et al.* (1985) showed that organic bound phosphorus was the most dynamic fraction in the sediment of shallow eutrophic lake Vallentunasjön.

In this paper we evaluate the role of the sediment in the phosphorus cycle of a peaty eutrophic system. Extraction procedures are introduced to estimate the potentially bioavailable pools of organic and inorganic phosphorus in the sediment. Results are used in a simple descriptive model for evaluating the impact of additional restoration measures. The different scenarios presented concern: a) further reduction of the external phosphorus load; b) sediment removal; c) Fe(III) addition to the sediment; and d) reduction of downward seepage by changing the hydrology of the lake area.

Lake description

The Loosdrecht lakes are shallow, interconnected lakes which are partly surrounded by polders. They were created by peat excavation during the past centuries (Van Liere, 1986). Because of continuous wind-induced turbulence (Gons *et al.*, 1986), the surface sediment remains aerobic throughout the year. General lake characteristics are presented in Table 1.

The sediment has an extremely high organic matter content, exceeding 50% on a dry weight basis (Boers *et al.*, 1984). The lakes receive water with a relatively high alkalinity from the surrounding area and the sediment contains 3–6% $CaCO_3$ despite its peaty origin. The pH of the lake water varies around 8.4.

The hydrological setting of the lakes induces an average downward seepage of 1.5 mm d^{-1} (Engelen *et al.*, 1992). Therefore, approximately 6.10^6 m^3 of supplemental water is needed during summer to maintain the water table at a fixed level. The high phosphorus concentration of this water has largely contributed to a serious deterioration in the water quality during the past decades (Van Liere *et al.*, 1989). Since 1984 the supplemental water has been dephosphorized. This measure reduced the external phosphorus

Table 1. General characteristics of the Loosdrecht lakes and data on the composition of the water column and the sediment.

Lake area	14.5 km^2
Average depth	1.8 m
Water inflow and precipitation	2.3 m^3 m^{-2} y^{-1}
Water outflow (surface)	1.3 m^3 m^{-2} y^{-1}
Water outflow (seepage)	0.65 m^3 m^{-2} y^{-1}
Average external P load	1.0 mg m^{-2} d^{-1}
Average P discharge	0.4 mg m^{-2} d^{-1}
LAKE WATER (1990) ($N = 13$)	
Total phosphorus	75 (±22) mg m^{-3}
Soluble reactive phosphorus	2 (±3) mg m^{-3}
Chlorophyll *a*	93 (±18) mg m^{-3}
pH	8.4 (±0.3)
LAKE SEDIMENT 0–10 cm (1989) ($N = 8$)	
Dry weight	9.0 (±1.0) %
Organic carbon content	53.6 (±3.0) %
Iron (total)	20.3 (±2.0) g kg^{-1}
Iron (acid extractable)	11.3 (±1.9) g kg^{-1}
Phosphorus total	0.67 (±0.08) g kg^{-1}
Phosphorus (acid extractable)	0.15 (±0.02) g kg^{-1}
Cacium (total)	40.2 (±3.4) g kg^{-1}
$CaCO_3$	46.1 (±13.8) g kg^{-1}

load from 3.3 to 1 mg m^{-2} d^{-1}. However, the rate of decrease in the phosphorus concentration of the lake of ca. 12 mg m^{-3} y^{-1} is less pronounced and amounts to <5% per year (Van Liere *et al.*, 1990). The lake is still highly eutrophic and in summer the chlorophyll *a* level often exceeds 100 mg m^{-3}. Because the average Sechhi depth of the Loosdrecht lakes is still limited to 0.3 m, application of additional restoration measures is considered desirable.

Materials and methods

Phosphorus release and uptake experiments

Sediment cores were sampled at the centre of the Loosdrecht lakes with a stainless steel sampler having a perspex inner tube of 0.40 m length and a diameter of 0.05 m. After sampling, intact sediment cores were transferred immediately to the laboratory. The overlying lake water was siphoned off and replaced with artificial lake water (ALW = 1.0 mM Ca^{2+}, 2 mM Na^{+}, 0.2 mM Mg^{2+}, 0.2 mM SO_4^{2-}, 2.4 mM HCO_3^- and 1.6 mM Cl^-; pH 8.3). The water level was made up to 15 cm and flushed with air for the aerobic incubations, or with a mixture of 99.7% N_2 and 0.3% CO_2 for the anaerobic incubations. The CO_2 was added to maintain the pH in the overlying water at 8.0 ± 0.3.

Aerobic phosphorus release was determined monthly in cores sampled in the period 1983–1991 according to Boers & van Hese (1988). The overlying water was replaced continuously with ALW and filtered over 0.45 μm membrane filters and acidified to pH = 1 before analyses. Anaerobic release of phosphorus was measured in 1990. To prevent contamination with oxygen, sampling was performed through needles in the rubber stoppers of the columns, using a peristaltic pump. The samples were acidified before exposure to the air in order to prevent oxidation of dissolved ferrous iron and subsequent precipitation of iron hydroxyphosphate.

The uptake of phosphorus by the sediment was measured under aerobic and anaerobic conditions. 10 ml of 1 mM NaH_2PO_4 was added to the overlying water (final volume 290 ml) of the aerobic cores at time zero and once again seven days later. The anaerobic cores received only one addition after 20 days incubation. One week later, the remaining overlying water was removed as much as possible and the water level made up to a height of 15 cm with ALW. Thereafter, the water was sampled by replacing half of the overlying water with fresh ALW once or twice a week (Keizer *et al.*, 1991).

The effect of iron addition on the phosphorus uptake was measured in anaerobic cores to which 10 ml of 0.01 M freshly-prepared $Fe(OH)_3$ (2.8 g Fe m^{-2}) was added prior to the phosphorus addition. After settling, the precipitate was gently mixed with the upper 5 mm of the sediment. The $Fe(OH)_3$ suspension was prepared by titrating a known volume of 0.2 M $FeCl_3$ in 2 h to pH 7 with 0.6 M NaOH, and diluted 10 times with ALW just before use.

Sediment analysis and phosphorus fractionation

Total phosphorus and iron contents of the top layer (0–2 cm) of the lake sediment were determined in cores sampled at four different locations in 1982, 1985 and 1991. After determining dry weight (24 h at 105 °C) and loss on ignition (4 h at 550 °C), the remaining ash was digested for 24 h with a 3:1 mixture of concentrated HCl and HNO_3 at 150 °C. After filtration (0.45 μm) phosphorus and iron in the samples were analyzed.

Total inorganic phosphorus in the sediment was determined by a 24 h extraction with 0.24 N H_2SO_4 (1% w/w) of slices of the upper 10 cm of the sediment. After that, the slurry was filtered (0.45 μm) and reactive phosphorus was measured in the filtrate. In mixtures of sediment and iron hydroxyphosphate, this extraction method gave a >95% recovery of the added iron and phosphorus.

To distinguish between adsorbed phosphorus and phosphorus included in minerals, 0.15% w/w sediment suspension in ALW were incubated for two months in an anaerobic glove box (Coy

temperatures and temperature dependent oxygen consumption rates reported by Sweerts (1990). With these input variables the missing rates for P_{seep}, P_{sed} and P_{int} can be estimated under the assumption of steady state.

According to Van Liere *et al.* (1990) total phosphorus in the water column decreased between 1984 and 1989 at an average rate of 12 mg m^{-3} y^{-1}, which corresponds to 0.054 mg m^{-2} d^{-1} on an areal basis. The total phosphorus content of the 0–2 cm sediment has decreased since 1982 with an average rate of 34 mg kg^{-1} y^{-1}, which corresponds to 0.15 mg m^{-2} d^{-1}. It seems reasonable to assume that the present decrease rate is smaller than immediately after the reduction of the external phosphorus load in 1984. So the average value over the past years presumably overestimates the present decrease. Although it cannot be ruled out that a decrease still proceeds at a slow rate, it is obvious that the present average net input of phosphorus to the sediment approximates the difference between P_{ext} and P_{out}, viz. 0.6 mg m^{-2} d^{-1}. In addition, the input of phosphorus on top of the sediment must be almost offset by downward seepage. Based on an average seepage of 1.5 mm d^{-1} and an annual average phosphorus concentration of 0.5 mg l^{-1} at 10 cm depth (A. Sinke, pers. comm.), downward transport of phosphorus due to seepage is estimated at 0.8 mg m^{-2} day^{-1}. Thus removal by downward seepage may indeed balance the major part of the phosphorus input onto the top the sediment. Moreover, phosphorus retention by sediment burial or diagenesis must be very small. However, because the difference between annual phosphorus retention and the estimated phosphorus removal by seepage is only a rough estimate, it cannot be said whether the phosphorus content of the upper 10 cm is still decreasing or not. Although spatial variability of seepage rates and phosphorus concentrations in pore water have to be examined in more detail, the results emphasize the importance of downward seepage as a limiting process for the accumulation of potentially available phosphorus near the sediment-water interface.

The estimated rate for the phosphorus transport from the sediment to the overlying water (P_{int}) is much higher than the diffusive release of 0.3 mg m^{-2} d^{-1} measured with the intact sediment cores. The limited validity of laboratory release experiments to determine the total flux of phosphorus from the sediment to the overlying water in shallow lakes, was earlier suggested by Ryding & Forsberg (1977) and confirmed recently by Sinke *et al.* (1990).

The presented sedimentation of phosphorus of 1.5 mg m^{-2} d^{-1} corresponds to a settling velocity of 2 cm d^{-1}, which is a factor two smaller than measured by Gons & van Keulen (1987) in a laboratory set up. This discrepancy is partly a result of wind-induced turbulence in the field and partly of using annual averages rather than seasonal values. Actual summer values for P_{min}, P_{sed} and P_{int} may be approximately twice as high as indicated in Fig. 5, but in winter they may be close to zero.

In order to predict future trends after application of various additional management measures, relations between rates and pool sizes were defined according to Equations 1–3.

$$dPWAT/dt = P_{ext} - (K_{sed} + K_{out})\,PWAT + K_{int}PADS \quad (1)$$

$$dPORG/dt = K_{sed}PWAT - K_{min}\sqrt{PORG} \quad (2)$$

$$dPADS/dt = K_{min}\sqrt{PORG} - K_{int}PADS - K_{seep}PADS \quad (3)$$

It is assumed that the total flux from the sediment to the overlying water (P_{int}) and the removal by downward seepage (P_{seep}) are linearly proportional with factors K_{int} and K_{seep} to PADS. Sedimentation (P_{sed}) and the discharge of phosphorus (P_{out}) are assumed to be linearly proportional with factors K_{sed} and K_{out} to the total amount of phosphorus in the overlying water (PWAT). P_{min} is assumed to be proportional to the square root of the degradable organic pool size, similar to the relation between the sediment oxygen demand and the volumetric oxygen demand reported by Bouldin (1968). The values of the proportionality constants were calculated from the steady state values presented in Fig. 5.

With Equations 1–3 the trends are calculated after the following treatments: A) 50% reduction of the external phosphorus load; B) 50% reduction of the external phosphorus load and complete reduction of downward seepage by inundation of surrounding polders; C) sediment dredging of the sediment top layer which removes 80% of the easily degradable organic matter (PORG); and D) combination of dredging and a 50% reduction of the external phosphorus load.

Due to the boundary condition, that at present the system is in steady state, no changes occur without any additional management measurements. According to the model, a reduction of the external phosphorus load will result in a gradual reduction of PWAT (Fig. 6). The new equilibrium will be reached much faster if this measure is accompanied by sediment dredging. Dredging without a reduction of the external phosphorus load will initially improve the water quality, but eventually the system will return to its former equilibrium. These predictions reflect the role of the mobile sediment fractions (PORG and PADS) in the internal nutrient cycle. The rates at which these pools decrease determine largely the changes in PWAT. A reduction of the external load is followed by a gradual decrease of these

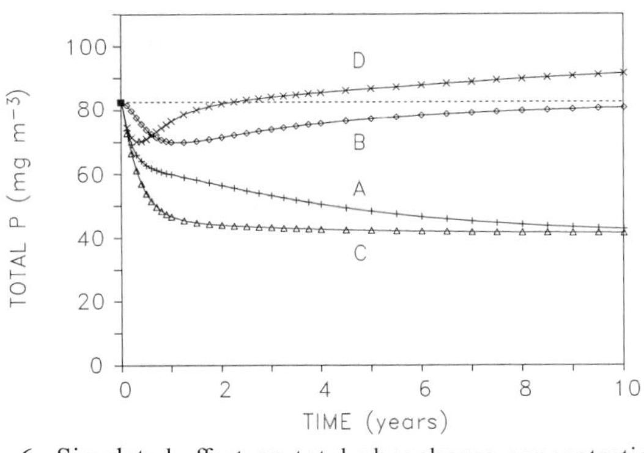

Fig. 6. Simulated effect on total phosphorus concentrations in the water column after various additional management measures. Dotted line: no additional measurements; A) 50% reduction of the external phosphorus load; B) sediment removal; C) sediment removal and 50% reduction of external phosphorus load and D) 50% reduction of external phosphorus load without downward seepage.

pools as long as the total phosphorus input P_{ext} is lower than the total output ($P_{out} + P_{seep}$). Sediment dredging of the top layer results in an instantaneous reduction of the mobile phosphorus pool. Therefore, even without a reduction of the external load, dredging results in an initial reduction of PWAT. However, after dredging the total output of phosphorus from the system is lower than the total input. Under these conditions the mobile fractions of phosphorus in the sediment are likely to increase in size till the original steady state conditions are reached again. If a 50% reduction of the external phosphorus load is combined with a complete elimination of the downward seepage, the changes in PWAT are relatively small, because the difference between input and output hardly differs from the present situation.

The trends revealed by the model calculations demonstrate that the difference between phosphorus input and output, rather than the external phosphorus load alone, determines the final concentration of phosphorus in the water. In lakes in which the phosphorus output by either discharge or seepage is less favourable, restoration perspectives are worse. A reduction of the external phosphorus load has a limited effect in case of a large potentially available pool of phosphorus and a small discharge of phosphorus after the measurement. In such cases dredging may contribute significantly to the improvement of the water quality. Such a behaviour was, indeed, observed for Lake Trummen (Bengtsson *et al.*, 1975). Limited success of sediment dredging has been, however, reported by Ryding (1982) for lake Trehorningen.

The effects of application of 100 gram m^{-2} iron as $Fe(OH)_3$ were calculated with and without a simultaneous 50% reduction of the present external phosphorus load. Four additional assumptions were made: 1) the added iron is reduced at a rate of 0.05 $g\ m^{-2}\ d^{-1}$ (Fig. 2), i.e. after six years all the added iron is converted into reduced precipitates; 2) initially, the addition of iron stops the removal of phosphorus by downward seepage completely (K_{seep} is set at zero); 3) to simulate the effects of improved coagulation in the sediment top layer and suppressed resuspension of sediment material, the internal phosphorus load decreases (K_{int} is reduced by a factor five) immediately after the iron addition; 4) K_{seep} and K_{int} return to their original values after complete reduction of the added iron.

After the addition of iron an initial decrease of PWAT occurs, because of the improved adsorption of phosphorus in the sediment (Fig. 7). However, the adsorption capacity of the iron decreases because of chemical and microbial reduction processes. At the same time PADS increases as mineralization proceeds at its original rate, while downward seepage and transport to the water are suppressed because of the strong adsorption to the added iron. The iron treatment initially improved the retention of phosphorus in the sediment, but this results eventually in a relatively large internal load as the reduction of the iron proceeds. Finally, the gradual increase of PADS and K_{int} results in an internal load (P_{int}) at a somewhat higher level than in the absence of the iron treatment.

The time scale on the abscissae in Fig. 6 and Fig. 7 must be considered as a rough indication because of the limited accuracy of the model. The scaling is in particular sensitive to the difference between input and output relative to the pool sizes in the sediment. Because of its simplicity, further adaptations will be needed to predict the behavior of the lake if the structure of the nutrient cycle changes e.g. after the introduction of submersed marcophytes or after biomanipulation. Part of these processes are included in more sophisticated models like PCLOOS (Janse *et al.*, 1992) which contains a description of sediment phosphorus dynamics that is largely comparable to the approach presented here.

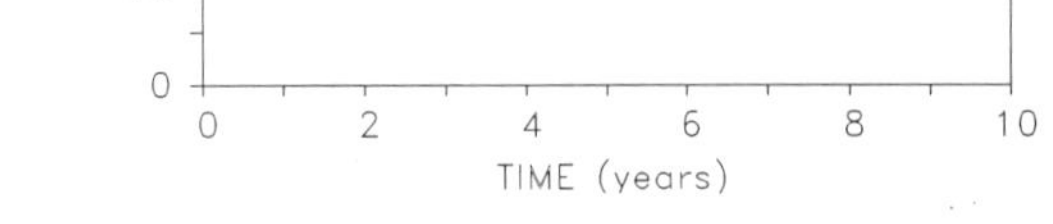

Fig. 7. Simulated effects of iron additions on total phosphorus concentrations in the Lake water. Dotted line: no treatment; A) iron addition, no reduction of external phosphorus load; B) 50% reduction of external load and C) combination of iron addition and 50% reduction of external load.

Conclusions

The diffusive release of phosphorus from the sediment of the Loosdrecht lakes is small, both under aerobic and anaerobic conditions. Inorganic phosphorus generated by mineralization processes is removed from the sediment by downward seepage and uptake in the nutrient cycle during resuspension. The phosphorus cycle in the lake, including the upper 10 cm of the sediment is close to steady state. In the Loosdrecht lakes phosphorus removal by downward seepage is responsible for the withdrawal of phosphorus from the nutrient cycle. It seems that diagenesis and sediment burial hardly act as a permanent sink for phosphorus. The total phosphorus content of the sediment top layer has decreased significantly since 1982. The amount of potentially bioavailable phosphorus in the upper 10 cm equals $<15\%$ of the total phosphorus content of the sediment, but is still three times larger than the annual external phosphorus load.

The simulations indicate that in the Loosdrecht lakes a further reduction of the external load will reduce the total phosphorus content of the lake water since the historical phosphorus load in the sediment will be reduced relatively fast by downward seepage. In comparable lakes without downward seepage, restoration perspectives are less favourable, because the bioavailable phosphorus load accumulated in the sediment can only be reduced by flushing through the water column. Sediment dredging and iron addition without reduction of the external load may be successful on a short term, but not on a longer time scale. Elimination of downward seepage by hydrological measures will counteract the effect of a reduction of the external load.

Acknowledgements

This study is part of the 'Water Quality Research Loosdrecht Lakes'. It was financially supported by the Provincial Waterboard of Utrecht and the Netherlands Integrated Soil Research Programme. The authors with to thank M. Buysman, P. de Joode, F. Cottaar and S. Visser for technical assistance.

References

Bengtsson, L., S. Fleisher, G. Lindmark & W. Ripl, 1975. Lake Trummen restoration project I. Water and sediment chemistry. Verh. int. Ver. Limnol. 19: 1080–1087.

Boers, P. C. M. & O. van Hese, 1988. Phosphorus release from the peaty sediments of the Loosdrecht Lakes (The Netherlands). Wat. Res. 22: 355–363.

Boers, P. C. M., J. W. Th. Bongers, A. G. Wisselo & Th. E. Cappenberg, 1984. Loosdrecht Lakes Restoration Project: Sediment phosphorus distribution and release from the sediments. Verh. int. Ver. Limnol. 22: 842–847.

Boström, B., M. Jansson & C. Forsberg, 1982. Phosphorus release from lake sediments. Arch. Hydrobiol. Beih. 18: 5–59.

Boström, B., I. Ahlgren & R. Bell, 1985. Internal nutrient loading in a eutrophic lake, reflected in seasonal variations some sediment parameters. Verh. int. Ver. Limnol. 22: 3335–3339.

Bouldin, D. R., 1968. Models for describing the diffusion of oxygen and other mobile constituents across the mud-water interface. J. Ecol. 56: 77–87.

Cooke, G. D., E. B. Welch, S. A. Peterson & P. R. Newroth, 1986. Lake and reservoir restoration. Butterworth Publ. Boston. pp. 392.

Cullen, P. & C. Forsberg, 1988. Experiences with reducing point sources. Hydrobiologia 170: 321–336.

Einsele, W., 1936. Uber die Beziehungen des Eisenkreislaufs zum Phosphatkreislauf im eutrophen See. Arch. Hydrobiol. 29: 664–686.

Engelen, G. B., B. F. M. Kal, J. J. Buyse & F. G. M. Van Pruissen. The hydrology of the Loosdrecht lakes area. Hydrobiologia 233: 21–38.

Fries, J. & H. Getrost, 1977. Organische Reagenzien fur die Spurenanalyse. E. Merck, Darmstadt.

Froelich, P. N., 1988. Kinetic control of dissolved phosphate in natural rivers and estuaries: A primer on the phosphate buffer mechanism. Limnol. Oceanogr. 33: 649–668.

Gons, H. J. & R. Van Keulen, 1987. The relatie tussen doorzicht en slib in the Loosdrechtse plassen in verband met de zwemwaternorm. WQL-rapport 1989–4. Limnological Institute Nieuwersluis. 35 pp.

Gons, H. J., R. Veeningen & R. Van Keulen, 1986. Effects of wind on a shallow lake ecosystem: resuspension of particles in the Loosdrecht Lakes. Hydrobiol. Bull. 20: 109–120.

Gulati, R. D., L. Van Liere & K. Siewertsen, 1991. The Loosdrecht lake system: Man's role in its creation, perturbation and rehabilitation. In O. Ravera (ed.), Terrestrial and aquatic ecosystems: perturbation and recovery. Ellis Horwood Ltd: 593–606.

Holdren Jr, G. C. & D. E. Armstrong, 1980. Factors affecting phosphorus release from intact lake sediment cores. Envir. Sci. Technol. 14: 79–87.

Hosomi, M., M. Okada & R. Sudo, 1982. Release of phosphorus from lake sediments. Envir. Int. 7: 93–98.

Janse, J. H., T. Aldenberg & R. P. G. Kramer, 1992. A mathematical model of the phosphorus cycle in Lake Loosdrecht and simulation of additional measures. Hydrobiologia 233: 119–136.

Kamp-Nielsen, L., 1974. Mud-water exchange of phosphate and other ions in undisturbed sediment cores and factors affecting the exchange rates. Arch. Hydrobiol. 73: 218–237.

Keizer, P., M. C. N. P. Buysman & Th. E. Cappenberg, 1991. Sorption and release of phosphorus in a peaty sediment. Verh. int. Ver. Limnol. 24: 722–725.

Lijklema, L., J. H. Janse, R. M. M. Royackers & M.-L. Meijer, 1988. Eutrofiëring in Nederland. H_2O 21: 462–467.

Marsden, M. W., 1989. Lake restoration by reducing the external phosphorus loading: the influence of sediment phosphorus release. Freshwat. Biol. 21: 139–162.

Mortimer, C. H., 1941. The exchange of dissolved substances between mud and water in lakes I. J. Ecol. 29: 280–329.

Mortimer, C. H., 1942. The exchange of dissolved substances between mud and water in lakes II. J. Ecol. 30: 147–201.

Murphy, J. & J. P. Riley, 1962. A modified single solution method for the determination of phosphate in natural waters. Analyt. Chim. Acta 27: 31–36.

Nürnberg, G. K., 1988. Prediction of phosphorus release rates from total and reductant soluble phosphorus in anoxic lake sediments. Can. J. Fish. aquat. Sci. 45: 453–462.

Redshaw, C. J., C. F. Mason, C. R. Hayes & R. D. Roberts, 1990. Factors influencing phosphate exchange across the sediment-water interface of eutrophic reservoirs. Hydrobiologia 192: 233–245.

Ryding, S. O., 1982. Lake Trehorningen restoration project. Changes in water quality after sediment dredging. Hydrobiologia 92: 549–558.

Ryding, S. O., 1985. Chemical and microbiological processes as regulators of exchange of substances between sediments and water in shallow eutrophic lakes. Int. Revue. ges. Hydrobiol. 70: 657–702.

Ryding, S. O. & C. Forsberg, 1977. Sediments as a nutrient source in shallow polluted lakes. In H. L. Golterman (ed.), Interactions between sediments and freshwater. Dr. W. Junk, The Hague: 227–234.

Sas, H. (ed.), 1989. Lake restoration by reduction of nutrient loading: expectations, experiences, extrapolations. Academia Verlag Richarz, Sankt Augustin, 497 pp.

Shaw, J. H. F. & E. E. Prepas, 1990. Exchange of phosphorus from shallow sediments at nine Alberta lakes. J. envir. Qual. 19: 249–256.

Sinke, A. J. C. & Th. E. Cappenberg, 1988. Influence of bacterial processes on the phosphorus release in the eutrophic Loosdrecht lakes, The Netherlands. Arch. Hydrobiol. Beih. 30: 5–13.

Sinke, A. J. C., A. A. Cornelese & Th. E. Cappenberg, 1991. Phosphatase activity in sediments of the Loosdrecht lakes. Verh. int. Ver. Limnol. 24: 719–721.

Sinke, A. J. C., A. A. Cornelese, P. Keizer, O. F. R. Van Tongeren & Th. E. Cappenberg, 1990. Mineralization, porewater chemistry and phosphorus release from peaty sediments in the eutrophic Loosdrecht lakes. Freshwat. Biol. 23: 587–599.

Søndergaard, M., 1989. Phosphorus release from a hypertrophic lake sediment: Experiments with intact sediment cores in a continuous flow system. Arch. Hydrobiol. 116: 45–59.

Søndergaard, M., P. Kristensen & E. Jeppesen, 1992. Phosphorus release from resuspended sediment in the shallow and wind-exposed Lake Arresø, Denmark. Hydrobiologia 228: 91–99.

Sweerts, J-P. R. A., 1990. Oxygen consumption, mineralization and nitrogen cycling at the sediment-water interface of north temperate lakes. Ph.D. Thesis University of Groningen.

Sweerts, J-P. R. A. & Th. E. Cappenberg, 1988. Use of micro-electrodes for measuring oxygen profiles in lake sediments. Arch. Hydrobiol. Beih. 31: 365–371.

Tessenow, U., 1972. Losungs- Diffusions- und Sorptionsprozesse in der Oberschicht von Seesedimenten. 1. Ein Langzeit-experiment unter aeroben und anaeroben Bedinungen im Fliessgleichgewicht. Arch. Hydrobiol. 4, Suppl. 38: 353–398.

Van Liere, L., 1986. Loosdrecht lakes, origin, eutrophication, restoration and research programme. Hydrobiol. Bull. 20: 9–15.

Van Liere, L., L. Breebaart, W. Kats & J. J. Buyse, 1989. De waterkwaliteit in het Loosdrechtse plassengebied. CHO/TNO rapport 22: 265–279.

Van Liere, L., O. F. R. Van Tongeren, L. Breebaart & W. Kats, 1990. Trends in chlorophyll and total phosphorus in the Loosdrecht lakes, The Netherlands. Verh. int. Ver. Limnol. 24: 707–710.

Hydrobiologia **233**: 51–59, 1992.
L. Van Liere & R.D. Gulati (eds), Restoration and Recovery of Shallow Eutrophic Lake Ecosystems in The Netherlands.

Coupling of phytoplankton and detritus in a shallow, eutrophic lake (Lake Loosdrecht, The Netherlands)

Herman J. Gons,[1] Tineke Burger-Wiersma,[2] Jerko H. Otten[1] & Machteld Rijkeboer[1]
[1] *Limnological Institute, 'Vijverhof' Laboratory, Rijksstraatweg 6, 3631 AC Nieuwersluis, The Netherlands;*
[2] *Laboratory for Microbiology, University of Amsterdam, Nieuwe Achtergracht 127, 1018 WS Amsterdam, The Netherlands*

Key words: phytoplankton, *Oscillatoria limnetica*, *Prochlorothrix hollandica*, detritus, shallow lakes, lake restoration

Abstract

An oscillating steady state is described of phytoplankton, dominated by *Prochlorothrix hollandica* and *Oscillatoria limnetica*, and sestonic detritus in shallow, eutrophic Lake Loosdrecht (The Netherlands). A steady-state model for the coupling of the phytoplankton and detritus is discussed in relation to field and experimental data on phytoplankton growth and decomposition. According to model predictions, the phytoplankton to detritus ratio decreases hyperbolically at increasing phytoplankton growth rate and is independent of a lake's trophic state. The seston in L. Loosdrecht contains more detritus than phytoplankton as will apply to many other lakes. The model provides a basis for estimating the loss rate of the detritus, including decomposition, sedimentation and hydraulic loss. In a shallow lake like L. Loosdrecht detritus will continue to influence the water quality for years.

Introduction

Reductions of the external phosphorus loading did not significantly improve the water quality of the shallow and highly eutrophic Lake Loosdrecht (Loogman & Van Liere, 1986; Van Liere *et al.*, 1990). The lake is still characterized by predominance of filamentous prokaryotic species, including *Oscillatoria* spp. and *Prochlorothrix hollandica* (De Kloet *et al.*, 1990; Van Liere *et al.*, 1989), absence of submerged vegetation (Malthus *et al.*, 1990), high stocks of bream (*Abramis brama*), lack of large-bodied grazing zooplankton (Gulati *et al.*, 1992), very low Secchi-disc transparency (almost always <0.5 m), and high annual mean concentrations of chlorophyll *a* and total phosphorus (both *ca.* 100 mg m^{-3}).

The low transparency is due to phytoplankton and particulate detritus. The detritus consists of peat fragments and fine-sized ($<15\ \mu m$) particles. The peat fragments appear in the water column after resuspension in periods of high wind-speed, but rapidly resettle in calm conditions. The fine-sized detritus is likely to derive mainly from the phytoplankton, has higher dry weight concentrations and a greater partial attenuation coefficient for downwelling light than the phytoplankton (Gons, 1987; Gons *et al.*, 1991; Otten *et al.*, 1992; Rijkeboer & Gons, 1990).

Each of the phytoplankton and detritus components alone would prevent the Secchi-depth from reaching 1 m. Because the lake shows frequent resuspension of sediments (Gons *et al.*, 1986b; 1991) and the uniformity of depth almost entirely precludes the permanent deposition of particles, it might be inferred that even a drastic reduction in phosphorus loading would not achieve this water-quality goal. However, the history of L. Loosdrecht (see further) and data on other lakes of various trophic states (Saunders, 1972; Wetzel, 1983; Wetzel *et al.*, 1972), demonstrate that the detritus concentration in lakewater is coupled to phytoplankton production. Furthermore, in having a higher concentration of particulate detritus than of phytoplankton, L. Loosdrecht is no exception but rather represents the rule (Wetzel, 1983). It can thus be surmised that a decrease in phytoplankton production will inevitably involve decrease of the detritus as well. The question remains how long the refractory, fine-sized detritus that accumulated in the past may prevent a clear state from developing.

Lake Loosdrecht exhibits only a weak seasonality in the concentrations of chlorophyll and seston dry weight, and their dynamics during summer can be characterized as an oscillating steady state (Gons *et al.*, 1986a; Gons & Rijkeboer, 1990). The validity of applying this concept is illustrated on a basis of a time series of chlorophyll *a* and seston dry weight concentrations in summer 1988. The question of how change in detritus concentration is related to the phytoplankton growth in this very shallow lake is addressed with a simple steady-state model. Model predictions are discussed in relation to concentrations of chlorophyll *a* and seston dry weight, and experimental data on phytoplankton growth and mortality.

Theory

We presume that the phytoplankton growth in the lake in summer is very nearly at steady state. The specific growth rate of the phytoplankton (μ) varies slightly around a mean value set by the specific rate of loss (D) due to grazing, sinking, and other causes, hence $\mu = D$ (time^{-1}). The phytoplankton biomass depends on the availability of light and nutrients. Steady-state phytoplankton growth may also involve constant concentrations of detritus and seston.

Besides the steady-state condition, three further assumptions are made:

(1) The seston dry weight concentration (S g m^{-3}) consists exclusively of phytoplankton (PH g m^{-3}) and fine-sized detritus (FD g m^{-3}):

$$[S] = [PH] + [FD] \qquad \text{(Eq. 1)}$$

For defined phytoplankton this partitioning can be solved using chlorophyll (CHL g m^{-3}) and seston dry weight measurements as in Eq. (2), using β (g PH g^{-1} CHL) as a symbol for a constant of conversion of chlorophyll into phytoplankton dry weight:

$$[S] = \beta[CHL] + [FD] \qquad \text{(Eq. 2)}$$

(2) Regardless of the cause of phytoplankton mortality, the only product is a fixed refractory fraction (r) of the biomass, which is instantaneously released into the pool of fine-sized detritus.

(3) Decrease of the fine-sized detritus has a constant over-all specific loss rate σ (time^{-1}), which includes decomposition, permanent deposition and hydraulic wash-out.

Change in the concentration of fine-sized detritus can now be given as:

$$d[FD]/dt = D\,[PH]r - \sigma[FD] \qquad \text{(Eq. 3)}$$

We presume [FD] to be also at steady state: $d[FD]/dt = 0$, hence:

$$[FD] = D/\sigma\,[PH]\,r \qquad \text{(Eq. 4)}$$

Consequently, the concentration of fine-sized detritus associated with a given concentration of phytoplankton with defined r-value depends on the ratio of the loss rates of the two components. We note that also [S] will be constant (Eq. 1).

After re-arrangement of Eq. (4) the phytoplankton to detritus ratio $R_{p/d}$ is given by:

$$R_{p/d} = [PH]/[FD] = \sigma/D1/r \qquad \text{(Eq. 5)}$$

and the value of σ:

$$\sigma = R_{p/d}Dr \qquad \text{(Eq. 6)}$$

Equation (5) implies that, in principle, the shares

of phytoplankton and detritus in seston are independent of a lake's trophic state.

The value of σ indicates the time needed for restoration of L. Loosdrecht's transparency once the phytoplankton growth has been minimized. For phytoplankton in L. Loosdrecht values of D ($=\mu$) have been estimated using field and experimental studies. Values of β and r have been obtained for *O. limnetica* and *P. hollandica*, the dominating species in summer, at defined conditions of light and phosphorus limitation.

The lake

Lake Loosdrecht (situated *ca.* 20 km SE of Amsterdam, The Netherlands; $A = 9.8\ \mathrm{km}^2$; $\bar{z} = 1.9$ m) has its origin in peat dredging and subsequent erosion by wind and waves. Until the nineteen-fifties the lake was clear and had extensive benthic growth of characeans. Only storms might turn the water turbid for one or two days. Green algae and diatoms dominated the phytoplankton. Phytoplankton dry weight concentration, estimated from the biovolume, was *ca.* $0.2\ \mathrm{g\ m^{-3}}$. Studies in the early sixties showed disappearance of the characeans and a drastic decrease in water transparency and changes in plankton composition symptomatic of eutrophication.

Lake Loosdrecht is a seepage lake needing substantial supplementary water supply to maintain its water table in summer. It is thought that phosphorus loading due to inlet of water from the highly polluted River Vecht has been the main cause of the eutrophication. Since 1984 onwards water from the Amsterdam-Rhine Canal has replaced the supply of the River Vecht water. Additional measures included sewerage diversion and phosphate removal of the inlet water by coagulation with $FeCl_3$.

Throughout the nineteen-eighties the phytoplankton showed perennial predomination by filamentous prokaryotic species, mainly *O. limnetica* (including incompletely identified filaments), *O. redekei*, and *Prochlorothrix hollandica* (Burger-Wiersma *et al.*, 1986, 1989; De Kloet *et al.*, 1990; Van Tongeren *et al.*, 1992). Unlike in other shallow, eutrophic Dutch lakes (Berger, 1987; Hosper & Meijer, 1986; Moed & Hoogveld, 1982) *Oscillatoria agardhii* is not predominant. Many eukaryotic species co-exist, among which nanoflagellates may have high cell densities, but only the diatoms *Diatoma elongatum* and *Melosira italica* reach a considerable biomass in early spring and autumn (Gons, 1991).

For details on the lake's morphometry, and hydrological and biological history, see Engelen *et al.* (1992), Gulati *et al.* (1991), Hofstra & Van Liere (1992) and Loogman & Van Liere (1986).

Materials and methods

In summer 1988 turbidity and the concentrations of seston dry weight and chlorophyll were monitored at a small island in the northern part of L. Loosdrecht (Gons *et al.*, 1991). From July onward, in the morning and evening, water was collected and stored using a refrigerating sampler (Ibuk Abwasser-technik, Königsbronn, Germany). The storage time varied from two hours to longer than three days, but all samples were processed within four days after collection. The influence of storage time on the analyses appeared to be insignificant.

Lakewater was prefiltered through 144 μm mesh screen. Seston dry weight was determined by filtering through weighed Whatman GF/F filters. Chlorophyll *a* and phaeopigment were measured after extraction in hot ethanol (Moed & Hallegraeff, 1978). Phaeopigment generally constituted about 20% and always less than 30% of the uncorrected chlorophyll values.

Growth of the natural phytoplankton was studied using bags of dialysis membrane *in situ* (Burger-Wiersma & Mur, 1986) and laboratory scale enclosures (LSE; Rijkeboer *et al.*, 1990b).

The enclosure in dialysis bags aimed at isolating the seston from large-bodied zooplankton and exchange of particles, while preserving the physical and chemical conditions *in situ*. After filling with prefiltered (144 μm) lakewater, the dialysis bags (length 0.6 m; pore size 2.4 nm) were suspended vertically in the 0.1–0.7 m depth layer. Change in protein was directly converted into specific growth rate of the phytoplankton. Growth rate estimates from carbohydrate decrease in the dark were made according to results with cultures of *P. hollandica* (Burger-Wiersma & Mur, 1986; Burger-Wiersma & Baard, 1987).

The LSE (Rijkeboer *et al.*, 1990ab) is a continuous flow system simulating the natural water column under light and temperature conditions *in situ*. Growth rates of phytoplankton in prefiltered (144 μm) lakewater were estimated by comparison of the chlorophyll-specific column-integrated oxygen production with the values for cultures of *P. hollandica* and *O. limnetica* in LSE (Rijkeboer *et al.*, 1990b).

Light- and phosphorus-limited growth, and chemical characteristics of *P. hollandica* were studied using turbidostat cultures (Burger-Wiersma & Baard, 1987), and those of *O. limnetica* and *P. hollandica* using chemostat cultures in LSE (Rijkeboer *et al.*, 1990b).

Fractions of refractory particulate matter of *O. limnetica* were determined after three weeks of incubation in the dark in aerobic conditions (Otten *et al.*, 1992).

Results and discussion

Seston dynamics

In a system at steady state the concentrations of all components are constant, a situation made feasible by finely tuned feed-back control. Environmental perturbations may temporarily cause great oscillations, but the feed-back mechanisms force the concentrations to return to the values imposed by the steady-state relationships.

From May until September in 1988 the concentrations of chlorophyll *a* (CHL, including phaeopigment) and seston dry weight (S) ranged from 103 to 202 mg m^{-3} and from 16 to 67 g m^{-3}, respectively. Though the short-term fluctuations were marked, seasonal trends were absent. The averages of CHL and S in these months remained close to those of the whole period, i.e. 160 mg m^{-3} and 34 g m^{-3}, respectively.

Stability of the phytoplankton and detritus relationship is reflected by the chlorophyll content of seston (Fig. 1). From May until September the values fluctuated around 5 mg CHL g^{-1}, within the range of 3 to 8 mg g^{-1}. This range is quite narrow considering the variability of pigment content in phytoplankton cultures alone (see further). Meteorological variability appears to be the main cause of these fluctuations (Gons & Rijkeboer, 1990; Gons *et al.*, 1991). For example, the lowest record was due to resuspension of sediment during a severe storm in September. Thereafter more resuspended phytoplankton relative to other particles remained in suspension, hence the values increased to 7 mg g^{-1}, but later decreased to 5 mg g^{-1} again.

In 1988 prolonged stability of the phytoplankton growth was aided by evenness of insolation and temperature. A warm, sunny spring was followed by a remarkably cloudy summer. Throughout May–September the water temperature was invariably between 15 and 20 °C.

However, it needs to be emphasized that evidently there is feed-back control of the chlorophyll content of the seston. In the theory section it was explained that the seston concentration may be constant at given steady-state phytoplank-

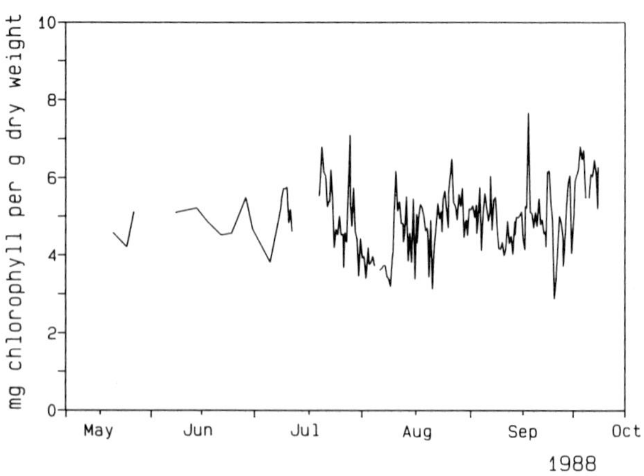

Fig. 1. Time series of chlorophyll *a* content (including phaeopigment) of seston dry weight in Lake Loosdrecht.

ton production, but not how such a steady state is controlled. The answer must be sought in the functioning of the detritus derived from the phytoplankton itself. It has been recognized for a long time (Hutchinson, 1957) that detritus plays an important role in the metabolism of lakes. Saunders (1972) explicitly showed how detritus may function as a buffer as regards flows of energy and nutrients. With respect to phytoplankton production in L. Loosdrecht, this capacity of detritus has been found to control the biomass under both light and phosphorus limitation (Rijkeboer & Gons, 1990; Rijkeboer *et al.*, 1990b; in press).

Furthermore, there is evidence that in the case of L. Loosdrecht the detritus precludes the development of the large-bodied zooplankton able to graze the filamentous prokaryotic species (Gulati *et al.*, 1992). In this way the range of the phytoplankton loss rate will be restricted to low values. The deposition of these organisms may be due to entrainment in flocs of detritus resettling after resuspension rather than sinking of the individual trichomes (Gons *et al.*, 1991). This mechanism might set the lower limit to the loss rate.

We conclude that the conditions for coupling of phytoplankton and detritus concentrations (Eq. 4) do exist. The seston dynamics in L. Loosdrecht in summer 1988 possesses the features of an oscillating steady state. The described theory is applicable to the mean concentrations of CHL and S measured in this period.

Phytoplankton to detritus ratios

The maximum value of the specific growth rate (μ_{max}) of cyanobacteria at about 20 °C is approximately 1.0 d^{-1} (Van Liere & Mur, 1980; Reynolds, 1987). *P. hollandica* has a μ_{max} values of 0.52 d^{-1} at 20 °C (Burger-Wiersma *et al.*, 1989). Refractory fractions, estimated as the particulate matter remaining after 21 days of aerobic decomposition of *O. limnetica*, averaged 0.18 (Otten *et al.*, 1992). This information suffices for exploring the phytoplankton to detritus ratio ($R_{p/d}$) as a function of μ ($= D$) for hypothetical values of σ (Eq. 6).

The relationship (Fig. 2) has $\mu = 0$ as an asymptote, because there is no detritus formation when there is neither growth nor loss of the phytoplankton population. As μ increases, $R_{p/d}$ decreases hyperbolically and only for values of $\sigma > 0.02$ d^{-1} there can be more phytoplankton than detritus over a wide range of growth rates. For $\sigma = 0.01$ and 0.005 d^{-1}, the concentration of phytoplankton will be lower than of detritus at $\mu > 0.1$ and > 0.05 d^{-1}, respectively. For $\sigma = 0.001$ d^{-1} there will be much more detritus than phytoplankton, except at extremely low growth rates.

The value of σ, for fine-sized particles discussed here, will be lower than 0.02 d^{-1} in most lakes. First, reported decomposition rates of refractory detritus are invariably one to two orders of magnitude lower (Otten *et al.*, 1992; Saunders *et al.*, 1980). Secondly, although the sedimentation rates, i.e. intrinsic settling velocities over depth in mixed layers (Smith, 1982), of the fine-sized detritus in a shallow lake like Loosdrecht can be about 0.02 d^{-1}, the net loss rate will be much lower due to resuspension (Gons, 1991; Gons *et al.*, 1986b; 1991). In epilimnia of all but very

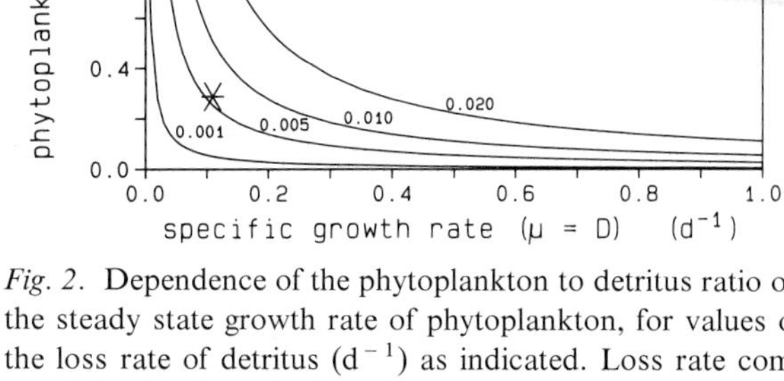

Fig. 2. Dependence of the phytoplankton to detritus ratio on the steady state growth rate of phytoplankton, for values of the loss rate of detritus (d^{-1}) as indicated. Loss rate computed for fraction of refractory particulate matter = 0.18, see text. Asterisk represents position of Lake Loosdrecht in summer 1988.

small deep lakes the sedimentation rates will be much lower than $0.02\ d^{-1}$, and also in this case the loss will be reduced by resuspension of litoral deposits. Thirdly, water retention times are generally too long for hydraulic loss rates as high as $0.02\ d^{-1}$. In a seepage lake like Loosdrecht, this loss will probably be lower than indicated by the value of $0.006\ d^{-1}$ corresponding to the water retention time of *ca.* half a year, because the particles may be retained at the sediment/water interface.

Even though refractory fractions may greatly depend on phytoplankton species and growth conditions (Jewell & McCarty, 1971; Fallon & Brock, 1979; Otten *et al.*, 1992) it can be concluded that in lakes in general and L. Loosdrecht in particular the mean $R_{p/d}$ value will be less than unity over almost the whole range of phytoplankton growth rates. The proportions of phytoplankton and detritus may greatly vary seasonally in non-steady-state situations. In Frains Lake (Michigan) $R_{p/d}$ ranged from 0.06 to 0.77 (Saunders, 1972).

The specific loss rate σ for Lake Loosdrecht

The parameter β is needed for partitioning the seston dry weight into phytoplankton and detritus based on measured values of CHL and S (Eq. 2). For known specific growth rate of the phytoplankton the value of σ is fixed (Eq. 4). The values of β and μ are not independent of each other.

The chlorophyll content of phytoplankton is highly dependent on both energy and nutrient status of the cells. Depending on mean irradiance and growth rate, chlorophyll *a* contents of *O. agardhii* in phosphorus- and light-limited chemostats varied from 3 to 8 and 5 to 17 $mg\ g^{-1}$ dry weight, respectively (Zevenboom *et al.*, 1982). Thus, for this organism, β may range 60 to 330. The high value would imply that phytoplankton constituted >100% of the seston in L. Loosdrecht. In light-limited turbidostat cultures of *P. hollandica*, β-values were 28 and 142 at mean irradiances of 10 and 240 $\mu E\ m^{-2}\ s^{-1}$ PAR (= photosynthetically available radiation), respectively (Burger-Wiersma & Post, 1989). Should the high β-value of these cultures apply to the phytoplankton in L. Loosdrecht in summer 1988, two-thirds of the seston was phytoplankton.

However, vertical light attenuation coefficients in the lake in summer are so high ($\geq 5\ m^{-1}$), that the euphotic depth is much less than 1 m, and the depth- and time-averaged irradiance is less than 50 $\mu E\ m^{-2}\ s^{-1}$ of PAR. For steady-state growth in the 'natural' light conditions in the laboratory scale enclosures (LSE), receiving 550 $\mu E\ m^{-2}\ s^{-1}$ as averaged incident daylight (PAR) in summer, the range of β (including phaeopigment) was relatively narrow. The minimum was 42 – for *P. hollandica* under both light and phosphorus limitation – and the maximum was 77 – for *O. limnetica* under phosphorus limitation (Rijkeboer *et al.*, 1990b). The growth rate ranged 0.04 to $0.17\ d^{-1}$. The highest β-value from these results would imply that the phytoplankton constituted one-third of the seston dry weight.

In summer 1988, the phytoplankton growth rate estimates were made using only the LSE method (Rijkeboer *et al.*, 1990b). For both light- and phosphorus-limited growth the derived values of μ were $0.09–0.12\ d^{-1}$. These μ-values were about half of those obtained for 1987 and 1989 using the same method. The reasons for annual variability in μ remain to be investigated. In the summers of 1984–1986, both the protein change over 24 h and the carbohydrate decrease in the dark gave about $0.1\ d^{-1}$ as mean value of μ (Burger-Wiersma & Baard, 1987).

In summary, the mean value of μ (= D) in summer 1988 probably was close to $0.10\ d^{-1}$. At this growth rate, the variation in β for light- and phosphorus-limited steady state cultures of *P. hollandica* and *O. limnetica* was 42 to 50 (Rijkeboer *et al.*, 1990b). A mean value of 46 implies that the mean contribution of phytoplankton to the seston dry weight in summer 1988 was 22%. Assuming the validity of Eq. (1), the remaining 78% of the seston dry weight was refractory fine-sized detritus. Thus, the value of $R_{p/d}$ would be 0.28.

The validity of Eq. (1) is questionable because

part of the seston may be constituted by peat fragments and mineral particles not associated with phytoplankton decomposition. However, remembering the former clear state of the lake, with limitation of the Secchi-depth by the bottom only, it is reasonably safe to conclude that these materials cannot represent more than 10% of the present mean of seston dry weight. Even a dry weight concentration as low as 5 g m^{-3} would allow a Secchi-depth of only 1 m (Gons, 1987). No data are available to substantiate that the former submerged vegetation dampened the wind-resuspension of sediment particles.

The share of the refractory detritus could also be over-estimated because the labile particulate detritus has been ignored in Eq. (1). For steady-state conditions in L. Loosdrecht the labile detritus can be neglected, because the decomposition rate of this fraction often will be higher than the phytoplankton growth rate (Otten *et al.*, 1992) and, consequently, its concentration must be very low.

Thus, a phytoplankton to detritus ratio of 0.3 appears reasonable, and, from the value of μ, the position of L. Loosdrecht in summer 1988 can be located (Fig. 2). The value of σ can be precised using Eq. (6). For D $(=\mu)=0.1$ d^{-1}, $r=0.18$ and $R_{p/d}=0.3$, the value of σ is 0.0054 d^{-1}. This value equals the decomposition rate of the refractory matter remaining after three weeks of incubation of *O. limnetica* in the dark (Otten *et al.*, 1992). It is too early, however, to conclude that other causes of decline of the detritus *in situ* are insignificant.

The fate of the refractory detritus

Besides D and σ, the refractory fraction r controls the coupling of phytoplankton and detritus (Eq. 4). Both the formation of the detritus $(=D[PH]r)$ and the value of σ (Eq. 6) are linearly equated to r. Consequently, values of r greatly influence predictions about formation and fate of the fine-sized detritus. The value of r is likely to reflect metabolic adaptation to environmental conditions, and seems to be related to the growth rate in particular (Otten *et al.*, 1992). However, the data needed to predict values of r in relation to the growth conditions of the organisms are too fragmentary to allow discussion.

For D $(=\mu)=0.10$ d^{-1} and $r=0.18$, the dry weight of detritus produced $(=D[PH]r)$ during May–September 1988 was 22 g m^{-3}. Extrapolation of $\sigma=0.0054$ d^{-1} to a whole year involves that 3 g m^{-3} would remain in the water column one year later, which does not seem to be a bad perspective for the lake's restoration. However, this annual decrease of 'old' detritus, will be over-estimated for two reasons. First, the decomposition component of σ is likely to gradually decline with ageing of the particles, and also will decrease with temperature in winter (Saunders *et al.*, 1980). Secondly, the deposition component of σ will vary seasonally. Sinking rates of the particles decrease with temperature due to change in viscosity, and wind-resuspension occurs more frequently and in larger areas of the lake from September through April than in summer (Gons *et al.*, 1986b; Gons, 1991). Therefore, the deposition rate in summer can be considerably higher than the annual mean, which has the effect of substantially lowering the value of σ on an annual basis.

A general conclusion is that refractory detritus may continue to significantly influence the water transparency in L. Loosdrecht for years after its origin, even after the dominance of the cyanobacteria and *P. hollandica* has ceased. Apparently, the situation is very similar to the shallow L. Veluwe (Hosper & Meijer, 1986). It remains to be seen which mechanisms may accelerate the elimination of the phytoplankton remnants in shallow, wind-exposed lakes.

Acknowledgements

We wish to thank Louis van Liere for the pleasant co-operation. Acknowledgements are due to Ronny van Keulen for technical assistance. This research has been financially supported by the Netherlands Ministry of Housing, Physical Planning and Environment, and the Commission of the European Communities (EV/4V-0137-NL).

The majority of the detritus occurs in particle aggregates in the sediment/water interface, the epipelon. It includes fragments of eroded peat, settled algae and fine-sized detritus. The detritus is resuspended frequently over large areas of the lake (Gons *et al.*, 1991; 1986b). The peat fragments settle again in a few hours but the fine-sized detritus remains in the water column for several weeks after resuspension.

Algal mortality is caused by different factors. Besides the zooplankton grazing in the water layer (Gulati *et al.*, 1992), sedimented algae on the lake bottom die after a few days (Otten & Gons, 1991a) due to lack of light. Microfauna, bacteria and viruses may also contribute to algal mortality in the epipelon (Daft & Steward, 1973; Granhall & Berg, 1972).

Little is known about the inputs of dissolved detritus in L. Loosdrecht. The main sources are likely to be decomposing phytoplankton and aquatic plants. Allochthonous inputs include humic substances leaching from the lake's peaty substratum and those formed during decomposition of terrestrial plants (Christman & Gjessing, 1983; Wetzel, 1983; De Haan, 1982), entering the lake via run-off.

Several stages can be distinguished in the decomposition of phytoplankton (Fig. 1; Otten & Gons, 1991a, b). After the death of cells, part of the biomass is released into the DOM pool. The

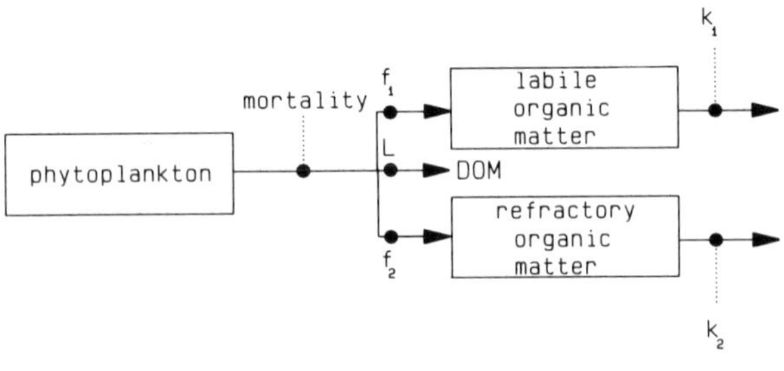

Fig. 1. Model of dynamics of particulate detritus (Otten & Gons, 1991b). Mortality = phytoplankton death rate coefficient (time^{-1}); L = fraction of biomass released as dissolved substances at cell lysis; f_1 = labile fraction of biomass; f_2 = refractory fraction of phytoplankton biomass; k_1 = decomposition rate coefficient of labile detritus (time^{-1}); k_2 = decomposition rate coefficient of refractory detritus (time^{-1}); *DOM* = dissolved organic matter.

larger fraction of the remaining particulate matter is labile (i.e. readily utilized) whereas the rest is refractory (i.e. hardly utilized by heterotrophic microorganisms). Not included in Fig. 1 is the origin of refractory dissolved compounds in the decomposition of the particulate phases.

Refractory material appears to constitute the major fraction of detritus in the lake, and the accumulation of refractory detritus in the epipelon has been experimentally confirmed (Otten & Gons, 1991a). Due to its slow decomposition rate, this detritus might retard the lake's restoration (Gons *et al.*, 1992). Therefore, it is of special interest to obtain data on the refractory fraction of phytoplankton, its decomposition rate and its distribution in the lake. For this study, on the origin and fate of phytoplankton detritus, an experimental approach was chosen, using cultures of *O. limnetica*.

Materials and methods

Growth of Oscillatoria limnetica

Oscillatoria limnetica (isolate Limnological Institute) was grown in synthetic medium in a laboratory scale enclosure (LSE) operated as a chemostat (Rijkeboer *et al.*, 1990a). The conditions simulated averaged *in situ* summer conditions with respect to incident light (120 W m^{-2} PAR), pH (8–8.5) and water temperature (20 °C). The cultures were non-axenic. Phosphorus was supplied to the algal cultures at a rate of 105 μg P l^{-1} d^{-1} (light-limited algal growth in steady state), or at a rate of 7.5 μg P l^{-1} d^{-1} (phosphorus limited algal growth in steady state).

Refractory fractions

Refractory fractions of algal biomass were estimated in batch experiments after transferring phosphorus- or light-limited *O. limnetica* suspensions to 5 l vessels. The suspensions were kept in the dark for three weeks at 20 °C. Each experiment was performed in duplicate. Conditions were kept aerobic by aeration with water-saturated air. The pH remained close to 8 during these experiments. After three weeks, the particulate detritus appeared to be decomposed at a very low rate. The remaining particulate material was defined as the refractory part of the particulate matter initially present. Refractory fractions of seston of L. Loosdrecht were estimated after incubating lake water in duplicate as described above.

Dissolved detritus

A nutrient-saturated batch culture of *O. limnetica* was kept in the dark at 20 °C during three weeks. Absorbance spectra of pre-filtered samples (0.45 μm Millipore HA) were measured at 2 nm intervals using a Philips PU 8600 UV/VIS spectrophotometer. Dry weight concentrations and absorbance spectra of non-filtered suspensions were also measured. In the latter case particle scattering caused part of the absorbance.

Decomposition of refractory detritus

A 12 l continuous flow system, designated epipelon trap (ET), was fed with effluent from an *O. limnetica* culture at about 400 ml h^{-1} (Otten & Gons, 1991a, b). The ET was kept in the dark at 20 °C and aerated to maintain the dissolved oxygen content of the water close to saturation. Respiration was measured daily by monitoring the dissolved oxygen change during interruption of the aeration and the dilution of the system (Otten & Gons, 1991b), and the layer of detritus was sampled weekly. After six weeks the inflow of culture effluent was stopped and the ET diluted with synthetic medium instead. The labile detritus formed during these first six weeks decomposed within two to three weeks, thus only refractory detritus remained. From then on, the measured oxygen consumption was due to respiration by microorganisms decomposing refractory detritus. Decomposition rate coefficients (k_2, see Fig. 1) were calculated assuming the stoichiometry for complete oxidation of $C_6H_{12}O_6$ (Otten & Gons, 1991b).

Results and discussion

Refractory fractions of phytoplankton and seston

After incubating a suspension of steady-state light-limited *O. limnetica* (specific growth rate $\mu = 0.083\ d^{-1}$) in the dark at 20 °C, almost all algae died within twelve days. Absorbance spectra and dry weight concentrations of the suspension were measured to establish the refractory fraction of the particulate matter and to study the transformation of materials from the particulate into the dissolved phase.

Absorbance of the particles (Fig. 2) was obtained by subtracting the absorbance of 0.45 μm filtered water from the total absorbance. The absorbance above 700 nm is mainly due to particle scattering. In the spectra, until day 9, not only the absorption peaks of chlorophyll *a* (around 440 and 680 nm) but also that of phycoerythrin (around 570 nm) are pronounced. Absorption by C-phycocyanin (around 630 nm) was rather low. The sudden decrease of absorbance between day 9 and 12 was caused by mortality and lysis of most *O. limnetica* cells. Except around 680 nm, the presence of photosynthetic pigments was no longer indicated. From day 15 onwards, the ab-

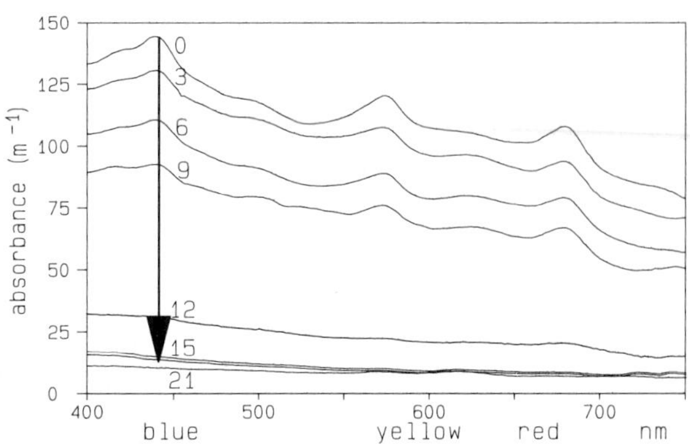

Fig. 2. Absorbance spectra of particulate matter during decay of *Oscillatoria limnetica*. Time course indicated by arrow and day numbers.

sorbance decreased only slightly: mainly refractory detritus was left and it decomposed at a much lower rate.

The decomposition process shows a sigmoid course (Fig. 3; upright triangles), considering the dry weight concentration vs. time data. The data points until day 9 represent the stage characterized by respiration and fragmentation of *O. limnetica* preceeding death (Otten & Gons, 1991b). Cell lysis and rapid decomposition of the labile fraction occurred mainly around day 11. After another ten days of aerobic decomposition the refractory fraction left was 0.24 of the initial biomass. The experiment was repeated with phosphorus-limited algae at the same growth rate ($\mu = 0.083\ d^{-1}$), the refractory fraction being again one-quarter of the initial biomass (Fig. 3; upside-down triangles).

Refractory fractions of freshwater algae vary from 0 to 0.86 (Fallon & Brock, 1979; Jewell & McCarthy, 1971), depending on algal species (Foree & McCarty, 1970; Jewell & McCarty, 1971; Tezuka, 1989), temperature and the prevailing communities of bacteria and zooplankton (Jewell & McCarty, 1971). Earlier, we found the low value of 0.07 for light-limited *O. limnetica* at $\mu = 0.17\ d^{-1}$ (Otten & Gons, 1991a). Our study indicates that the proportion of the refractory fraction of phytoplankton also depends on the growth rate, but is quite insensitive to the type of growth limitation.

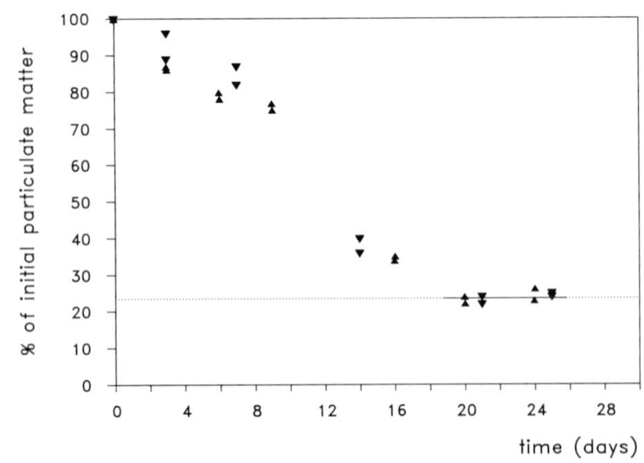

Fig. 3. Time course of dry weight concentration during decay of light-limited (upright triangles) and phosphorus-limited (upside-down triangles) *Oscillatoria limnetica*. Dotted line indicates the refractory matter as % of the initial biomass.

Seston from L. Loosdrecht was incubated in the dark at 20 °C under aerobic conditions and the dry weight concentration was measured over three weeks. This experiment was repeated five times in 1990 and 1991 (Table 1). The particulate refractory fraction of the sestonic matter ranged from 0.50 to 0.63: no seasonality has been found so far. The biomass of non-blue-green algal species in the lake was negligible (P. J. Boesewinkel-de Bruyn, pers. comm.). Thus the phytoplankton biomass constituted maximally 37 to 50% of the total seston dry weight during the study period 1990/1991. These percentages indicate that the relative share of phytoplankton was higher than that estimated for summer 1988 (Gons *et al.*, 1992).

Dissolved detritus

Organic matter mainly enters the dissolved phase during lysis of the phytoplankton cells (Fig. 1) but there is also some contribution during further decomposition of the particles. Part of the dissolved organic matter is available to free-living and attached bacteria, whereas the rest can be considered to be refractory humic compounds (Geller, 1986).

Absorbance spectra of filtrates in the experiment with light-limited *O. limnetica*, described above, showed a marked shift towards the blue

Table 1. Refractory detritus fractions of *O. limnetica* and seston from L. Loosdrecht. Steady-state specific growth rate and the growth-limiting factor are given in parenthesis.

Source	Fraction
O. limnetica (0.083 d^{-1}; phosphorus)	0.24
O. limnetica (0.083 d^{-1}; light)	0.24
O. limnetica (0.17 d^{-1}; light)	0.07
Seston (Aug. 1990)	0.50
Seston (Oct. 1990)	0.55
Seston (Nov. 1990)	0.53
Seston (July 1991)	0.63
Seston (Aug. 1991)	0.50

(Fig. 4) in comparison with the spectra of the particles (Fig. 2). The chlorophyll *a* peak in the red was still present but not that in the blue wavelength band. The absorption peak of phycoerythrin was also absent. Differences with absorption spectra of filtered lake water (Gons *et al.*, 1992; Malthus & Dekker, 1990) are the absorption peak of chlorophyll *a* in the red band and the flattening near 400 nm in the spectra in Fig. 4. These characteristics indicate that the filtrates in the experiment contained very small particles besides true dissolved compounds.

During the experiment the mainly in blue-light absorbing dissolved substances increased (Fig. 4). This increase was greatest in the period of lysis of the organisms but continued until the end of the experiment. The absorption coefficient of filtered lake water at 440 nm is 1–1.5 m^{-1}. In the experiment an absorbance at 440 nm of 8 was reached, which corresponds to an absorbtion coefficient of about 3. The initial dry weight concentration of the *O. limnetica* suspension was 45 g m^{-3}, which is equivalent to the phytoplankton production during 15 to 30 days *in situ* (Rijkeboer *et al.*, 1990b).

Further kinetic studies, including fluxes of allochthonous humic substances, are needed to establish whether a causal relationship underlies the positive correlation between the growth of *O. limnetica* and the concentration of dissolved organic carbon *in situ* (Van Tongeren *et al.*, 1992). Also needed is a qualitative comparison of the dissolved detritus derived in the laboratory with that in the lake (De Haan *et al.*, 1990; De Haan, 1983).

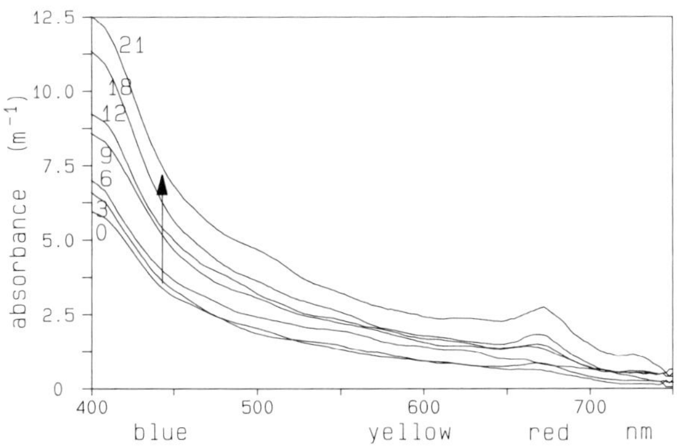

Fig. 4. Absorbance spectra of dissolved matter during decay of *Oscillatoria limnetica*. Time course indicated by arrow and day numbers.

Decomposition rate of particulate detritus

The oxygen uptake of refractory detritus derived from *O. limnetica* was measured using the epipelon trap (see Methods). From these results, the decomposition rate coefficient (k_2; see Fig. 1) was calculated. After three weeks of incubation of *O. limnetica* in the dark the remaining particulate matter was decomposed very slowly. The decomposition rate of this refractory detritus was 0.005 d^{-1} at 20 °C, implying that 16% could still be present after one year. Because the decomposition is likely to slow down later (Saunders *et al.*, 1980) and because decomposition is slower at lower temperatures, these particles may influence the water quality for several years after their origin.

As has been shown above, the major part of the biomass after mortality flows into the pool of labile detritus, which has a quite high decomposition rate (k_1; see Fig. 1): varying from 0.06 to 0.19 d^{-1}, depending on algal growth conditions (Otten & Gons, 1991a, b). Because these decomposition rates are more than an order of magnitude higher than those of the refractory detritus, the labile detritus in general will be a small part of the total of phytoplankton detritus in L. Loosdrecht, like in many other lakes (Wetzel, 1983).

Dynamics of particulate detritus in L. Loosdrecht

To evaluate the impact of the detritus on the underwater light climate and nutrient cycling, it is important to estimate the areal concentration of the detritus in both seston and epipelon ([D_A] gm^{-2}). According to the model for dynamics of detritus (Fig. 1, and Otten & Gons, 1991b), the steady-state value of D_A in L. Loosdrecht is given approximately by the quotient of formation

Hydrobiologia **233**: 69–86, 1992.
L. Van Liere & R.D. Gulati (eds), Restoration and Recovery of Shallow Eutrophic Lake Ecosystems in The Netherlands.

The dynamics and role of limnetic zooplankton in Loosdrecht lakes (The Netherlands)

R. D. Gulati, A. L. Ooms-Wilms, O. F. R. Van Tongeren, G. Postema & K. Siewertsen
Limnological Institute, Vijverhof Laboratory, Rijksstraatweg 6, 3631 AC Nieuwersluis, The Netherlands

Key words: eutrophication, lake restoration, zooplankton grazing, rotifers, crustaceans, P-loading, P-excretion

Abstract

The paper summarizes the results of a ten-year (1981–1991) zooplankton research on the Lake Loosdrecht, a highly eutrophic lake. The main cause of the lake's eutrophication and deteriorating water quality was supply up to mid 1984 of water from the River Vecht. This supply was replaced by dephosphorized water from the Amsterdam-Rhine Canal in 1984. The effects of this and other restoration measures on the lake's ecosystem were studied. Despite a reduction in the external P-load from *ca.* 1.0 g P m^{-2} y^{-1} to *ca.* 0.35 g m^{-2} y^{-1} now, the filamentous prokaryotes, including cyanobacteria and *Prochlorothrix*, continue to dominate the phytoplankton.

Among the crustacean plankton *Bosmina* spp, *Chydorus* sp. and three species of cyclopoid copepods and their nauplii are quite common. Though there was no major change in the composition of abundant species, *Daphnia cucullata*, which is the only daphnid in these lakes, became virtually extinct since 1989. Among about 20 genera and 40 species of rotifers the important ones are: *Anuraeopsis fissa, Keratella cochlearis, Filinia longiseta* and *Polyarthra*. The rotifers usually peak in mid-summer following the crustacean peak in spring. The mean annual densities of crustaceans decreased during 1988–1991. Whereas seston (< 150 μm) mean mass in the lake increased since 1983 by 20–60%, zooplankton (> 150 μm) mass decreased by 15–35%.

The grazing by crustacean community, which was attributable mainly to *Bosmina*, had mean rates between 10 and 25% d^{-1}. Between 42 and 47% of the food ingested was assimilated. In spring and early summer when both rotifers and crustaceans have their maximal densities the clearance rates of the rotifers were much higher. Based on C/P ratios, the zooplankton (> 150 μm) mass contained 2.5 times more phosphorus than seston (< 150 μm) mass so that the zooplankton comprised 12.5% of the total-P in total particulate matter in the open water, compared with only 4.5% of the total particulate C. The mean excretion rates of P by zooplankton varied narrowly between 1.5 and 1.8 μg P l^{-1} d^{-1}, which equalled between 14 and 28% d^{-1} of the P needed for phytoplankton production.

The lack of response to restoration measures cannot be ascribed to one single factor. Apparently, the external P-loading is still not low enough and internal P-loading, though low, may be still high enough to sustain high seston levels. Intensive predation by bream is perhaps more important than food quality (high concentrations of filamentous cyanobacteria) in depressing the development of large-bodied zooplankton grazers, e.g. *Daphnia*. This may also contribute to resistance of the lake's ecosystem to respond to rehabilitation measures.

Introduction

There are only a few long-term studies in Europe dealing with the response of zooplankton to restoration measures (e.g. see in Faafeng & Nilssen, 1981; Gulati, 1990; Ewald, 1991). The present studies on zooplankton in Loosdrecht lakes were started in July 1981 in anticipation of the planned restoration of the lakes (Van Liere *et al.*, 1992). A five-year (1982–1986) study on zooplankton in three shallow and highly eutrophic lakes in the Loosdrecht area did not reveal any significant changes following the restoration measures, which have resulted in considerable reduction in external phosphorus loading since mid-1984 (Gulati, 1990). The research work on the lakes was continued uninterrupted until 1991 when the results were presented at the concluding 'Water Quality research Loosdrecht lakes' (WQL) symposium (Van Liere & Gulati, 1992).

The major aim of the zooplankton research in Loosdrecht lakes has been: 1) to provide baseline data needed for understanding the structure and functioning of these lake ecosystems, and 2) to examine changes in zooplankton dynamics, especially those in response to the restoration measures. Several aspects of zooplankton in the lake have been investigated and published. These included changes in zooplankton densities and grazing, lake's trophic status and food quality (Gulati, 1983, 1984; Gulati *et al.*, 1985); food selection among cladocerans (Irvine, 1986; Gulati *et al.*, 1991a); feeding in rotifers (Gulati *et al.*, 1987, 1992; Ooms-Wilms, 1991; Ooms-Wilms *et al.*, 1992; Wilms *et al.*, 1991); carbon and phosphorus contents and relationships with food, and phosphorus regeneration by zooplankton (Den Oude & Gulati, 1988; Gulati *et al.*, 1989, 1991c).

The main objective of this paper is to present an integrated picture of all aspects of zooplankton studied in the Lake Loosdrecht between 1981 and 1991, with special remarks on changes.

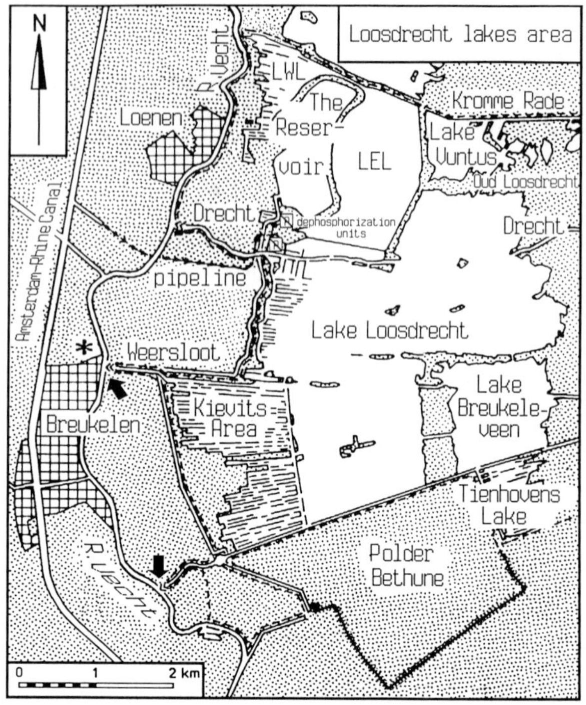

Fig. 1. Location and surroundings of the Lake Loosdrecht.

The lake

The zooplankton study in the Lake Loosdrecht, the largest among lakes of the Loosdrecht area (Fig. 1) was started in July 1981. Important morphometric and limnological features of the Lake Loosdrecht are summarized in Table 1. For a detailed account of the creation, history of formation, morphology and hydrology of Loosdrecht lakes the reader is referred to papers in Loogman & Van Liere (1986) and in Van Liere & Gulati

Table 1. Morphometry and general limnological characteristics of the Lake Loosdrecht (data from various sources).

Area	979 ha
Mean depth	1.85 m
Temperature	11.3 ± 6.1 °C
Water residence time	0.6 year
External P-loading	$0.35\ g\ P\ m^{-2}\ y^{-1}$
Total-P	$89 \pm 22\ mg\ m^{-3}$
Chlorophyll	$93 \pm 18\ mg\ m^{-3}$
Secchi disc	0.3–0.5 m

(1992). Recently, Gulati *et al.* (1991b) have provided a concise account of man's role in the creation and perturbation and the recent attempts to rehabilitate these lakes.

The lakes are situated about 20 km southeast of Amsterdam (Lat. 52° 20′ N; Long. 5° 5′ E) in the province of Utrecht (Fig. 1). They are stripped *Sphagnum* bogs. The first indications of the existence of the Loosdrecht lakes date back to early seventeenth century when dredging of peat under the water surface was formalized in legislation. Precipitation and water pumped from the Polder Bethune on the south and from the River Vecht (up to 1983) on the west formed up to two-thirds of the lake's water input. Evaporation and infiltration (seepage), mainly to the Polder Bethune, are the main sources of water loss from the Loosdrecht lakes.

The water, generally supplied during short periods in summer, from the River Vecht, was the main cause of the lakes' eutrophication and deteriorating water quality. This supply was stopped in 1984 and replaced by water from the Amsterdam-Rhine Canal. This water was first stripped of its phosphorus content by treating with $FeCl_3 \cdot 6H_2O$ in a treatment basin (Fig. 1). This main restoration measure has led to a considerable reduction of the external P-loading of the Lake Loosdrecht since 1984, *viz.* from about 1.1 g P m^{-2} y^{-1} to about 0.35 g P m^{-2} y^{-1} now (Van Liere & Janse, 1992).

The Lake Loosdrecht is dominated by filamentous prokaryotes, cyanobacteria and *Prochlorothrix hollandica*; the former persist throughout the year reaching concentrations of 200×10^3 filaments l^{-1} or more in summer (De Kloet *et al.*, 1991). The mean seston (< 150 μm) mass varies between 8 and 12 mg C l^{-1} (Gulati, 1990), about two-thirds being detritus (Gons & Rijkeboer, 1990). The seasonal means of chlorophyll concentration ranged from 100 to 155 μg l^{-1}, and the mean primary production rates, based on ^{14}C-assimilation, from 1.7 to 2.5 mg C m^{-2} d^{-1} (De Kloet *et al.*,1991).

Methods

The details of sampling procedures and counting techniques used are given in Gulati (1990). The lakes were sampled routinely every four weeks but fortnightly during 1983, 1984 and 1985. A 30-L composite sample was made up of five-litre samples taken from a drifting boat at six stations.

The changes in the zooplankton densities and environmental variables were investigated applying Principal Component Analysis (PCA) to the log-transformed zooplankton species counts, computing linear correlations of the environmental (independent) variables, with PCA axes (for details see in Van Tongeren *et al.*, 1992). The zooplankton densities in 1985 were compared with those measured by Geelen (1955) in 1955 in the Lake Loosdrecht.

The zooplankton (> 150 μm) mass measurements refer to mainly crustaceans. The 'contamination' of this mass by rotifers, coarser algae and detritus was insignificant, except during the occasional stormy periods when resuspension of the peaty sediments did not allow such community biomass measurements. Both the zooplankton mass and seston (< 150 μm) mass were determined as COD and converted to carbon as outlined in Gulati (1990). Setting the mean values of the annual biomasses in 1983 at 100% the data of 1981–1991 were compared. The biomass data of 1985 were compared with those calculated from Geelen (1955) for 1955.

Carbon and phosphorus in the zooplankton and seston fractions were measured routinely starting from 1983 (Gulati *et al.*, 1991c). Phosphorus regeneration by zooplankton was measured in the laboratory (Den Oude & Gulati, 1988) to assess the role of zooplankton in the internal P-loading in the lake; this was done for the period 1987–1990. From the weight-specific P-excretion rates (0.17 μg P mg^{-1} DW h^{-1}) corrected for temperature, P-excretion rates (Den Oude & Gulati, 1988) for the entire study period (1981–1991) were computed.

Grazing rates were determined using the ^{14}C-technique (Gulati *et al.*, 1982), both for the zooplankton (> 150 μm) community in 1981-1984

(Gulati, 1984), and separately for the dominant filter-feeders (Irvine, 1986; Gulati *et al.*, 1991a). Also, rotifer grazing measurements have been attempted for the dominant species, *viz. Anuraeopsis fissa* (Wilms *et al.*, 1991; Ooms-Wilms, 1991) and *Euchlanis dilatata lucksiana* (Gulati *et al.*, 1987; Gulati *et al.*, in press). The clearance rates of the dominant species of both cladocerans and rotifers were measured simultaneously using latex spheres (Sanders *et al.*, 1989) in the diameter range 0.5–6.5 μm as tracers (Ooms-Wilms *et al.*, in press). This was done to assess and compare the grazing pressure of the two communities on food particles which will generally include most of the nanoplanktonic algae and flagellates, as well as bacterioplankton.

Results

Composition and densities

The zooplankton species composition did not change much during the study years. Virtual disappearance of *Daphnia cucullata* since 1981 and the absence of large-bodied *Daphnia* sp. (*D. galeata*) are quite conspicuous of the Lake Loosdrecht. Among the crustaceans, *Bosmina* spp. (*B. longirostris* and *B. coregoni*), *Chydorus sphaericus* (Fig. 2) and cyclopoid copepods (*Diacyclops bicuspidatus, Mesocyclops leuckarti, Acanthocyclops robustus*, and *Cyclops vicinus*) and their nauplli were important (for details see in Gulati, 1990). The annual means of crustacean densities varied more than two-fold, between 402 ± 395 ind l^{-1} in 1990 and 943 ± 883 ind l^{-1} in 1985 (Table 2). From about one-half to two-thirds (45–69%) of the crustaceans on the average were cyclopoid copepods, including nauplii. Virtually in all the study years the annual maximum for the crustaceans fell within a seven-week period from 23 April to 10 June (Fig. 3).

Among about 20 genera and 40 species of rotifers that were recorded in the Loosdrecht lakes (Gulati, 1990), the four most abundant species were *Anuraeopsis fissa*, *Keratella cochlearis*, *Filinia longiseta* and *Polyarthra* (see in Fig. 2). The size among the rotifers varied greatly: from about

Table 2. Mean annual densities (ind l^{-1} ± SD) of crustaceans and relative proportions of copepods and cladocerans in the total and *Bosmina* spp. (*B. longirostris* and *B. coregoni*) in the Lake Loosdrecht.

Year	Crustacean	Copepods %	Cladocerans %	*Bosmina* spp.		
				Total ind l^{-1}	*B. long.* %	*B. cor.* %
1981	447 ± 284	55	45	81 ± 33	–	–
1982	636 ± 612	45	55	257 ± 430	–	–
1983	559 ± 381	61	39	187 ± 213	–	–
1984	714 ± 527	62	38	215 ± 294	–	–
1985	943 ± 883	54	46	204 ± 459	–	–
1986	675 ± 474	62	38	190	19	81
1987	775 ± 647	55	45	192	67	33
1988	465 ± 281	65	35	113	1	99
1989	651 ± 452	69	31	157	27	73
1990	405 ± 418	56	46	119	14	86
1991	402 ± 395	62	38	129 ± 269	11	89

Fig. 2. Photo-micrographs of some important species of cladocerans and rotifers in the Loosdrecht lake; scale for each species is given by vertical bars (100 μm). The names of the species are: 1a, *Bosmina longirostris*; 1b, *B. coregoni*; 2, *Daphnia cucullata*; 3, *Chydorus sphaericus*; 4, *Anuraeopsis fissa*; 5, *Keratella cochlearis*; 6, *Filinia longiseta*; 7, *Polyarthra* sp.; and 8, *Euchlanis dilatata lucksiana*.

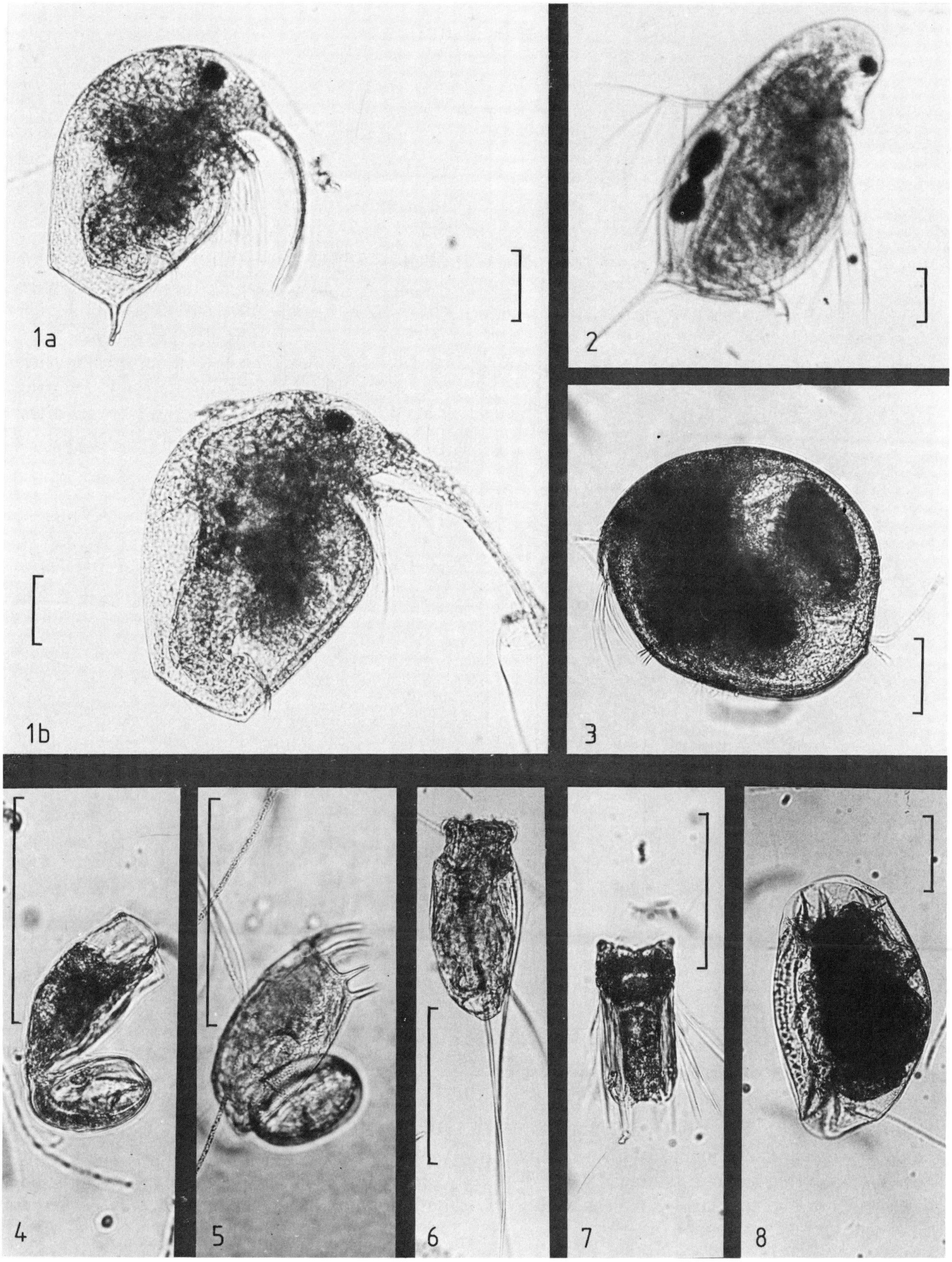
1a
1b
2
3
4
5
6
7
8

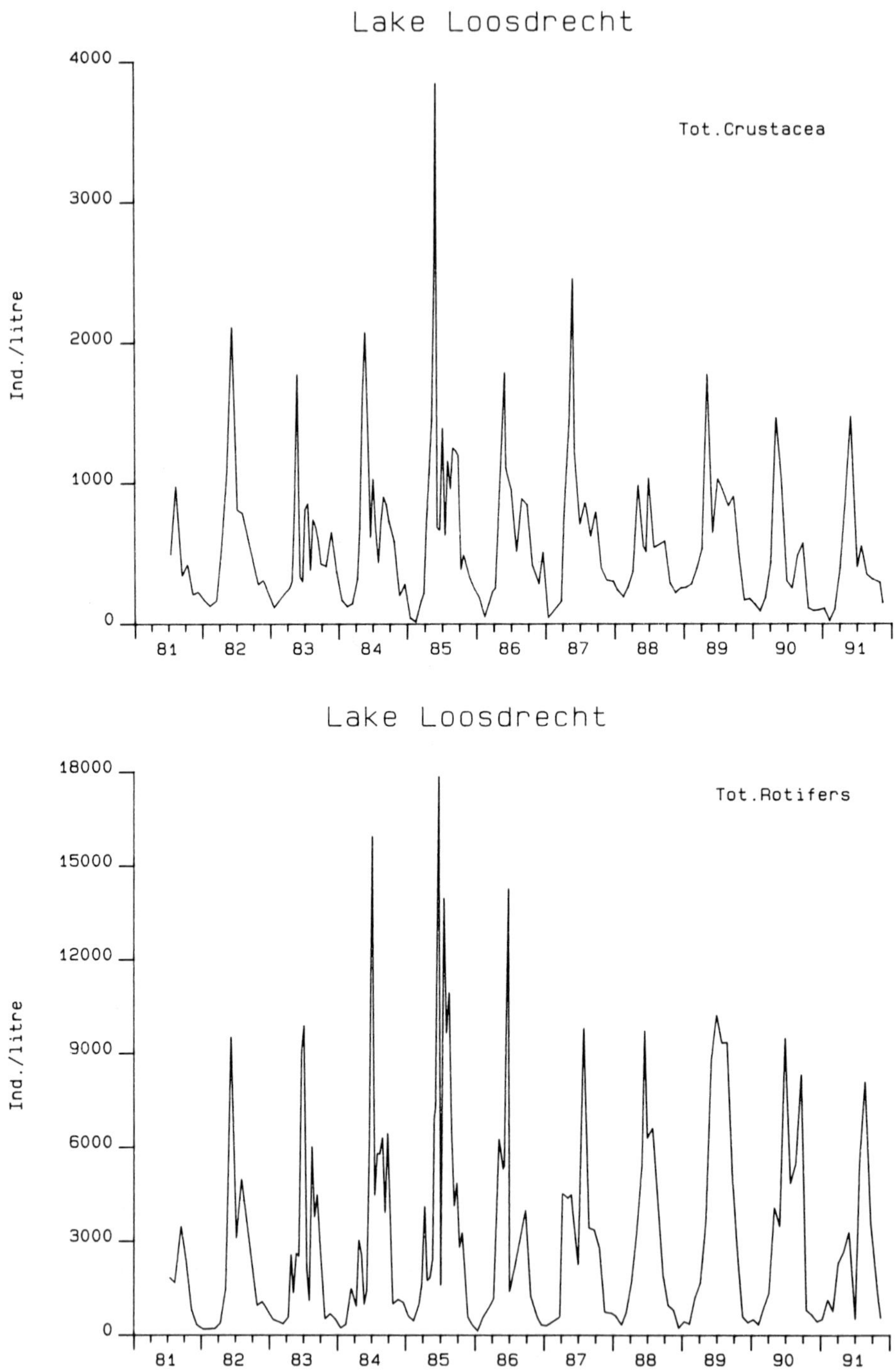

Fig. 3. Densities of pelagial *Crustacea* and *Rotifera* in the Lake Loosdrecht.

60 μm in *A. fissa* to *ca.* 350 μm in *Euchlanis dilatata lucksiana*. Their annual rotifer maximum generally occurred in mid summer, about 3–6 weeks after the main crustacean peak (Fig. 3). *Anuraeopsis* contributed mainly to the seasonal trends of total rotifers throughout the summer period. *Keratella cochlearis*, the second most important rotifer in the lakes, generally peaked in

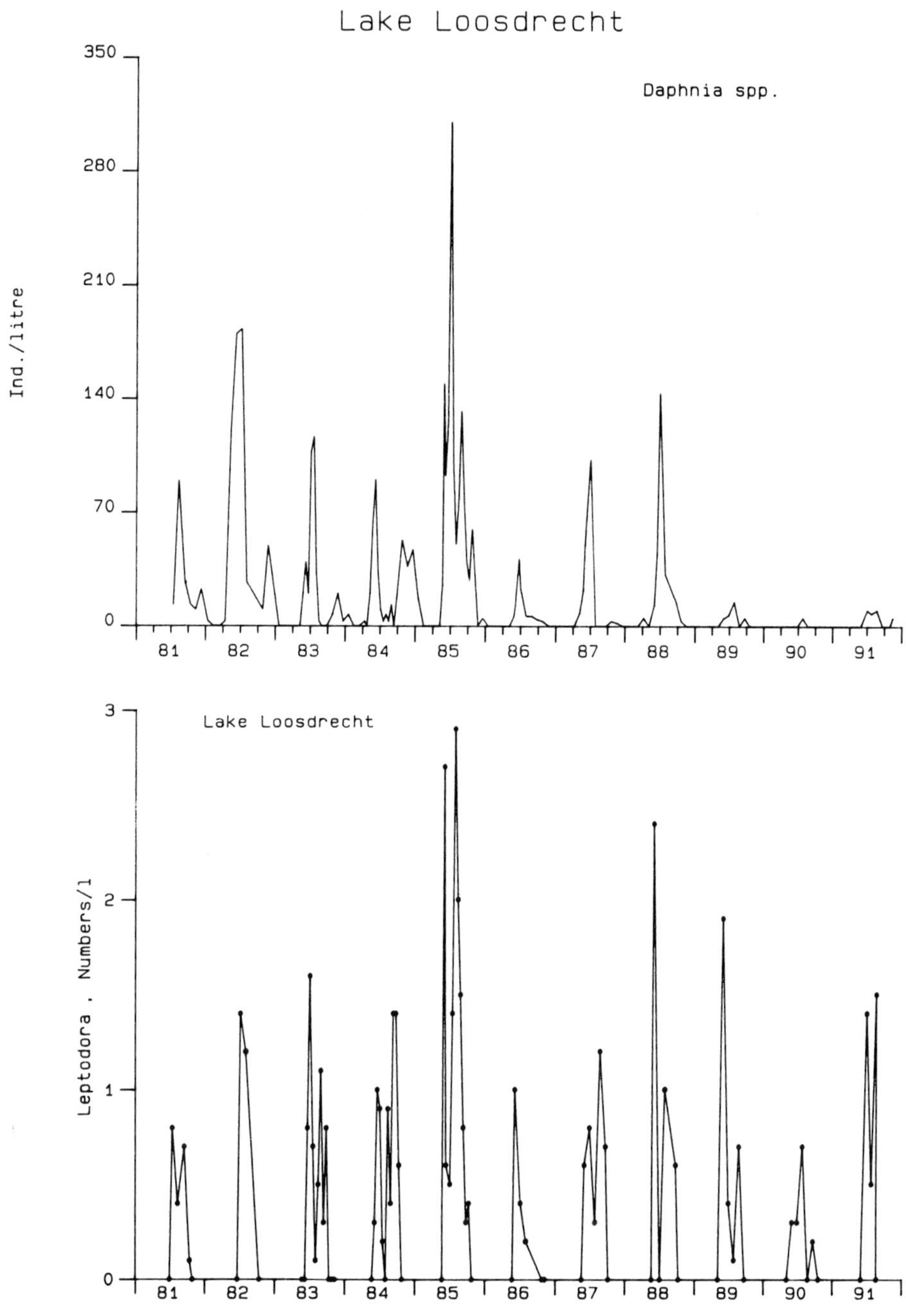

Fig. 4. Daphnia sp. and *Leptodora kindtii* in the Lake Loosdrecht.

spring concomitantly with the maximum of cladoceran filter-feeders (*Bosmina* spp.). But for an increase in the annual maxima of rotifers observed in the years 1984–1986, there was no other noteworthy trend in the rotifer dynamics. The ratio of mean annual densities of rotifers to those of the crustaceans, which was between 3.4 and 5 in the years from 1981 to 1987, increased to between 6 and 7.7 in the following four years.

Bosmina longirostris was generally the first cladoceran to start the spring increase; and it coexisted for a short period in late spring and early summer with *B. coregoni* which generally peaked in mid summer. The mean annual densities of *Bosmina* spp. during 1988–1991 (130 ind l^{-1}) were only about one-third lower than the mean densities in the four-years preceding 1988 (Table 2). The copepods exhibited a rather similar trend in this regard.

Daphnia cucullata was characterized by moderate densities and transient peaks up to 1987, became very sparse in 1989 and was nearly absent in 1990 and 1991 (Fig. 4, upper panel). Its maximal densities in the last three years ranged between 5 and 15 ind l^{-1} and its occurrence was generally restricted to a period during late spring – early summer.

Leptodora kindtii was the only invertebrate predator in the lakes. In the Lake Loosdrecht it was encountered in the limnetic zooplankton samples (> 150 μm) from June to September in densities that usually varied between 0.5 and 1.5 ind l^{-1}, but occasionally reached between 2 and 3 ind l^{-1} as in 1985 (Fig. 4, lower panel). The high numbers in 1985 coincided with those of zooplankton in general, but with those of *Daphnia* sp. and the rotifers in particular.

The PCA revealed that the within-year variation in the composition of zooplankton community is larger than that between the years. A trend over the years was lacking. The order in which the species exhibited their maxima can be read from the plot by viewing them clockwise, starting at the lower end of axis I (left end of the abscissa in Fig. 5). The changes in densities of the species located in the centre of the plot are relatively

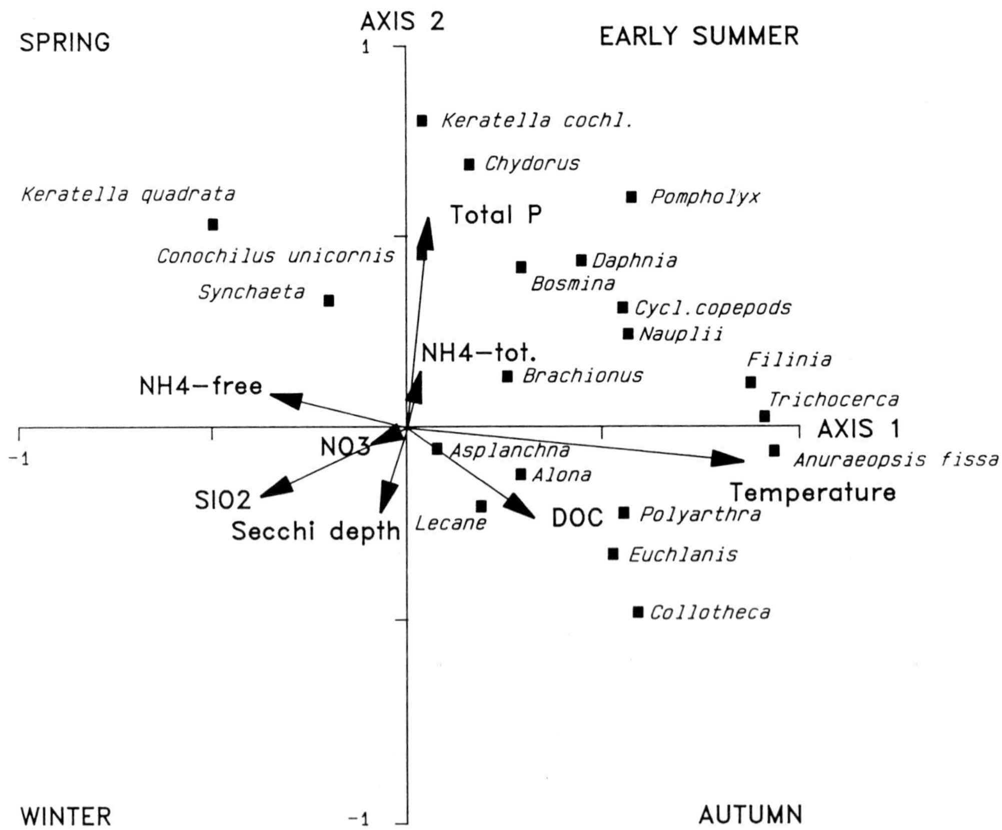

Fig. 5. PCA correlation biplot of limnetic zooplankton and environmental variables; species vectors are not drawn. Correlations between species, between environmental factors, and between species and environmental factors can be derived by projecting the vectors onto each other.

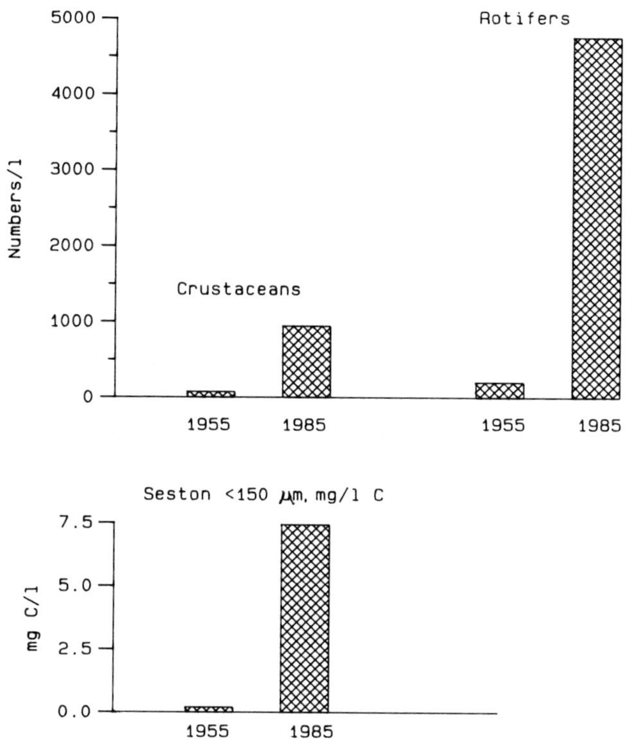

Fig. 6. A comparison of mean annual zooplankton densities and seston mass in 1985 with those in 1955 in the Loosdrecht lake. The data of 1955 are derived from Geelen (1955).

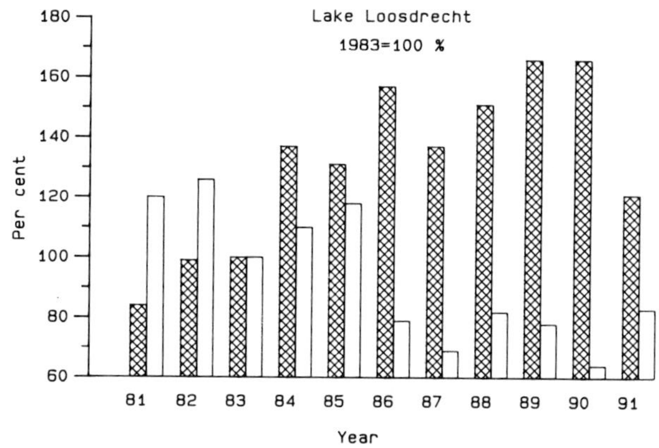

Fig. 7. Changes in the annual means of seston ($< 150\ \mu m$) and zooplankton masses in the Loosdrecht lake expressed as per cent of the respective biomasses in 1983, which were set at 100%; hatched bars, seston; unshade bars, zooplankton.

poorly described, and those towards the periphery are better described. The general pattern of the seasonal succession and deviation from it can be derived from the PCA diagram and from data for separate species. The succession starts in spring with the rotifer species *Keratella cochlearis*.

The correlation of the PCA axis with environmental variables suggests that temperature is the main controlling factor (Fig. 5). This is demonstrated, e.g., by the very high positive correlation between *A. fissa* and temperature on one hand and the high negative correlation between temperature and *K. cochlearis*, a spring form, on the other. The contrast between the late-spring-early-summer period and autumn period lies in the concentrations of Total-P which are higher in the former period than in the latter. The other environmental variables depicted in the figure were rather weakly correlated with the species abundances.

Temporal changes in densities and mass

The long-term comparison indicates 13-fold and 5-fold increases in the crustacean and rotifer densities between 1955 (Geelen, 1955) and 1985, respectively (Fig. 6). The most pronounced increase was, however, noted for seston ($< 150\ \mu m$) which has increased by about forty times since 1955.

The trends in mass of seston and zooplankton during 1981–1991 appeared to be roughly opposite (Fig. 7). The seston mass has increased since 1983 and was between 20 and 60% higher in the years 1984–1991. The zooplankton mass increased during 1984 and 1985 only. Thereafter, it slumped, in the years 1986–1991, to between 65 and 84% of the mean value in 1983.

Grazing

The grazing by the crustacean community had generally a spring maximum caused mainly by *Bosmina* spp. and *Daphnia* sp. (Fig. 8). In 1983, however, both the spring and late autumnal peaks were attributable to *Bosmina* alone. Most other grazing values were less than 25% d^{-1}. The means for daily grazing rates in 1982, 1983 and 1984 were 25%, 10.6% and 11.7%, respectively, and the mean net assimilation rates measured

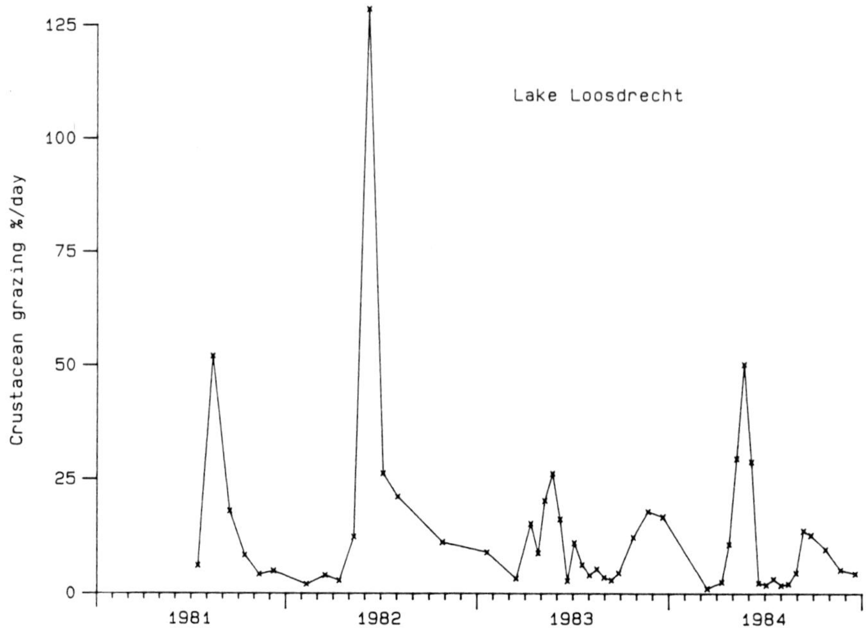

Fig. 8. Daily grazing (%) by zooplankton community (> 150 μm) in the Lake Loosdrecht (1981–1984), measured using ^{14}C-technique (see in Gulati, 1984).

simultaneously in these years were 12%, 4.7% and 4.8% d^{-1}, respectively.

Based on the measured clearance rates of *Bosmina* species in the Loosdrecht lakes during 1983 and 1984 (0.4–0.6 ml ind^{-1} d^{-1}: Gulati, 1984; Gulati *et al.*, 1991a) and *Bosmina* densities, the calculated consumption of *Bosmina* population as % of the community grazing, was found to be *ca.* 61% in 1983 and 72% in 1984. Thus, bosminids, which accounted for about two-thirds of the community grazing, were the most important filter-feeding crustaceans in the Lake Loosdrecht.

However, these community grazing rates do not represent total zooplankton grazing since we measured grazing for the zooplankton fraction > 150 μm, excluding a major part of rotifers. A comparison of the clearance rates (CR) of the most abundant rotifer grazers and important cladocerans shows marked differences in CR between the two taxonomic groups. The mean CR values for rotifers except for *Brachionus* decreased to nearly zero for tracer particles of 6.3 μm; but those for cladocerans increased by one to two orders of magnitude as particle size increased from 0.5 μm to 2.4 μm and 6.3 μm (Ooms-Wilms, unpubl. data). The specific clearance rates (SCR) derived from the respective CR values and weights were inversely related ($P < 0.05$) with the filter-feeders' weight for all the three tracer types (Fig. 9). In the range of weights of our animals, *viz.* decreasing from *Daphnia* (2.7 μg C ind^{-1}) to *Anuraeopsis* (0.009 μg C ind^{-1}), whereas CR decreased by about 30 times, the SCR increased between 80 and 100 times, depending on the size of tracer particles (Fig. 9).

Based on the CR, using latex spheres and literature values, for the cladocerans and rotifers and their respective densities on four dates in May-June 1990, we observed the following daily grazing rates for the two communities: crustaceans, 29%, 8% 18% and 5%; and rotifers, 66%, 32%, 26% and 60%, respectively. This clearly shows that in spring-early-summer, when both communities have their maximal densities, the CR of rotifers may far exceed those of crustaceans. However, this observation is tentative and restricted to food particles in the range 0.5–6 μm.

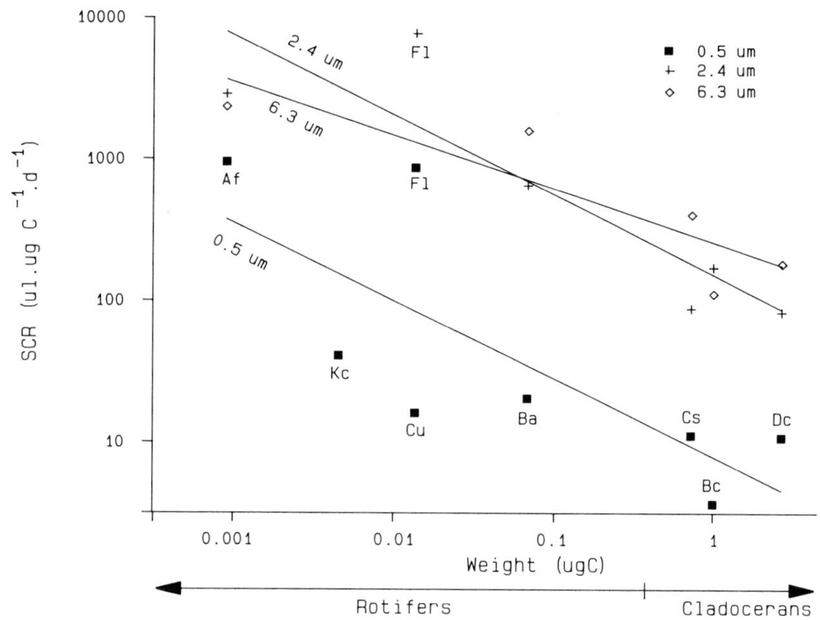

Fig. 9. A comparison of the regression relationships between specific clearance rates (SCR) and weights of some important species of crustaceans and rotifers based on three sizes (0.5 μm; 2.4 μm and 6.0 μm) of microspheres. The SCR was derived from unpublished data of A. Ooms-Wilms. The abbreviations for filterfeeders species are given along the SCR for only the 0.5 μm particles, but unless mentioned refer also to the SCR for the other two particle sizes (see along the upper two regression lines): Af, *Anuraeopsis fissa*; Kc, *Keratella cochlearis*; Cu, *Conochilus unicornis*; Fl, *Filinia longiseta*; Ba, *Brachionus angularis*; Cs, *C. sphaericus*; Bc, *B. coregoni*; and Dc, *D. cucullata*.

Carbon and phosphorus in seston and zooplankton

Carbon and phosphorus relationships of zooplankton and its seston (< 150 μm) food in Loosdrecht lakes revealed pronounced seasonal differences in the C/P ratios of seston, except in the Lake Loosdrecht (Gulati *et al.*, 1991c). In this lake the ratios were generally concentrated around 115 during 1983–1985 but increased in the subsequent years and had a mean annual level of *ca.* 150 in 1990. Also seasonal trends, which were lacking earlier, have become evident (Fig. 10, upper panel), because of increased P-limitation in summers. In contrast to the C/P ratios of seston, the C/P ratios of zooplankton fluctuated more narrowly and were constant around 45. For the six-year period (1983–1988), in which both C and P were measured simultaneously, the zooplankton comprised *ca.* 4.5% C but 12.7% P of the total contents of these elements, respectively, in the particulate suspended matter. Thus, changes in the C and P contents of zooplankton were strongly correlated ($P < 0.001$) such that the C/P ratio remained unaffected.

One main effect of relatively more P in zooplankton than in their sestonic food is that zooplankton contributes more to the total particulate-P (seston-P + zooplankton-P) in the lake, than to the total particulate-C (Fig. 10, lower panel). During the peak densities and biomass of zooplankton as in spring periods, the share of zooplankton-P in the total particulate-P increased markedly and reached up to 20% or more for short periods.

Phosphorus excretion by zooplankton

The computed rates of P excretion by zooplankton (Fig. 11) show that but for two high values in 1982, all others are less than 5 μg PO_4-P l^{-1} d^{-1}.

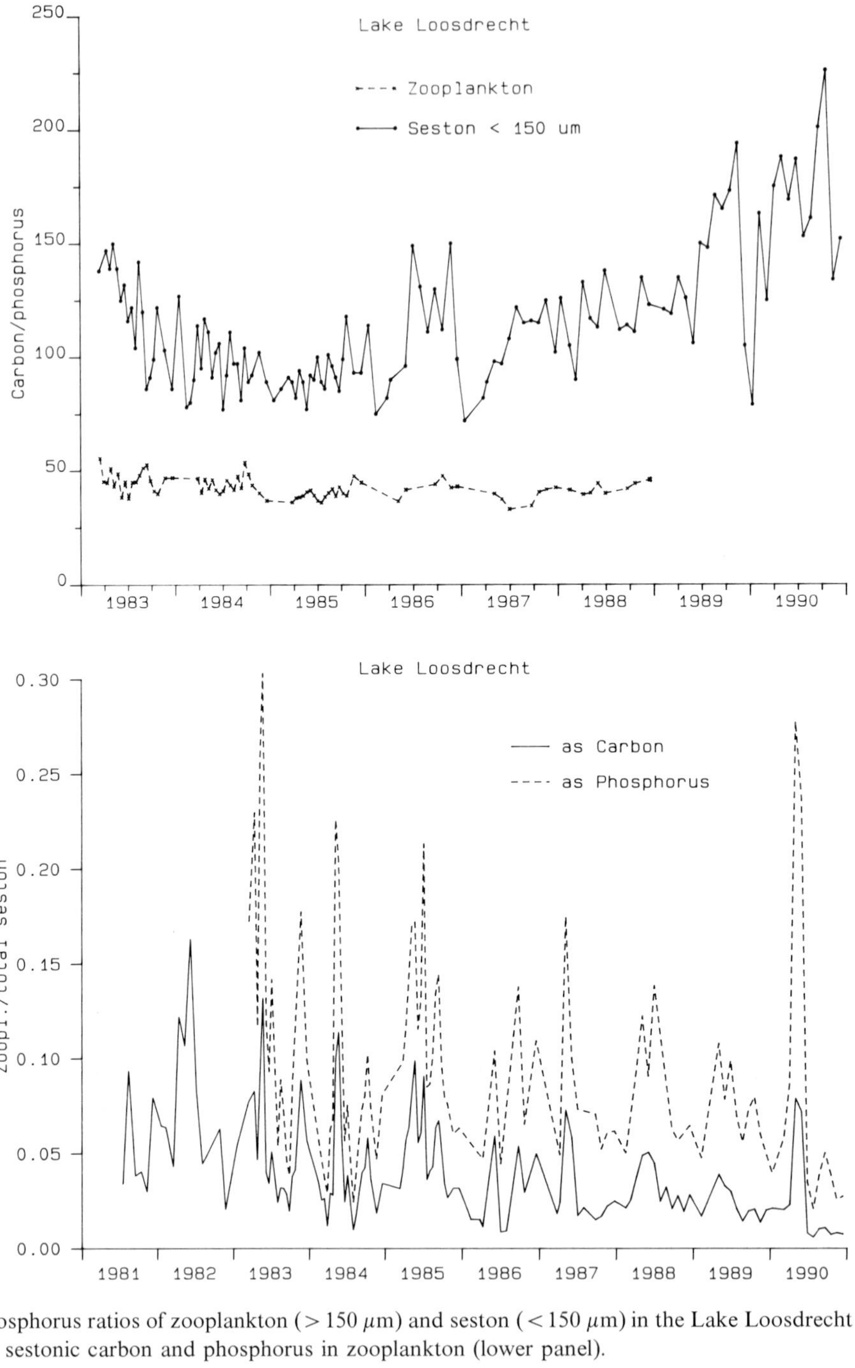

Fig. 10. Carbon-phosphorus ratios of zooplankton (> 150 μm) and seston (< 150 μm) in the Lake Loosdrecht (upper panel); and relative proportions sestonic carbon and phosphorus in zooplankton (lower panel).

Generally, the primary annual peak which occurred in spring was caused mainly by the increase in crustacean biomass, especially *Bosmina* spp. The second peak, if recorded, was in mid summer when zooplankton was generally present in modest densities, but the temperature was at its annual maximum (18–20 °C). Mean excretion rates in May-September varied narrowly between 1.5 and 1.8 μg P l^{-1} d^{-1}.

The measured excretion rates in *Euchlanis di-*

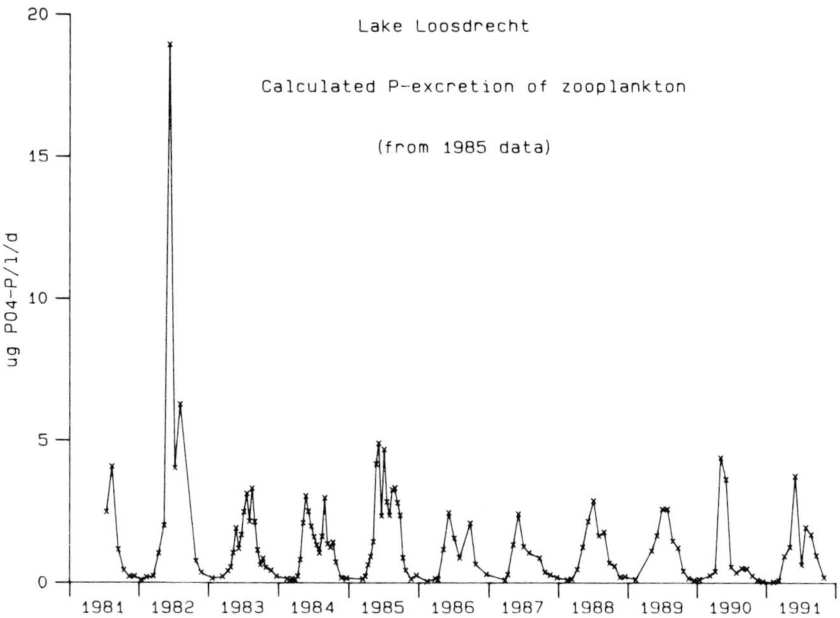

Fig. 11. Daily excretion of phosphorus by the zooplankton (> 150 μm) community in the Lake Loosdrecht. The data are based on routine measurements on the *in situ* community during 1985–1986 in the ambient lake water, and the regression model ($Y = 0.436e^{0.145X}$) for relationship between water temperature (X) and specific P excretion (Y): μg P mg^{-1} C zoop d^{-1} (Den Oude & Gulati, 1988).

latata lucksiana (isolated from and cultured in the lake water) of 0.06-0.18 μg P mg^{-1} DW h^{-1} (Gulati *et al.*, 1989), were similar to the mean weight-specific rates for crustacean community. However, these rates were not representative for the main rotifer species in the Lake Loosdrecht, because these species (*A. fissa* and *K. cochlearis*) were much smaller, weighing about an order-and-one-half less than *Euchlanis*. Moreover, this large euchlanid is relatively sparse and restricted to late summer and early autumn periods.

Role of zooplankton in P-recycling

The calculated P-recycling in the Lake Loosdrecht, for the years 1987–1990, is based on phosphorus in seston and zooplankton and its recycling *via* zooplankton grazing and excretion and phytoplankton production rates (Table 3). Zooplankton, *via* its grazing, removed daily between 4 and 6% of the P stock in seston (item 3, Table 3) but returned more than half of this gross

Table 3. Phosphorus recycling by crustacean zooplankton in the Lake Loosdrecht, based on means of May-September during 1987–1990. The ingestion (I), assimilation (A), and egestion of sestonic P by zooplankton were measured as carbon but converted to P using C/P ratios of seston (Gulati *et al.*, 1991c); the zooplankton P-excretion (E) is based on measurements by Den Oude & Gulati (1988). The zooplankton P losses other than *via* excretion (E) i.e. due to predation and natural mortality, are derived as the difference between A and E. Uptake of P during phytoplankton production (item 8) was assumed to be in the same ratio as C/P ratios measured in seston, and item 6 divided by item 8 gives item 9, i.e., the phytoplankton daily P requirements covered by daily zooplankton excretion. All the biomass data are expressed as μg P l^{-1} and rates as μg Pl^{-1} d^{-1}.

Parameter	1987	1988	1989	1990
1. Seston mass	86	84	74	59
2. Zooplankton mass	8.7	8.8	6.5	8.3
3. Ingestion (I)	5.5	5.9	3.4	3.5
4. Assimilation (A)	2.2	2.6	1.5	1.5
5. Egestion (I–A)	3.3	2.3	1.9	2.0
6. Excretion (E)	1.4	1.5	1.8	1.6
7. Zooplankton loss (A–E)	0.8	2.1	− 0.3	− 0.1
8. Primary produktion (PP)	10.2	9.1	7.1	5.8
9. E/PP × 100 (% d^{-1})	13.7	16.1	25.3	27.6

P-uptake as egestion. The fate of P in the egestion products was difficult to quantify since these products will partly lose some $P0_4$-P due to leaching but most of it in particulate form will re-enter the seston as detrital-P. If we consider leaching of P from the egestion products as insignificant, the daily P-input to seston (item 1) pool *via* zooplankton egestion (item 5), will be roughly 25–33% (item 5/item 8) of the P input *via* the phytoplankton production. Simultaneously, zooplankton excretion equalled between 14 and 28% of the phytoplankton production demand of P (see item 9 in Table 3).

The importance of zooplankton P-excretion, as a potential P-source for phytoplankton production, increased markedly in 1989 and 1990 compared with the preceding two years. This is due to a decrease in primary production in 1989–1990, rather than to increased zooplankton excretion rates.

The zooplankton recycling of its body-P *via* excretion (item 6/item 2, in Table 3) was about 0.20 d^{-1}, the annual means ranging between 0.19 and 0.28 d^{-1}.

Discussion

Here we attempt an integrated discussion of the results of 10-year monitoring of zooplankton densities and biomass, and studies on zooplankton feeding and excretion rates, including the more recent work on rotifer feeding.

There is practically no change since 1986 in composition of abundant species of either rotifers or crustaceans (Gulati, 1990). Moreover, because of the normal year-to-year oscillations, the amplitude of changes in annual mean densities do not warrant clear-cut conclusions on the likely trends, either natural or in response to restoration measures. This has been also confirmed by analysis of zooplankton data in the light of physicochemical variables, using Canonical Ordination Techniques (Van Tongeren *et al.*, 1992). The question arises: 'have the basic conditions that should bring about a change in zooplankton composition or abundance been brought about or met by the rehabilitation measures'?. A cursory answer to this is: no! In this regard the important aspects to be examined critically are the zooplankton food and predation. We have noted that, contrary to the expected decreases, the seston mass (this study) and the filamentous cyanobacteria have, in fact, increased (see e.g. in Van Tongeren *et al.*, 1992). Moreover, this latter has occurred despite the increasing evidence of P-limitation derived both from laboratory data (Rijkeboer *et al.*, 1991) and increase in C/P ratios of seston (Gulati *et al.*, 1991c). The reduction in P-load to *ca.* 0.3 g P m^{-2} y^{-1} is apparently less than expected (Van Liere & Janse, 1992) or needed. Therefore, filamentous cyanobacteria still continue to dominate this lake system. There are no recent data on predation by planktivorous fish, nor do we know if the standing crop of bream (*Abramis brama*) has changed. It may be mentioned that a major fish kill involving bream', besides management and P-reduction measures in the Schlachtensee, resulted within three years in important shifts in zooplankton structure (Ewald, 1991). The changes relate to the appearance of large-bodied *Daphnia* (*D. galeata*) and reduction in the numbers of *Bosmina* sp.

The Loosdrecht lakes differ markedly from other eutrophic lakes in The Netherlands, e.g. Lake Tjeukemeer which has a net P-retention of about 10% of the external-P load of 25–50 mg P m^{-2} d^{-1} (Van Huet, 1991). This is more than an order-of-magnitude higher than the net daily retention calculated for the Lake Loosdrecht by Janse *et al.* (1992). Moreover, the total-P concentration in the Tjeukemeer is also much higher (Lammens *et al.*, 1992). However, despite these differences in phosphorus concentrations, the mean chlorophyll contents during spring and summer in the two lakes (130–150 μg l^{-1}) are similar. Nevertheless, in the Lake Loosdrecht throughout the year more than 90% of the chlorophyll comes from filamentous cyanobacteria and the prochlorophyte *Prochlorothrix hollandica*. On the other hand, in Tjeukemeer green algae and diatoms contribute nearly as much to the chlorophyll as the filamentous prokaryotes, at least in spring period. Therefore, the food situation in the

Lake Loosdrecht during early spring is less optimal for cladocerans than in Tjeukemeer. Even though the total fish stock in the two lakes is comparable, namely *ca.* 300 kg ha^{-1}, and both have bream as the most dominant planktivorous fish, essential differences can be observed in the size composition of fish (Lammens *et al.*, 1992). Whereas in the Tjeukemeer standing stock of bream <25 cm is only about 50 kg ha^{-1}, in the Lake Loosdrecht it is three times as high –150 kg ha^{-1}. On the other hand, whereas the Tjeukemeer contains about 150 kg ha^{-1} bream >25 cm, the Loosdrecht has about 80 kg ha^{-1} of this. These are, however, very rough estimates based on annual means of several years since differences in individual years may be pronounced but will be offset by averaging over the years. Undoubtedly, the effects of the differences between the lakes in their nutrient status, producers' composition and relative abundance and of fish size composition and standing crop are likely to be reflected in the producer-consumer interphase, the zooplankton.

We do not know to what extent the absence in the Lake Loosdrecht of *Daphnia galeata* (which is quite common in many other eutrophic lakes in the Netherlands) can be attributed to food quality or fish predation. From the lakes undergoing biomanipulation (see e.g. Lyche *et al.*, 1990; and other references in Gulati *et al.*, 1990) it is clear that with a decrease in the standing crop of planktivorous fish, the share of larger grazers in zooplankton generally rises and remains high. An intriguing factor also in the Lake Loosdrecht is the virtual extinction since 1989 of *D. cucullata*, which is the smallest among the *Daphnia* species in the Dutch lakes, and which generally co-occurs with *D. galeata*, irrespective of the trophic status. At present one can only conjecture that predation by bream has become even more intense and has markedly reduced the *Daphnia* population in Loosdrecht lakes.

In the absence of large-bodied *Daphnia* species and because of the disappearance of *D. cucullata*, one would expect the planktivores, particularly bream <25 cm, to exert a high pressure on cyclopoid copepods and bosminids. But because of relatively smaller size, more so for copepods with their much smaller juvenile stages, can escape fish predation. The bosminids are apparently well adapted to the food dominated by cyanobacteria (Gulati *et al.*, 1991a), which are abundant even in early spring. The bosminids can generally quickly build-up their populations, in the absence of daphnid competitors, which may cause quick and severe depression of food resources (see e.g. in Romanovsky, 1985).

The grazer-induced net mortality (assimilation) of cyanobacteria is essentially low (4.7–12% d^{-1}) as observed from the grazing studies, and due mainly to *B. coregoni*. In view of the assimilation efficiencies of around 45%, the amount egested will be about 55% of that ingested. The role of bosminids in transforming cyanobacteria into detritus may not be unimportant since in the Lake Loosdrecht constantly a high proportion of seston (<150 μm) exists as detritus (Gons *et al.*, 1990). Also, model studies on phosphorus flow in the lake's ecosystem (Janse *et al.*, 1992: Fig. 2) reveal that, based on carbon, the proportion of detritus consumed will be even higher because of the expected higher C/P ratios of detritus than cyanobacteria and other planktonic food-items. Besides, lack of competition among cladocerans to feed on cyanobacteria may be related to differential use of food particles by *Bosmina* and *Daphnia* species (Irvine, 1986). Moreover, the bosminids exert a much stronger effect due to a more marked effect of body length on clearance rates, than *Daphnia*, feeding on particles both in the size range 0–7 μm and 7–15 μm. In laboratory studies using the Lake Loosdrecht water and large-bodied *Daphnia* (*D. magna*) Dawidowicz *et al.* (1988) demonstrated that the fecundity of the reared daphnids markedly decreased as the cyanobacterial concentration was increased to 80000 filaments l^{-1}. Though the measured ingestion rates were higher than the animals reared on 'diluted' lake water, the advantage of higher consumption rates at the high food levels appears to be more than off set by the expected very high respiratory cost of food gathering/rejection. Thus, besides the intense predation by bream, cyanobacterial dominance in the lake is not conducive to the establishment of large-bodied forms. *Leptodora kindtii*, despite its large size, can escape predation by fish, both because of its transparent body and tendency to migrate to the bottom layers in the day-time (J. Vijverberg, pers. comm.). It has persisted, but we do know its role as a predator on microzooplankton on which it is most likely to feed.

Though the planktivore predators, and perhaps in some situations food, adversely affect the optimal development of crustacean zooplankton, especially the large-bodied filter-feeding forms, the rotifers, in contrast, are abundant and have high diversity indicating favourable conditions. Two important features of rotifers need to be emphasized: 1) they appear to be an important factor in the mortality of nanoplankton ($<10\ \mu m$), including the unidentified flagellates, diatoms and bacterioplankton; and 2) they have much higher specific clearance rates on the small food types than cladocerans, and because of their much higher densities appear to be an important link in the microbial food chain. Besides, the small size confers upon the rotifers an added advantage in escaping predation better than the crustaceans. These characteristics make rotifers superior to crustaceans under the existing conditions, as well as have probably allowed them to virtually outcompete the latter during the process of eutrophication of the Loosdrecht lakes. Their success appears to have gone hand in hand with the increased success of bream in this lake ecosystem. The same factors which adversely the affect the crustaceans in Loosdrecht lakes perhaps also prevent the large-bodied rotifers, namely the brachionids (*Euchlanis* sp. and *Brachionus* sp.), from establishing. Among these, *Euchlanis dilatata lucksiana* represents both a unique and extreme organism: being among the largest of the rotifer taxa, it is able to consume prokaryotic filamentous plankton, cyanobacteria and *Prochlorothrix hollandica*, of varying lengths (Gulati *et al.*, in press). However, the failure of this rotifer to establish and build up a sizeable population, in mid summer, can be explained by its vulnerability to predation (Ejsmont-Karabin *et al.*, 1988). Also, *Brachionus* species which are known to attain densities as high as 17000 ind l^{-1} in eutrophic lakes, as in Lake Søbygard (Jeppesen *et al.*, 1990), were never abundant in the Lake Loosdrecht. Their numbers are generally high if green algae (e.g. *Scenedesmus* spp.) on which they feed (Gilbert & Starkweather, 1978) are available in high concentrations.

The role of zooplankton in P-recycling in the lake will be grossly underestimated if the excretion rates of rotifers are not included. This point has been earlier emphasized by Gulati (1990). The computed excretion rate of rotifers for the summer of 1985, using the rotifer densities and the relationships of excretion rate with size and temperature given by Ejsmont-Karabin (1983) was $33 \pm 19.7\ \mu g\ P\ l^{-1}\ d^{-1}$ which is thirteen times higher than that measured for the crustaceans by Den Oude & Gulati (1988), *viz* $2.4\ \mu g\ P\ l^{-1}\ d^{-1}$. Though these rates are very high, as well as their extrapolation to lakes is questionable, it does, nevertheless, indicate that zooplankton excretion, especially that of rotifers, can be an extremely important factor in the lake's internal P-loading.

Lastly, the zooplankton research as well as other studies on Loosdrecht lakes have gone a long way in increasing our insight into the functioning as well as the resilience of a hypertrophic lake ecosystem. The expected changes, or, taking a retrospective view, the lack thereof, have been the driving force of the long-term, integrated studies on the Lake Loosdrecht.

Acknowledgements

The first author is highly indebted to Dr. Sikko Parma, the Director of the Limnological Institute, who encouraged zooplankton research in the Loosdrecht lakes from its start in 1981; he provided the modest funds at his disposal to support the stay at the Institute of several foreign scientists to cooperate in the zoological research. Our colleague Dr. Koos Vijverberg made several useful suggestions to improve the manuscript. The European Community funded the investigations on zooplankton excretion and rotifer grazing. We are grateful to Dr. Louis van Liere, the WQL coordinator, who contributed to the success of zooplankton studies silently but surely. We also thank Mrs. Cecilia Janssen-Kroon for taking care of the lay-out of this typescript and of publications and reports on zooplankton that have appeared since 1984.

References

Dawidowicz, P., Z. M. Gliwicz & R. D. Gulati, 1988. Can *Daphnia* prevent a blue-green algal bloom in hypertrophic lakes? A laboratory test. Limnologica 19: 21–26.

De Kloet, W. A., P. J. Boesewinkel-de Bruyn & L. Breebaart, 1991. Phytoplankton and its production rates in the Loosdrecht lakes before and after P-load reduction. Verh. int. Ver. Limnol. 24: 715–718.

Den Oude, P. J. & R. D. Gulati, 1988. Phosphorus and nitrogen excretion rates of zooplankton from the eutrophic Loosdrecht lakes, with notes on other P sources for phytoplankton requirements. Hydrobiologia 169: 379–390.

Ejsmont-Karabin, J., 1983. Ammonia nitrogen and phosphorus excretion by planktonic rotifers. Hydrobiologia 104: 231–236.

Ejsmont-Karabin, J., R. D. Gulati & J. Rooth, 1989. Is food availability the main factor controlling the abundance of *Euchlanis dilatata lucksiana* Hauer in a shallow, hypertrophic lake? Hydrobiologia 186/187: 29–34.

Ewald, S., 1991. Long-term changes of crustacean plankton during successful restoration of Lake Schlachtensee (Berlin-West). Verh. int. Ver. Limnol. 24: 866–872.

Faafeng, B. A. & Jens Petter Nilssen, 1981. A twenty-year study of eutrophication in a deep, soft-water lake. Verh. int. Ver. Limnol. 21: 412–424.

Geelen, J. F. M., 1955. The plankton of the lakes. Unpublished report no. 38. Amsterdam Municipal Waterworks (in Dutch).

Gilbert, J. J. & P. L. Starkweather, 1978. Feeding in the rotifer *Brachionus calyciflorus* III. Direct observations on the effect of food type, food density, change in food type and starvation on the incidence of pseudotrochal screening. Verh. int. Ver. Limnol. 20: 2382–2388.

Gons, H. J. & M. Rijkeboer, 1990. Algal growth and loss rates in Lake Loosdrecht: first evaluation of the roles of light and wind on a basis of steady state kinetics. Hydrobiologia 191: 129–138.

Gulati, R. D., 1983. Zooplankton and its grazing as indicators of trophic status in Dutch lakes. Envir. Monit. Ass. 3: 343–354.

Gulati, R. D., 1984. Zooplankton and its grazing as measures of trophy in the Loosdrecht lakes. Verh. int. Ver. Limnol. 22: 863–867.

Gulati, R. D., 1990. Zooplankton structure in the Loosdrecht lakes in relation to trophic status and recent restoration measures. Hydrobiologia 191: 173–188.

Gulati, R. D., K. Siewertsen & G. Postema, 1982. The zooplankton: its community structure, food and feeding and role in the ecosystem of Lake Vechten. Hydrobiologia 95: 127–163.

Gulati, R. D., K. Siewertsen & G. Postema, 1985. Zooplankton structure and grazing activities in relation to food quality and concentration in Dutch lakes. Arch. Hydrobiol. Beih. 21: 91–102.

Gulati, R. D., J. Rooth & J. Ejsmont-Karabin, 1987. A laboratory study of feeding and assimilation in *Euchlanis dilatata lucksiana*. Hydrobiologia 147: 289–296.

Gulati, R. D., J. Ejsmont-Karabin, J. Rooth & K. Siewertsen, 1989. A laboratory study of phosphorus and nitrogen excretion of *Euchlanis dilatata lucksiana*. Hydrobiologia 186/187: 347–354.

Gulati, R. D., E. H. R. R. Lammens, M.-L. Meijer & E. Van Donk (eds), 1990. Biomanipulation, tool for water management (First International Conference 8-11 August, 1989, Amsterdam). Developments in Hydrobiology 61: Kluwer Academic Publishers, Dordrecht, 628 pp.

Gulati, R. D., C. Vuik, K. Siewertsen & G. Postema, 1991a. Clearance rates of *Bosmina* species in response to changes in trophy and food concentration. Verh. int. Ver. Limnol. 24: 745–750.

Gulati, R. D., L. Van Liere & K. Siewertsen, 1991b. The Loosdrecht lake system: Man's role in its creation, perturbation and rehabilitation. In: Terrestrial and Aquatic Ecosystems: Perturbation and Recovery; ed. O. Ravera. Chichester, Ellis Horwood. pp. 593–606.

Gulati, R. D., K. Siewertsen & L. Van Liere, 1991c. Carbon and phosphorus relationships of zooplankton and its seston food in Loosdrecht Lakes. Mem. Ist. ital. Idrobiol. 48: 279–298.

Gulati, R. D., J. Ejsmont-Karabin & G. Postema, in press. Feeding in *Euchlanis dilatata lucksiana* Hauer on filamentous cyanobacteria and a prochlorophyte. Hydrobiologia.

Irvine, K., 1986. Differential feeding behaviour of the dominant Cladocera as an explanation of zooplankton community structure in the Loosdrecht Lakes. Hydrobiol. Bull. 20: 121–134.

Janse, J. H., T. Aldenberg & P. R. G. Kramer, 1992. A mathematical model of the phosphorus cycle in Lake Loosdrecht and simulation of additional measures. Hydrobiologia 233: 119–136.

Jeppesen, E., M. Søndergaard, O. Sortkjaer, E. Mortensen & P. Kristensen, 1990. Interactions between phytoplankton, zooplankton and fish in a shallow, hypertrophic lake: a study of phytoplankton collapses in Lake Søbygard, Denmark. Hydrobiologia 191: 149–164.

Lammens, E. H. R. R., P. J. Boesewinkel-de Bruyn, H. Hoogveld & E. Van Donk, 1992. Phosphorus-load, phytoplankton, zooplankton and fish stock in Loosdrecht Lake and Tjeukemeer: confounding effects of predation and food availability. Hydrobiologia 233: 87–94.

Loogman, J. G. & L. van Liere (eds), 1986. Proceedings of the Waterquality Research Loosdrecht Lakes Symposium on:

important exception to the rule was the application of biomanipulation in lake Breukeleveen, one of the Loosdrecht lakes, where the reduction in fish biomass had hardly any effect (Van Donk *et al.*, 1990). The high density of filamentous cyanobacteria and the large impact of turbulence causing wind seemed to be the most important counteracting effects. Enclosure experiments in Tjeukemeer did not result in clear water either, even though no fish were present (Richter *et al.*, 1987).

Because of the difficulties experienced in improving the transparency in Lake Breukeleveen and in Tjeukemeer enclosures by biomanipulation, we were interested in comparing the food chains in these lakes. We used data concerning external P load, total-P concentration, concentrations of the main phytoplankton groups (Cyanobacteria, Chlorophyceae and Diatomeae), the density and average size of main zooplankton species or groups and fish species density and composition for the period 1984–1990. We were mainly interested in the difference between the lakes rather than the seasonal or yearly trends (see Van Tongeren *et al.*, this volume) and therefore we compared only the averages of the May–October period.

Fig. 1. Location of the Loosdrecht lakes (Lo) and Tjeukemeer (Tj) in the Netherlands.

Study area

Tjeukemeer is a shallow (average depth 1.5 m), eutrophic and turbid (Secchi-disc 25–35 cm) freshwater lake with a surface area of 21.5 km^2. Vegetation is rare, and is represented by only a limited amount of reed. As the lake is shallow and exposed to wind, summer stratification does not occur and in consequence the phytoplankton and zooplankton are almost homogeneously distributed (De Nie *et al.*, 1978). The lake is part of an interconnected system of water bodies, which receives water from the nearby lake IJsselmeer during summer and from the surrounding polders in winter, when precipitation exceeds evaporation (Fig. 1). The management of the hydrological regime is designed to control the water level which is required for agricultural purposes. The Loosdrecht lake (980 ha) and Lake Breukeleveen (180 ha and a part of the Loosdrecht system) are similar to Tjeukemeer regarding depth and turbidity, although the external P-load is different. Phosphorus reduction measures have reduced the yearly external P-load in Loosdrecht lakes to *ca.* 0.37 g m^{-2} (Van Liere *et al.*, 1990a; Van Liere & Janse, this volume), whereas the yearly external P-load in Tjeukemeer is 25–50 g m^{-2}, mainly because the lake receives drainage from the agricultural area (Van Huet, 1991). However, P-retention in the latter lake is only 10%, while in the Loosdrecht lakes it is about 60%.

Methods and materials

Fish were collected in the period May–October in the years 1984–1990. In Tjeukemeer the fish were caught once a month, while in Loosdrecht Lake they were taken three times per year. In both lakes a small-mesh trawl (5.5 mm cod end) was used at five stations and standard hauls were made at a speed of 1 m s^{-1} for 10 min. Length-frequency distributions (fork length) and total numbers of each species were determined. Catches of young-of-the-year were only used for the months September–October, because before that period the fish were still too small to be retained by the trawl or confined to the inshore area.

Zooplankton was sampled at the same time and place of trawling. Using a 5-L Friedinger sampler with a cylinder of 60 cm height, two samples were taken at each of the five stations, one just below the water surface and the other just above the bottom. The samples were concentrated by filtering through a 120 μm mesh and preserved in 4% formalin. The combined sample was subsampled with a whirling vessel after Kott (1953), and organisms were identified and their lengths measured. The length of the cladocerans was measured from the top of the head or, if a helmet was present, from the base of the helmet to the base of the tail spine.

In Tjeukemeer phytoplankton was sampled fortnightly, while in the Loosdrecht lakes it was sampled every four weeks. Further details of the sampling procedure and the method used for counting are given in Moed & Hoogveld (1982), Gulati (1990) and Van Tongeren *et al.* (1992). For sampling and determination of chlorophyll-*a* and total P concentration see Moed (1973), De Haan & Moed (1984) and Van Liere *et al.* (1990b).

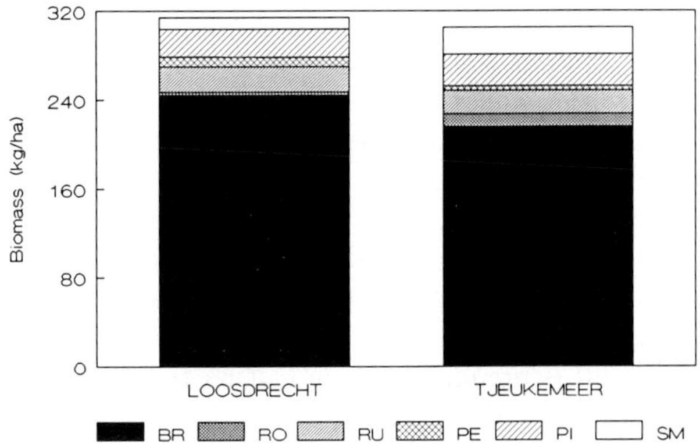

Fig. 2. Biomass per ha of bream (BR), roach (RO), ruffe (RU), perch (PE), pikeperch (PI) and smelt (SM) in the Loosdrecht lakes and Tjeukemeer as an average for the periods May–September in 1984–1990.

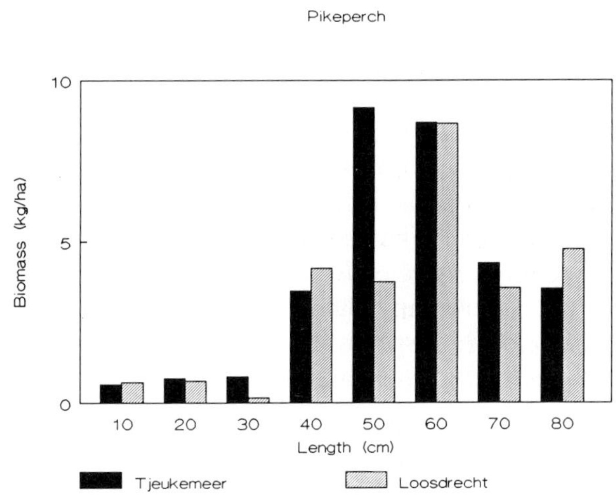

Fig. 3. Size distribution of pikeperch in Loosdrecht Lake and Tjeukemeer expressed in biomass per 10 cm length class averaged for the periods May–September in 1984–1990.

Results

The biomass and species composition of the fish communities in Loosdrecht Lakes and Tjeukemeer are similar. The total biomass in both lakes is *ca.* 300 kg ha^{-1} and is dominated by bream (68% in Tjeukemeer, 77% in Loosdrecht Lake) (Fig. 2). Pikeperch is the most abundant predator (9% in Tjeukemeer, 8% in the Loosdrecht Lake) and in both lakes its population is dominated by large specimens (45–85 cm forklength) (Fig. 3). The other fish species present are smelt, perch (*Perca fluviatilis*), roach, ruffe and white bream (*Blicca bjorkna*) varying from 7.6% to 0.5% of the total biomass (Fig. 2). These species are the main prey fish of pikeperch and selected in the order as indicated (Lammens, unpublished results). Smelt, white bream and roach are more abundant in Tjeukemeer and ruffe and perch occur in similar densities in both lakes.

The fish communities in Loosdrecht Lake and Tjeukemeer differ mainly in the size composition of the bream population. In Loosdrecht Lake the largest part of the fish biomass is comprised of bream <25 cm (*ca.* 65%), whereas in Tjeukemeer bream > 25 cm make up the largest part (*ca.* 75%, Fig. 4). Van Densen *et al.* (1986) and Lammens (1982) determined the growth rate of bream in Loosdrecht Lake and in Tjeukemeer respectively. Bream up to 25 cm grew very slowly in Loosdrecht Lake, but very fast in Tjeukemeer. During the study period the recruitment in 0+ bream did not significantly differ, but we did find a significant ($p < 0.005$), t-test) difference in size of the 0+ bream at the end of its first growing season (5.8 cm in Tjeukemeer, 5.2 cm in Loosdrecht; Fig. 5). Therefore the difference in length-frequency distribution is caused not by different recruitment but by different growth rates which result in year-classes accumulating in the length ranges up to 25 cm in Loosdrecht Lake.

Zooplankton

The lakes mainly differ in the absence of *Daphnia hyalina* in the Loosdrecht Lakes in most years (Fig. 6a), whereas it is an important daphnid in Tjeukemeer. The abundances of *Bosmina coregoni*, *B. longirostris*, and cyclopoid copepods is not significantly different ($p > 0.10$, $p > 0.35$, and $p > 0.30$ respectively, t-test). Only *Daphnia cucullata* was significantly ($p < 0.01$, t-test) more abundant in Loosdrecht Lake (Fig. 6a). Only the smallest species, *B. longirostris* and *C. sphaericus* did not differ significantly ($p > 0{,}35$ and $p > 0.39$ respectively, t-test) in size composition, but all the other species are significantly larger ($p < 0.01$ or smaller, t-test) in Tjeukemeer (Fig. 6b).

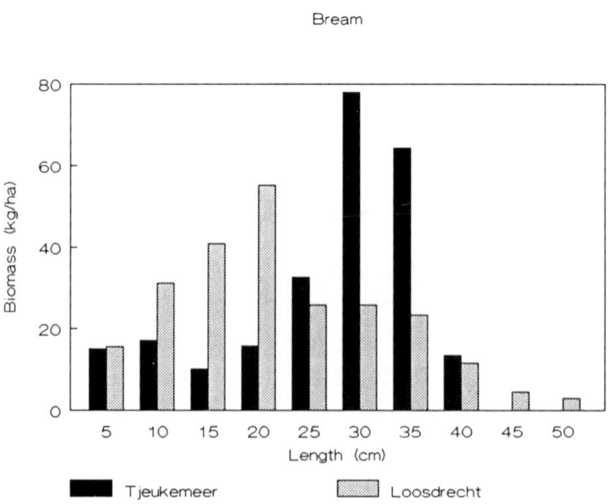

Fig. 4. Size distribution of bream in Loosdrecht Lake and Tjeukemeer expressed in biomass per 5 cm length class averaged for the periods May–September in 1984–1990.

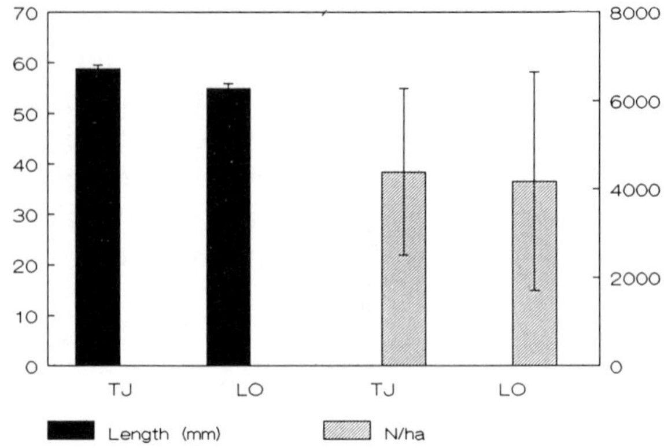

Fig. 5. Size and recruitment of 0+ bream in Loosdrecht Lake (LO) and Tjeukemeer (TJ) as determined in September averaged for 1984–1990. Bars indicate SE.

*Phytoplankton, chlorophyll-*a*, Secchi-depth and total-P*

Filamentous cyanobacteria constituted the bulk of the phytoplankton in both lakes. They were significantly ($p < 0.0001$, t-test) more abundant in the Loosdrecht Lakes, in which the average number of cells amounted to *ca.* 150 000 ml^{-1}. In contrast the average density in Tjeukemeer was *ca.* 30 000 ml^{-1} in Tjeukemeer (Fig. 7), but the filaments were much thicker and longer (H. Hoogveld, pers. comm.) than in Loosdrecht Lake and therefore the chlorophyll-*a* concentration hardly differed (Fig. 8). In both lakes the green algae had relatively low densities, but their numbers were in Tjeukemeer much higher than in Loosdrecht Lake (*ca.* 500 vs. 100 ml^{-1}). Diatoms formed the third relative abundant phytoplankton group and were on average more abundant ($p < 0{,}01$, t-test) in Loosdrecht Lake (2000 vs. 1000 ml^{-1}) (Fig. 7). Average chlorophyll-*a* concentrations in both lakes were around 150 $\mu g\,l^{-1}$ and the Secchi-depth (*ca.* 35 cm) did

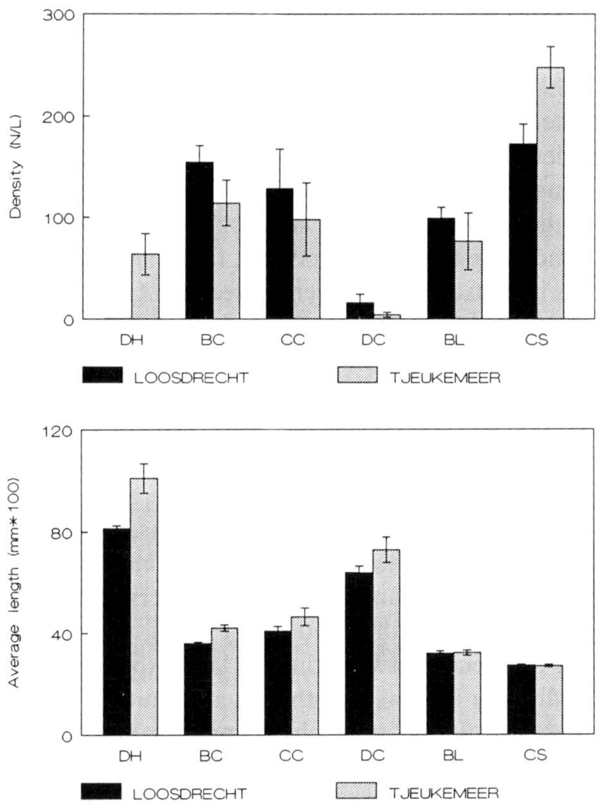

Fig. 6a. Average density of the most abundant zooplankton species (DH = *Daphnia hyalina*, BC = *Bosmina coregoni*, CC = cyclopoid copepods, DC = *Daphnia cucculata*, BL = *Bosmina longirostris*, CS = *Chydorus sphaericus*) for the periods May–September in 1984–1990. Bars indicate SE. *6b.* Average length of most abundant zooplankton species (see 6a for abbreviations) for the periods May–September in 1984–1990. Bars indicate SE.

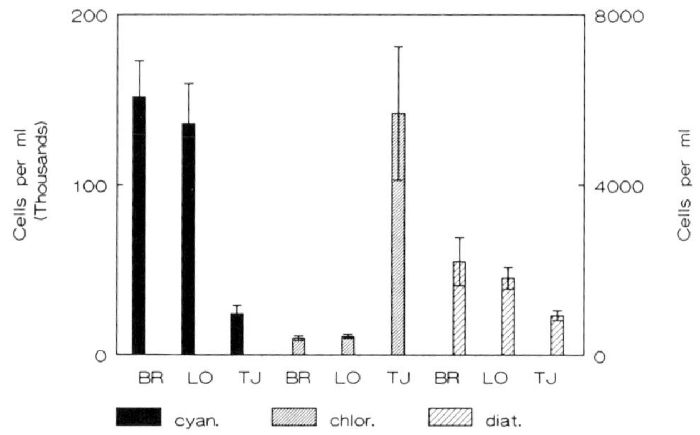

Fig. 7. Average density of the three main phytoplankton (cyan = Cyanobacteria, left scale; chlor. = chlorophyta, right scale; diat. = diatomeae, right scale) groups for the periods May–September in 1984–1990 in two Loosdrecht lakes (BR = Lake Breukeleveen, LO = Loosdrecht Lake) and in Tjeukemeer (TJ). Bars indicate SE.

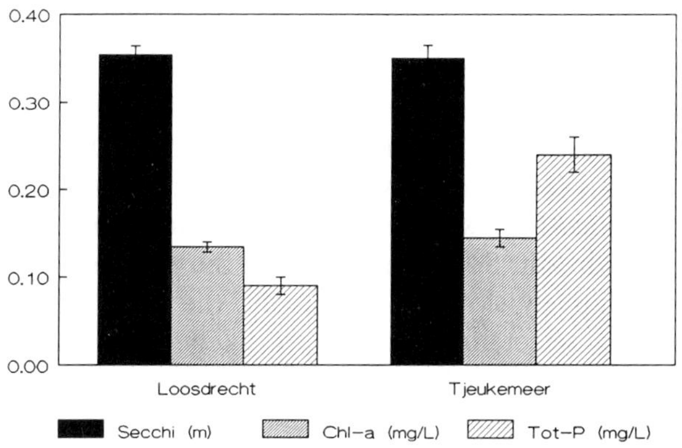

Fig. 8. Secchi-depth (m), chlorophyll-*a* (mg l^{-1}) and total-P (mg l^{-1}) as an average for the periods May–September in 1984–1990. Bars indicate SE.

not differ either (Fig. 8). Surprisingly the total-P concentration was 2–3 times higher ($p < 0.001$, t-test) in Tjeukemeer (*ca.* 250 $\mu g\, l^{-1}$), but the daily external P-load is even 50 times higher in Tjeukemeer (50–100 $\mu g\, l^{-1}$) (Van Huet, 1991).

Discussion

Considering the potential predation pressure on zooplankton the main difference between the fish communities in Loosdrecht Lake and Tjeukemeer is the relatively small individual size of bream in Loosdrecht Lake. Because the bream population makes up almost 70–80% of the total fish biomass the predation effect on zooplankton will consequently be different in the two lakes. In Tjeukemeer the bream population can feed only on large zooplankton, i.e. *Daphnia hyalina* > 1 mm, because of the size composition of the population (Lammens, 1985). When the *Daphnia* density is lower than 20 l^{-1} large bream will switch to benthivorous feeding (Lammens & Hoogenboezem, 1991; Hoogenboezem *et al.*, 1991) and therefore unable to eradicate large zooplankton. This is probably the main reason why *D. hyalina* never disappears from Tjeukemeer. In Loos-

Van Liere, L., O. F. R. Van Tongeren, L. Breebaart & W. Kats, 1990b. Trends in chlorophyll and total phosphorus in Loosdrecht lakes. The Netherlands. Verh. int. Ver. Limnol. 24: 707–710.

Van Liere, L. & J. H. Janse 1997 Restoration and resilience to recovery of the Lake Loosdrecht ecosystem in relation to its phosphorus flow. Hydrobiologia 223: 95–104.

Van Tongeren, O. F. R., L. Van Liere, R. D. Gulati, G. Postema & N. Boesewinkel-de Bruyn 1992. Multivariate analysis of the plankton communities in the Loosdrecht lakes: relationship with the chemical and physical environment. Hydrobiologia 233/ 105–117.

Hydrobiologia **233**: 95–104, 1992.
L. Van Liere & R.D. Gulati (eds), Restoration and Recovery of Shallow Eutrophic Lake Ecosystems in The Netherlands.

Restoration and resilience to recovery of the Lake Loosdrecht ecosystem in relation to its phosphorus flow

Louis Van Liere [1,2,*] & Jan H. Janse[2]
[1] *Limnological Institute, Rijkstraatweg 6, NL-3631 AC Nieuwersluis, The Netherlands;* [2] *National Institute of Public Health and Environmental Protection, P.O. Box 1, NL-3720 Bilthoven, The Netherlands*
(author for correspondence)*

Key words: water management, food-web, phosphorus flow, lake restoration, recovery, resilience

Abstract

A reduction in external phosphorus loading since 1984 to Loosdrecht lakes system by the dephosphorization of the inlet water, yielded only minor effects in Lake Loosdrecht. This reduction measure turned out to have decreased the loading only by a factor of two. A conceptual model was constructed based on laboratory measurements to describe phosphorus flow in the lake ecosystem for the summer of 1987. The role of zooplankton and fish was more important in phosphorus recycling than diffusion at the sediment-water interface. The input and output of phosphorus of the lake were at equilibrium and therefore, further reduction in external loading was needed for recovery. The results of the conceptual model agreed well with the output of the mathematical model PCLOOS. Additional measures such as dredging, flushing, chemomanipulation, or biomanipulation would be ineffective at the present level of external loading. Only a significant further reduction in external input will restore Lake Loosdrecht's water quality over a long period of time.

Introduction

The Loosdrecht lakes system (Fig. 1) consists of the three shallow (average depth about 2 m) interconnected lakes, Loosdrecht, Vuntus, and Breukeleveen (for detailed description, see Gulati *et al.* 1991a; Hofstra & Van Liere, 1992). The banks and ditches of the Kievits Area also form a part of the system influencing its water and phosphorus balances. The eutrophication of these lakes started from the beginning of the 20th century from various sources (Engelen *et al.*, 1992). The trophic status of these lakes, passing through various intermediate stages, has become eutrophic, with cyanobacteria and detritus dominating the particulates in the water column (Hofstra & Van Liere, 1992). To restore the water quality, a series of measures have been taken to reduce phosphorus input. The most important step was the replacement of inflow to the lake in 1984; the inflow from River Vecht, with its phosphorus level at 1–2 mg P l^{-1}, was switched to Amsterdam-Rhine Canal by means of a pipeline (Fig. 1). The Canal water was dephosphorized before entering the lakes by coagulation with iron (III)-chloride followed by sedimentation of the resulting floc (Van der Veen *et al.*, 1987). L. Breukeleveen and L. Vuntus have shown little effect by the diversion, probably because they are situated far from the inlet. Thus, this paper will focus on L. Loosdrecht near the inlet.

To understand the persistence of trophic status

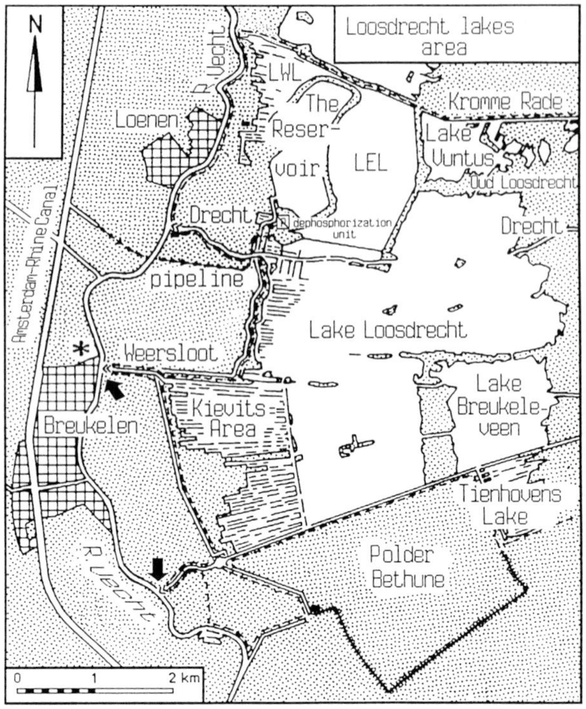

Fig. 1. The Loosdrecht lakes area. Arrows indicate the sluices used before 1984 to supply the lakes with water from the River Vecht. The Reservoir, L. Eastern Loenderveen (LEL) and L. Western Loenderveen (LWL) are hydrologically separated from the L. Loosdrecht and L. Vuntus. Asterix: Limnological Institute.

in the lake, an accurate phosphorus budget for the lake was first estimated (Engelen *et al.*, 1992). A conceptual model of phosphorus cycling through various ecosystem compartments was made as presented below. Using various parameters measured in water and various compartments of the ecosystem, a mathematical model was developed (Janse & Aldenberg, 1990). This model output was compared with empirical data, in order to test its possibilities to simulate further water management measures in L. Loosdrecht (Janse *et al.*, 1992). The needs for further (and additional) measures to restore Loosdrecht lakes are indicated.

Reduction of external phosphorus

Earlier calculations (Province of Utrecht, 1973, 1980) predicted that the replacement of inflow from R. Vecht with dephosphorized water from the Amsterdam-Rhine Canal would reduce the phosphorus load of the lake system (L. Loosdrecht, L. Vuntus and L. Breukeleveen) from 1.35 to 0.10 g P m^{-2} y^{-1}. Thus, the lake system was expected to react shortly to the replacement. However, no significant changes have been observed (Table 1).

In L. Vuntus and L. Breukeleveen, the external phosphorus load did not change. (Engelen *et al.*, 1992). For the whole lake area, the external phosphorus load after 1984 was 0.36–0.46 g P

Table 1. External phosphorus load (g P m^{-2} y^{-1}) of the whole Loosdrecht lakes' system and its separate compartments, excluding Kievits Area, before and after the restoration measures. Data according to Van Liere *et al.*, 1991.

Lake system	Before the measures (average for 1982 and 1983)	After measures (average for 1984 and 1985)	Increased load: leakage of sluices (average for 1986 and 1987)	Present estimated load (if no extra measures were to be taken)
Loosdrecht (excluding Kievits area)	0.69	0.31	0.44	0.35
Breukeleveen	0.57	0.59	0.59	0.60
Vuntus	0.24	0.26	0.25	0.25
Whole lake system (including Kievits area)	1.05	0.36	0.46	0.37

$m^{-2} y^{-1}$ (Table 1), which is much higher than the predicted 0.1 g P $m^{-2} y^{-1}$ by the earlier reports (Province of Utrecht, 1973, 1980), probably because they did not account for the load from the agricultural hinterland. The load to L. Loosdrecht was even higher in 1986 and 1987 because of a severe leakage of the Southernmost sluice. This leakage, however, did not have any effect on L. Vuntus and L. Breukeleveen.

Besides, the earlier calculation of external loading appeared to be an overestimate. Before 1984, water from R. Vecht was supplied mainly through two sluices (Fig. 1), passing through the Kievits Area with its banks and ditches where wind influence, and concomitant mixing, were negligible. The water lost a part of its phosphorus passing this area, resulting in a lower load for L. Loosdrecht. Phosphorus concentration in Kievits Area sediments was high when compared to other lakes (Boers *et al.*, 1984).

The phosphorus balance (Engelen *et al.*, 1992) revealed that the external load to the whole lake system had been reduced from 1.05 to 0.37 g P $m^{-2} y^{-1}$. The reduction of external load to L. Loosdrecht resulted in decrease only by a factor of about two (0.69–0.35 g P $m^{-2} y^{-1}$, Table 1).

However, certain parts of the system showed clear responses. Phosphorus release rate has decreased from 1 g P $m^{-2} d^{-1}$ in 1983 to 0.3 g P $m^{-2} d^{-1}$ in 1990 (Keizer & Sinke, 1992). The concentration of total phosphorus in L. Loosdrecht corrected for autocorrelation decreased from 1984 to 1988 by 12 ± 2 $\mu g\ l^{-1}$ P y^{-1}, but the concentration of chlorophyll *a* did not react (Van Liere *et al.*, 1990a; Van Tongeren *et al.*, 1992). Light energy was important as a growth-limiting factor before 1985, but later phosphorus has become more important (Rijkeboer *et al.*, 1991; Gons *et al.*, 1992). This shift in growth-limiting factor was also reflected in the C/P ratio of the seston (< 150 μm) which increased from 90 to 145 between 1985 and 1989 (Gulati *et al.*, 1991b). Other parts of the food chain showed little reaction (Gulati *et al.*, 1992; Lammens *et al.*, 1992).

The flow of phosphorus in Lake Loosdrecht

To gain more insight in the factors contributing to the persistence of the ecosystem's trophic status, a conceptual model for phosphorus flow was constructed. The model was based largely on the data from the scientific projects by the work group Water Quality Research Loosdrecht Lakes (Loogman & Van Liere, 1986; Van Liere & Gulati, 1992). The input and output of the system were derived from hydrological data (Buyse, 1988). Data on phosphorus-dynamics at the sediment-water interface were taken from Keizer & Sinke (1992) and Gons & Van Keulen (1989). Primary production was measured and calculated according to Van Liere *et al.* (1986a). A striking similarity was noted between the model that described primary productivity and the results derived from the mass balances obtained in Laboratory System Enclosures (Rijkeboer *et al.*, 1991).

Zooplankton grazing, consumption, assimilation, and egestion were determined according to Gulati (1984) and Gulati *et al.* (1992). Phosphorus flows within seston and zooplankton were calculated on the basis of the C/P ratios, of zooplankton and their seston-food, which were measured by the ^{14}C-tracer method in the laboratory. Zooplankton (> 150 μm) excretion was computed from zooplankton biomass and water temperature using the relationship between temperature and specific excretion rate (Den Oude & Gulati, 1989). Mineralization, which was measured only three times in 1986, was not included in the diagramme. The phosphorus mineralization rate of sestonic matter was in the same order of magnitude as phosphorus consumption rate (Van Liere *et al.*, 1986b). This suggests that the mineralization of seston occurs mainly in the water column and the top layer of sediments (epipelon) in contact with oxygenated water (see also Otten *et al.*, 1992).

Data of phosphorus flow to and from water plants were from calculations by Malthus *et al.* (1990).

Phosphorus recycling in the fish compartments was derived from Lammens *et al.* (1992), rates of consumption of zooplankton and benthic fauna

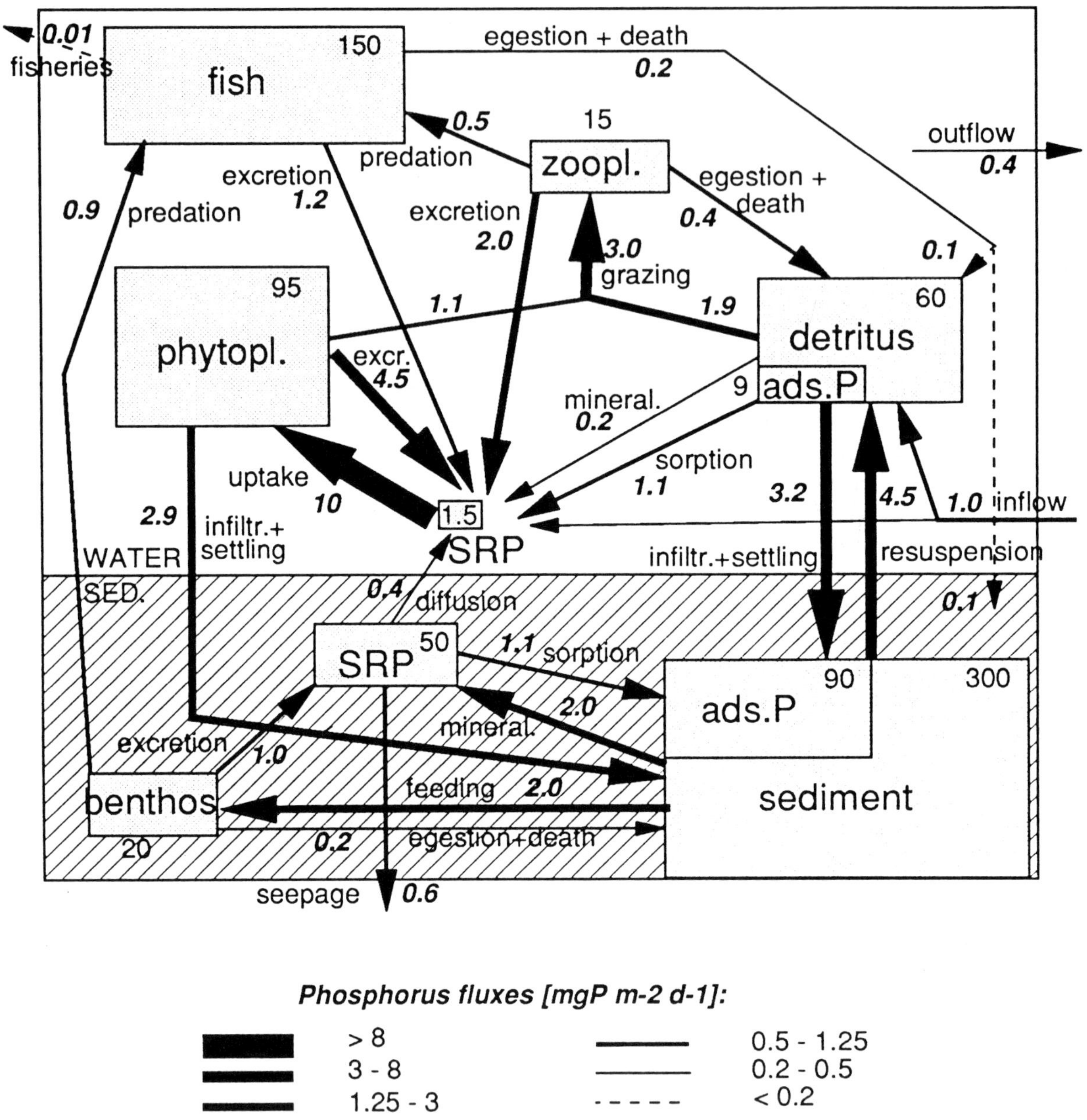

Fig. 3. Phosphorus content (mg P m^{-2}) and fluxes (mg P m^{-2} day^{-1}) and phosphorus amounts in various compartments of Lake Loosdrecht for the time period April–September 1987 as calculated by the mathematical model PCLOOS.

ematical model separated seston into two fractions, phytoplankton and detritus, whereas the flow diagramme used them as a single compartment. Calculated growth rates were different from the estimated values based on primary production measurements (Van Liere *et al.*, 1986). The calculated phosphorus flow from zooplankton deviated only a little from the measurements.

Overall, the model gave a good description of phosphorus flow through various compartments of the ecosystem. It predicted with a reasonable accuracy the outcome of further reduction of external phosphorus load over a short term, pointing out at the same time additional measurements to be taken to reduce primary production (Janse *et al.*, 1992).

Additional measures

If an ecosystem does not react instantaneously on the external phosphorus load additional measures are often called for. Such measures are reviewed here, especially in regard to their applicability or otherwise in the Loosdrecht lakes.

Further reduction of the external phosphorus load

For the Loosdrecht lakes, any further reduction of external phosphorus load is technically difficult, but possible. As mentioned earlier, a reduction of the external load by a factor of two was not sufficient to induce significant responses from the ecosystem. Engelen *et al.* (1992) have pointed out several possible hydrological measures which could decrease the load an additional 50%, including the diversion (or dephosphorization, De Ruiter, 1992) of the Eastern hinterland water with its high phosphorus concentrations.

Flushing

Hosper & Meijer (1986) and Jagtman *et al.* (1992) have shown positive results with flushing in the recovery of L. Veluwe and L. Wolderwijd. However, the chemical composition of the flushing water with high concentration of calcium and nitrate facilitated chemomanipulation (Jagtman *et al.*, 1992). In L. Loosdrecht the situation is more problematic. The inlet point of the dephosphorized Amsterdam-Rhine Canal water is close to the discharging pumping station. A mixing time of at least a week is needed and thus flushing has to be repeated several times. It also does not deal with the problem of organic deposits on the lake bottom. Therefore, it is not recommended.

Dredging

Keizer & Sinke (1992) have shown that a major phosphorus exchange at the sediment-water interface involves the diffusion of mineralized phosphorus to the overlaying water. Therefore, it is unlikely that phosphorus release will be reduced by removing the top layer of sediments. In fact Van Liere *et al.*, (1986) reported only 30% reduction in phosphorus release rate from Lake Loosdrecht sediments following the removal of its top layer. With the present high external load of phosphorus, organic matter in the top layer will be replenished rapidly. However, removing more than just the top layer will reduce resuspension (Gons & Van Keulen, 1989), which is an important factor in phosphorus exchange between seston and sediments (Rijkeboer *et al.*, 1990). Dredging may also increase downward seepage (Engelen *et al.*, 1992), which can readily be compensated for by an increased supply of dephosphorized water. Although dredging would increase external phosphorus load slightly per unit area, its load per unit volume would decrease relatively. The cost of dredging L. Loosdrecht is estimated to be 35 ± 10 10^6 ECU. Technical problems of dredging a large amount of organic deposits with its specific density close to that of water have not yet been solved (Van der Does *et al.*, 1992), not to speak of the disposal of the dredged material. Considering the present high external load and the equilibrium situation between its input and output, it is questionable whether dredging will significantly help the recovery of L. Loosdrecht.

Biomanipulation

As can be derived from Dawidowicz *et al.* (1988), the removal of a large amount of bream (*Abramis brama*) failed to improve L. Breukeleveen (Van Donk *et al.*, 1990), because this top-down control is effective only when the food quality for large *Daphnia* can sustain their growth. Apparently, this was not the case for L. Breukeleveen, one of the lakes in the Loosdrecht lakes' area. Furthermore, with the present high amount of organic deposits in the lake where wind action plays such a dominant role, any increase in water transparency resulting from biomanipulation may not be enough to sufficiently enhance the growth of submerged plants to remove any significant amount of phos-

Janse, J. H. & T. Aldenberg, 1990. A mathematical model of the P cycle in the Loosdrecht lakes. Hydrobiol. Bull. 24: 69–89.

Janse, J. H., T. Aldenberg & P. R. G. Kramer, 1992. A mathematical model of the phosphorus cycle in the Lake Loosdrecht, and simulation of additional measures. Hydrobiologia 233: 119–136.

Keizer, P. & A. J. C. Sinke, 1992. Phosphorus in the sediment of the Loosdrecht lakes and its implications for lake restoration persectives. Hydrobiologia 233: 39–50.

Lammens, E. H. R. R., N. Boesewinkel-De Bruyn, H. Hoogveld & E. Van Donk, 1992. Phosphorus load, phytoplankton, zooplankton and fish stock in Loosdrecht Lake and Tjeukemeer: confounding effects of predation and food availability. Hydrobiologia 233: 87–94.

Loogman, J. G. & L. Van Liere, 1986a. Restoration of shallow lake ecosystems with emphasis on Loosdrecht lakes. Proceedings of the Water Quality research Loosdrecht lakes Symposium. Hydrobiol. Bull. 20: 5–259.

Malthus, T. J., E. P. H. Best & A. G. Dekker, 1990. An assessment of the importance of emergent and floating-leaved macrophytes to trophic status in the Loosdrecht lakes (The Netherlands). Hydrobiologia 191: 257–263.

Otten, J. H., H. J. Gons & M. Rijkeboer, 1992. Dynamics of of phytoplankton detritus in a shallow, eutropic lake (Lake Loosdrecht, The Netherlands). Hydrobiologia 233: 61–68.

Province of Utrecht, 1973. Over de veiligstelling van de waterkwaliteit van de Loosdrechtse Plassen (Securing the Loosdrecht lakes water quality). Report of the Province of Utrecht, Utrecht. 63 pp.

Province of Utrecht, 1980. Over de keuze van een suppletiemodel ter beveiliging van de waterkwaliteit van de Loosdrechtse Plassen (The choice of a supply model to secure Loosdrecht lakes water quality. Report of the Province of Utrecht, Utrecht. 90 pp.

Rijkeboer, M., F. De Bles & H. J. Gons, 1991. The role of sestonic detritus as a P-buffer. Mem. Ist. ital. Idrobiol. 48: (in press).

Sinke, A. J. C. & Th. E. Cappenberg, 1988. Influence of bacterial processes on the phosphorus release from sediments in the eutrophic Loosdrecht lakes, The Netherlands. Arch. Hydrobiol. Beih. Ergebn. Limnol. 30: 5–13.

Van der Does, J., P. Verstraelen, P. Boers, J. Van Roestel, R. Roijackers & G. Moser, 1992. Lake restoration with and without dredging of phosphorus-enriched upper sediment layers. Hydrobiologia 233: 197–210.

Van Der Veen, C., A. Graveland & W. Kats, 1987. Coagulation of two different kinds of surface water before inlet into lakes to improve the self-purification process. Wat. Sci. Technol. 19: 803–812.

Van Der Vlugt, J. C., P. A. Walker, J. Van Der Does & A. J. P. Raat, 1992. Fisheries management as an additional lake restoration measure: biomanipulation scaling-up problems. Hydrobiologia 233: 213–224.

Van Donk, E., M. P. Grimm, R. D. Gulati, P. G. M. Heuts, W. A. De Kloet & L. Van Liere, 1990. First attempt to apply whole-lake food-web manipulation on a large scale in The Netherlands. Hydrobiologia 200/201: 291–301.

Van Liere, L., L. Van Ballegooijen, W. A. De Kloet, K. Siewertsen, P. Kouwenhoven & T. Aldenberg, 1986a. Primary production in the various parts of Loosdrecht lakes. Hydrobiol. Bull. 20: 77–85.

Van Liere, L., P. C. M. Boers, P. J. Den Oude, H. J. Gons, R. D. Gulati, M. Rijkeboer, A. J. C. Sinke, J. P. R. A. Sweerts, J. Bril & L. Postman, 1986b. Water Quality Research Loosdrecht lakes, studying and modelling the impact of water management measures on the internal nutrient cycle. Final report to the European Community ENV 839-NL(N). 155 pp.

Van Liere, L., O. F. R. Van Tongeren, L. Breebaart & W. Kats, 1990a. Trends in chlorophyll and total phosphorus in Loosdrecht lakes, The Netherlands. Verh. int. Ver. Limnol. 24: 707–712.

Van Liere, L., R. D. Gulati, F. G. Wortelboer & E. H. R. R. Lammens, 1990b. Phosphorus dynamics following restoration measures in the Loosdrecht lakes (The Netherlands). Hydrobiologia 191: 87–95.

Van Liere, L., J. Ebert, W. Kats & J. J. Buyse, 1991. The waterquality of Loosdrecht Lakes reviewed. Mem. Ist. ital. Idrobiol. 48: (in press).

Van Liere, L. & R. D. Gulati, 1992. Restoration and recovery of shallow eutrophic lake ecosystems in The Netherlands: epilogue, Hydrobiologia 233: 283–287.

Van Tongeren, O. F. R., L. Van Liere, R. D. Gulati, G. Postema & P. J. Boesewinkel-De Bruijn, 1992. Multivariate analysis of the plankton communities in the Loosdrecht lakes: relationship with the chemical and physical environment. Hydrobiologia 233: 105–117.

Zevenboom, W. & L. R. Mur, 1980. N_2-fixing cyanobacteria: why they do not become dominant in Dutch hypertrophic lakes. In: J. Barica & L. R. Mur (eds), Hypertrophic ecosystems pp. 123–130. Junk Publishers, The Hague.

Hydrobiologia **233**: 105–117, 1992.
L. Van Liere & R.D. Gulati (eds), Restoration and Recovery of Shallow Eutrophic Lake Ecosystems in The Netherlands.

Multivariate analysis of the plankton communities in the Loosdrecht lakes: relationship with the chemical and physical environment

O. F. R. Van Tongeren, L. Van Liere* R. D. Gulati, G. Postema & P. J. Boesewinkel-De Bruyn
*Limnological Institute, Rijksstraatweg 6, 3631 AC Nieuwersluis, The Netherlands; *present address: National Institute of Public Health and Environmental Protection, P.O. Box 1, 3720 BA Bilthoven, The Netherlands*

Key words: canonical ordination, lake restoration, phosphate, phytoplankton, redundancy analysis, succession, trends, zooplankton

Abstract

Data on the water temperature, Secchi depth, water chemistry and phytoplankton and zooplankton of the Loosdrecht lakes, collected during an eight-years monitoring study (1983–1990), were examined using Principal Components Analysis (PCA) and Redundancy Analysis (RDA). The Loosdrecht lakes are highly eutrophic lakes, in which several restoration measures were taken from 1970 to 1986. These measures included the diversion of sewage by the construction of sewerage and additional reduction of external phosphorus-loading by supplying from 1984 onwards water from the Amsterdam Rhine Canal, which had phosphorus removed from it, and which replaced the inlet of highly eutrophic water from the River Vecht. The descriptive statistical models based on this study show that, although the total P in the lake was decreasing, probably because of the reduced external loading, the sum of the densities of cyanobacteria and *Prochlorothrix hollandica* was increasing. The dominant species of the zooplankton community (rotifers and crustaceans) did not show changes related to the restoration measures. It is concluded that the available data are insufficient to predict future changes, but that the changes in the past can be described well by a simple linear ordination model.

Introduction

Eutrophication of lakes is a worldwide problem, resulting in systems dominated by a few species of algae in which the diversity of the fish and zooplankton communities is low and where macrophytes are virtually absent. Nowadays, the restoration of such systems is a major issue in scientific research, but only in a few occasions detailed information of the communities seems to be available. The Water Quality Loosdrecht (WQL) project was one of the intensive research projects of the last decade, planned to obtain more insight into the mechanisms responsible for the changes in such eutrophic systems. During this project detailed information was collected in the Loosdrecht Lakes. Information on overall process rates, e.g. primary production, grazing rates and phosphorus release from the sediments, as well as detailed information on the composition of the planktonic communities and physical and chemical environment were collected in a systematic way over a period of eight years. It was planned to use the available data for the construction of models describing the major changes in the system and predicting future changes.

The Loosdrecht lakes system consists of three shallow, mainly hypertrophic lakes situated in the

1973; Ter Braak, 1987a) and a hybrid of RDA and PCA were applied to the dependent variables (phytoplankton and zooplankton counts) and the independent variables. For all analyses program CANOCO (Ter Braak, 1987b, 1989, 1991) was used.

Environmental variables were analyzed using PCA after centering and standardization of the variables, i.e. PCA on the correlation matrix of the environmental variables. The correlation structure within the set of environmental variables was visualized by plotting the correlations of the variables with the extracted PCA axes. Because year and sampling station are not chemical or physical characteristics, these variables were analyzed passively: i.e. their correlations with the ordination axes were computed after the correlation matrix between the other variables was used to obtain the first four principal components (axes).

Plankton counts were log transformed ($\log(y + 1)$) and centered prior to the analysis. The results of the thus obtained RDA and PCA on the respective covariance matrices are presented as so-called correlation biplots (e.g. Ter Braak, 1987a) in which the correlations with the ordination axes for both the dependent and the independent variables are plotted. Besides, centroids (means) of the sample scores were computed for classes of samples and also plotted.

Forward selection of environmental variables was used in the analysis of plankton counts to obtain a small set of environmental variables maximally related to the species counts. An arbitrary threshold value of 3% of the variance in the species data 'explained' by the addition of the next environmental variable was used to decide whether an environmental variable should be included in the model as independent variable. This arbitrary percentage approximates the mean proportion of the total variance for each species, i.e. 100% divided by 38 for phytoplankton taxa or by 32 in case of zooplankton taxa. Variables like total phosphorus, chlorophyll and transparency (Secchi-depth) were considered not to be real independent variables, since their values and the densities and composition of zooplankton as well as phytoplankton are mutually dependent. Therefore, such variables were not entered into the model. The environmental variables not included in the model were analyzed passively, i.e. their correlations with the ordination axes were computed after the computations for the ordination. Phytoplankton and zooplankton data were analyzed separately. Additionally, the zooplankton counts were analyzed passively together with the phytoplankton counts, i.e. the correlations of zooplankton counts with the four ordination axes were computed after extraction of the axes. The CANOCO results were post-processed, mainly to compute centroids, using program CEDIT (Van Tongeren, 1991).

Results

The PCA analysis of the environmental variables describes 71% of the variance in the chemical and physical data in 4 axes. Passive analysis of the relationship with sampling station and year indicates that 54% of the relationship between chemical and physical data and sampling station and year is represented by this PCA solution. Temperature and year, which accounted for more than the arbitrary 3% of the variance in the transformed species counts, were selected to enter into the RDA for the phytoplankton. However, for zooplankton only temperature accounted for more than 3% of the total variance in the transformed counts. The hybrid (RDA and PCA) analyses of the dependent variables accounted for 34% and 64% of the variance in the phytoplankton and zooplankton counts, respectively, in 4 axes. The temperature constrained (RDA with temperature) axis of the zooplankton counts accounted for 31% of the variance. The temperature-and-year constrained axes accounted for only 11% of the variance in the phytoplankton counts. The unconstrained (PCA) axes describe 33% (64% minus 31%) and 23% (34% minus 11%) of the variance in the zooplankton (3 axes) and phytoplankton data (2 axes), respectively.

Environmental variables

The four PCA axes of the environmental variables account for 28.3%, 17.5%, 14.7% and 10.3% of the variance, respectively. The yearly temperature cycle was clearly related to the cycles of some of the environmental variables. Fig. 1 summarizes the results for the environmental variables as a plot of PCA axis 1 (abscissa) against axis 3 (ordinate) displaying most of the information on temperature. Temperature and concentrations of NH_4^+, NO_3^- and SiO_2 were negatively correlated (arrows pointing more or less into the opposite direction of the temperature arrow). The negative correlation of soluble total phosphate (STP) with temperature was weak, which is indicated by the short projection of the arrow on the temperature arrow. Rough estimates of the concentrations at different temperatures can be made from the means and standard deviations of the environmental variables (Table 1). In Fig. 2 the year arrow roughly points into the direction of axis 2 (ordinate), including that trends, displayed approximately along axis 2 (Fig. 2, ordinate), are very weak. Differences between stations are best displayed by axis 3 (contrast between P9 and P7: Fig. 1) combined with axis 4 (contrast between P6 and the other stations: Fig. 2). The arrow for Lake Breukeleveen (P6) roughly points into the same direction as that for dissolved organic carbon (DOC), whereas the arrow for Secchi depth points roughly into the opposite direction. Therefore, from Fig. 2 we derive that DOC was relatively highest and Secchi depth relatively lowest in Lake Breukeleveen (P6), and that STP and total phosphorus (TP) were decreasing in the course of years (arrows pointing into the opposite direction of the Year arrow). These observations are confirmed by inspection of the full correlation matrix among environmental variables. The correlations for soluble reactive phosphate (SRP) with the other environmental variables were low, probably because of the low accuracy of the SRP

Table 1. Environmental variables with their abbreviations, units, and means with standard deviations.

Variable	Variable name	Unit	Mean	s.d.
Year minus 1900	Year	Year	87.9	1.6
Temperature	Temp.	°C	11.3	6.1
Nitrate	NO_3	mg N l^{-1}	0.06	0.13
Free ammonium	NH_4-free	mg N l^{-1}	0.08	0.12
Total ammonium	NH_4-tot.	mg N l^{-1}	0.36	0.18
Secchi depth	Secchi	m	0.25	0.08
Dissolved organic carbon	DOC	mg C l^{-1}	12.4	3.0
SiO_2	Total solicate	mg SiO_2 l	2.0	1.5
Soluble reactive phospate	SRP	µg P l^1	3.2	3.0
Soluble total phosphate	STP	µg P l^{-1}	9.5	4.2
Total phospate	TP	µg P l^{-1}	89.0	22.0

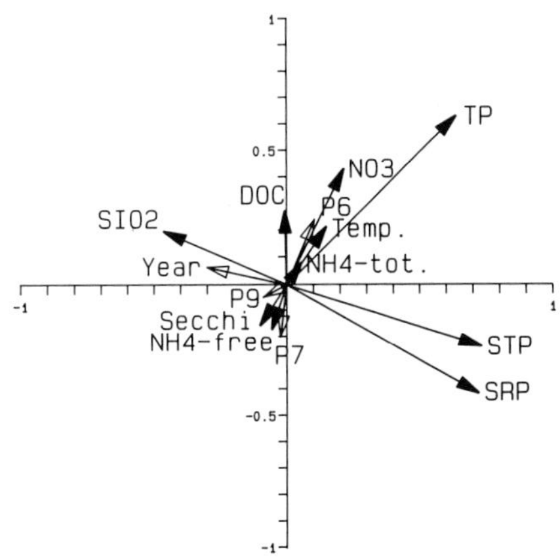

Fig. 2. PCA diagram of environmental variables, axis 2 (abscissa, 17.5% of the variance) against axis 4 (ordinate, 10.3% of the variance. Year, P6, P7 and P9 are passive variables. For abbreviations see Table 1.

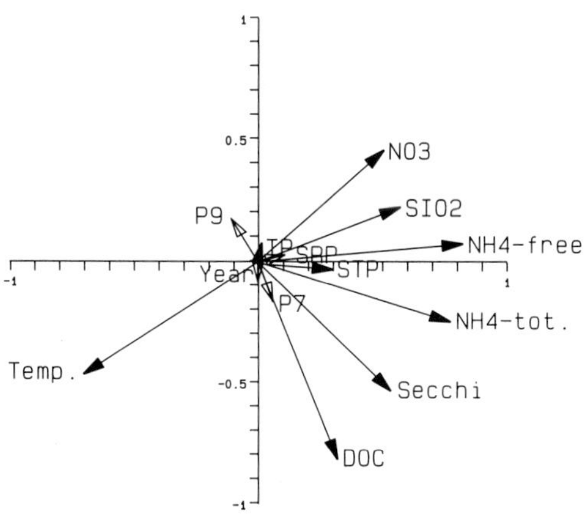

Fig. 1. PCA diagram of environmental variables, axis 1 (abscissa, 28.3% of the variance) against axis 3 (ordinate, 14.7% of the variance. Year, P6 (not visible), P7 and P9 are passive variables. For abbreviations see Table 1.

measurements, the concentrations being near the detection limit.

Phytoplankton

Temperature (Temp.) and the course of years (Year), accounted for 6% and 5% of the variance in the phytoplankton counts, respectively. This seemingly poor fit is caused mainly by the chosen transformation (log(y + 1)), which leads to a large weight (loading) for species with a high ratio between the highest and lowest numbers counted. These species are not necessarily the most abundant ones. However, the cumulative percentages fit for some important species indicate a fairly good description of these species by the model consisting of Temperature and Year (Table 2); however, only for the *Oscillatoria limnetica* group more than 50% of the variance was related to the independent variables. The two constrained axes (axes in which the relationship with the independent variables is expressed) account for 59.3 and 40.7% of the species environment relationship. The biplot of species and environmental variables (Fig. 3a) reveals several, fairly strong relationships. The length of the species vectors (not drawn, but being the arrows connecting the origin with the species points) reflects the proportion of the variance in the log-transformed counts related to the independent variables (Temp. and Year).

Some phytoplankton species were correlated negatively with temperature, indicating their peaks at lower temperatures and their relatively low numbers at high temperatures. Important among these are *Oscillatoria redekei* (max. 94000 cells ml^{-1} in June 1983), *Asterionella formosa* (max. 1040 cells ml^{-1} in April 1988, max. 160 cells ml^{-1} in August and September over the whole period) and *Dinobryon* (max. 1700 cells ml^{-1} in April 1984, max. in August and September over all years 203 cells ml^{-1}). Species correlated positively with temperature are: *Oscillatoria limnetica* groups (max. 388000 cells ml^{-1} in August 1983 and 350000 ml^{-1} in September 1990, max. spring value ever reached 203000 cells ml^{-1}), *Aphanizomenon flos-aquae* (16000 cells ml^{-1} in August 1983, 13600 cells ml^{-1} in the extremely warm month of June 1986, max. spring value over all other years 9500 cells ml^{-1}) and *Tetraedron* (200–400 cells ml^{-1}

Table 2. Cumulative fit per species for the first four ordination axes as fraction of the total variance and total variance per species (after log-transformation) for some important phytoplankton species. Fit is related to environmental data (temperature and year, axes 1 and 2) and latent variables (axes 3 and 4). Species (groups) not mentioned, but apearing in the figure 3: Actinast – *Actinastrum;* Anki/Mon – *Ankistrodesmus/Monoraphidium;* Chodatel – *Chodatella;* Cosmariu – *Cosmarium;* Cryptomo – *Cryptomonas;* Dinobryo – *Dinobryon;* Gymnodin – *Gymnodinium;* Lynglimn – *Lyngbya limnetica; Oocystis;* Peridini – *Peridinium; Phacus;* Restbitr – unidentified big trichomes; RestCent – unidentified centric diatoms; RestCyan – unidentified Cyanophyceae; Rhizosol- *Rhizosolenia;* and Tetraedr – *Tetraedron.* Note that *Prochlorothrix hollandica* is included in the *Oscillatoria limnetica* group.

Name	Abbreviation	Fraction of the variance 'explained'				
		Constrained axes		Unconstrained axes		Variance VAR (y)
		Axis 1	Axis 2	Axis 3	Axis 4	
Oscillatoria limnetica group	Osciligr	0.516	0.547	0.610	0.663	0.22
Aphanizomenon flos-aquae	Aphafl-aq	0.274	0.28	0.353	0.354	0.49
Oscillatoria redekei	Oscirede	0.242	0.242	0.356	0.370	1.36
Remaining Chlorophyceae	RestChlo	0.171	0.257	0.258	0.521	1.15
Pediastrum	Pediastr	0.040	0.139	0.148	0.396	1.36
Scenedesmus	Scenedes	0.019	0.059	0.067	0.428	0.72
Asterionella formose	Asteform	0.087	0.127	0.334	0.336	1.86
Melosira	Melosira	0.015	0.036	0.594	0.764	3.08
Diatoma elongatum	Diatelon	0.004	0.026	0.650	0.776	3.53

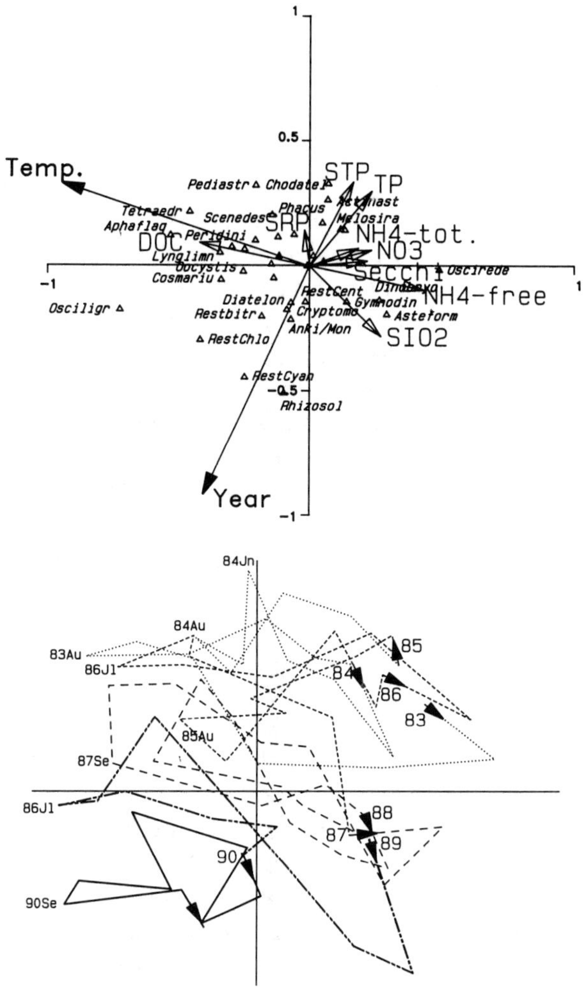

Fig. 3a. RDA correlation biplot of phytoplankton species and environmental variables, constrained by temperature and year: axes 1 (abscissa, 6.6% of the variance in the species counts) and 2 -ordinate, 4.5% of the variance in the species counts). Species vectors are represented as symbols at the position of the arrow-heads, vectors of environmental variables are drawn. For abbreviations see Tables 1 and 2.
b. Succession in the phytoplankton community represented by lines connecting the centroids (average scores) for all samples taken in each month. The beginning of each year is indicated by an arrow-head, for clarity some other points are labelled with year and month.

in summer, max. 150 cells ml^{-1} in spring) indicating summer peaks for these species.

An increase of numbers with time (Year) is inferred for the *Oscillatoria limnetica* group (minimum 770 cells ml^{-1} in 1983, gradually increasing to 51 500 cells ml^{-1} in 1990, maxima of the same order of magnitude in 1983 and 1990, being *ca* 400 000 cells ml^{-1}, gradually increasing from 1985, *ca* 200 000 cells ml^{-1} towards 1990), *Rhizosolenia* (reaching peaks of 1 000–4 000 cells ml^{-1} in 1987, 1989 and 1990, max. in the period 1983–1986 270 cells ml^{-1}) and *Ankistrodesmus/Monoraphidium* (max. 400 cells ml^{-1}), but *Pediastrum*, *Actinastrum*, *Phacus* and *Scenedesmus* (species of minor importance, which never reached counts of more than 400 cells ml^{-1}) decrease. For *Melosira* the decrease observed from the diagram is due to one extreme peak of 19 700 cells ml^{-1} in april 1984. Omitting 1984, no trend can be detected, maxima being invariably around 4 000 cells ml^{-1}.

The plot of centroids (mean scores) of the sample scores for the months within years (Fig. 3b) also shows the seasonal variation and the trend over years. The phytoplankton exhibited exceptional behaviour in 1985: the community in spring of 1985 was like that in the springs of the preceding years 1983, 1984 and 1986, but the community composition in summer 1985 closely resembled that during the summers of the following years 1987, 1988 and 1989. The extremely mild winter of 1989–1990 is noted from the triplot obtained by overlaying Fig. 3a on Fig. 3b. The abundance of species correlated positively with temperature remained high. None of the stations seems to have differed markedly from the other stations (first four axes).

Zooplankton

Only the water temperature was well related to the zooplankton numbers, accounting for 31% of the variance in the species counts. The second, unconstrained, axis accounts for another 18% of the variance, and it mainly describes the differences between spring and autumn. The biplot of species with environmental variables in Fig. 4a clearly shows the recurrent seasonal succession in the zooplankton community. The length of a species vector reflects the explained variance of the species (after log-transformation and centering), i.e. the square root of the proportion of the

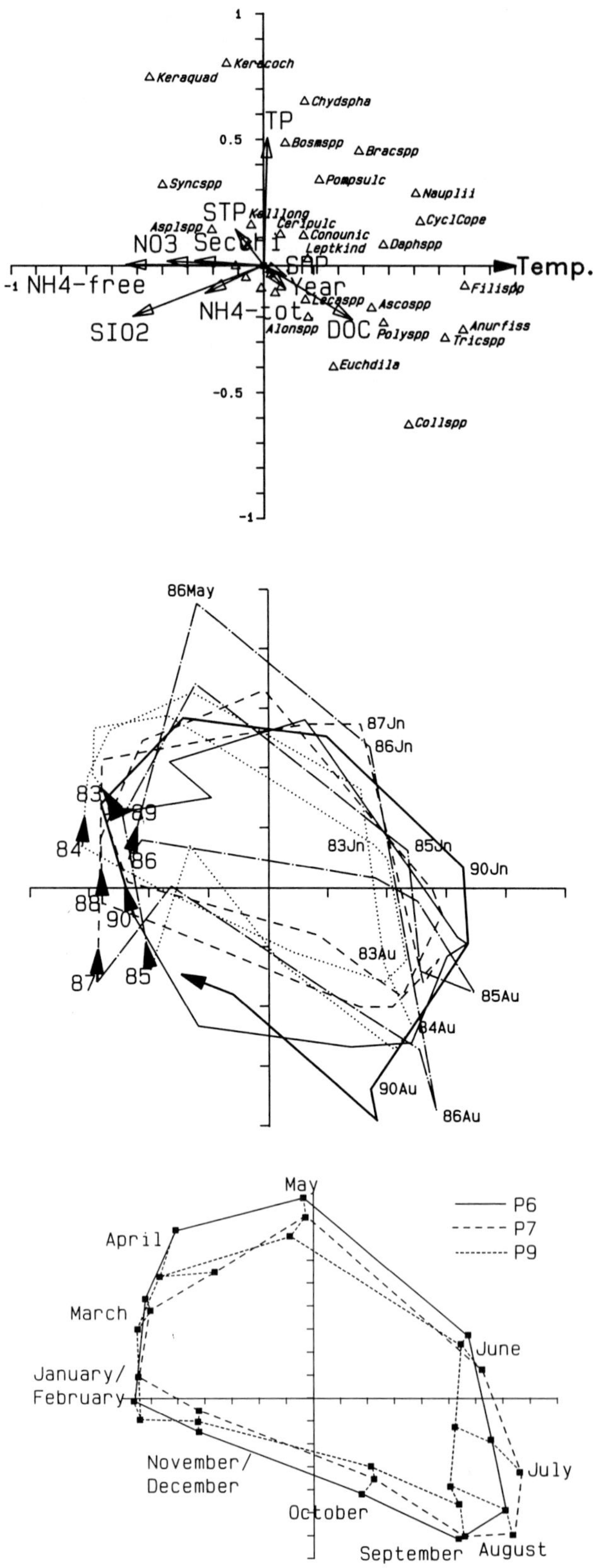

Fig. 4a. RDA correlation biplot of zooplankton species and environmental variables axes 1 (abscissa, constrained by temperature, 30.9% of the variance in the species counts) and 2 (ordinate, unconstrained, 17.5% of the variance in the species counts), samples are not shown. Vectors are indicated as in Fig. 3a. The direction of the species vectors reflects the timing of the maxima: turning clockwise around the origin, the successive peaks in the species densities can be read. For abbreviations see Tables 1 and 3.
b. Succession in the zooplankton community. For details see Fig. 3b.
c. Recurrent seasonal succession for the three stations represented by lines connecting the centroids (average scores) over 8 years for all zooplankton.

variance in the zooplankton counts related to seasonal variation (see Table 3 for the most important species). The general pattern in the cyclic succession thus derived is: in winter almost all species were nearly absent; in early spring relatively low peaks of *Asplanchna* (10–16 ind l^{-1}) and *Synchaeta* (500–600 ind l^{-1}) occurred, followed by higher peaks of *Keratella quadrata* (200–300 ind l^{-1}) and *K. cochlearis* (2000–3000 ind l^{-1}); late spring and early summer *Bosmina* species (*B. coregoni* and *B. longirostris* (*ca.* 600 ind l^{-1}) were followed by *Brachionus* (*ca.* 100 ind l^{-1}) and *Pompholyx sulcata* (*ca.* 250 ind l^{-1}) and later nauplii (*ca.* 500 ind l^{-1}), cyclopoid copepods (*ca.* 200 ind l^{-1}) and *Daphnia* species (*ca.* 40 ind l^{-1}); the carnivorous cladoceran *Leptodora kindtii* was most abundant (*ca.* 3 ind l^{-1}) at the same time, and later in summer the zooplankton community was dominated by rotifers, mainly *Filinia* spp. (*ca.* 500–800 ind l^{-1}), *Anuraeopsis fissa* (*ca.* 5000 ind l^{-1}), *Trichocerca* (*ca.* 300–600 ind l^{-1}), and *Polyarthra* (*ca.* 700 ind l^{-1}), the arrow of this latter species being shorter, because it was present during the whole year and only varied *ca.* tenfold in numbers). *Collotheca* spp. (*ca.* 500 ind l^{-1}) and *Euchlanis dilatata* (*ca.* 20 ind l^{-1}) were most abundant at the end of the season. Most species are well represented by the model: of the 32 species there are 10 for which more than 30% of the variance is described by the first two axes (Table 3). Species not recorded in Table 3 do not fit well in the four axes of the ordination, but are also of minor im-

Table 3. Cumulative fit per species as in Table 2, but for important zooplankton species. Fit is related to temperature (axis 1) and latent variables (axes 2, 3 and 4). Species not mentioned, but appearing with label in figures 4a and 5; Alonspp – *Alona;* Asplspp – *Asplanchna;* Cephspp – *Cephalodella;* Ceripulc – *Ceriodaphnia pulchella;* Conounic – *Conochilus unicornis;* Kellong – *Kellikottia longispina;* Lecaspp – *Lecane;* Leptkind – *Leptodora kindtii;* and Pompsulc – *Pompholyx sulcata.* The remaining species are of minor importance and indicated by symbols only.

Name	Abbreviation fraction of the variance 'explained'					
		Constrained	Unconstrained axes			VAR (y)
		Axis 1	Axis 2	Axis 3	Axis 4	
Daphnia	Daphspp	0.238	0.247	0.395	0.517	0.89
Bosmina	Bosmspp	0.009	0.260	0.262	0.482	1.1
Chydorus sphaericus	Chydspha	0.030	0.474	0.479	0.483	1.65
Cyclopoid copepods	CyclCope	0.402	0.439	0.42	0.486	0.50
Nauplii	Nauplii	0.386	0.474	0.476	0.556	0.38
Synchaeta	Syncspp	0.157	0.270	0.323	0.664	1.87
Polyarthra	Polyspp	0.240	0.288	0.291	0.338	0.59
Keratella quadrata	Keraquad	0.198	0.778	0.795	0.833	1.83
Keratella cochlearis	Keracoch	0.019	0.688	0.747	0.748	1.23
Filinia	Filispp	0.661	0.665	0.850	0.850	3.03
Brachionus	Bracspp	0.156	0.374	0.416	0.535	1.45
Anuraeopsis fissA	Anurfiss	0.654	0.714	0.896	0.896	4.49
Trichocerca	Tricspp	0.543	0.618	0.890	0.891	2.80
Collotheca	Colspp	0.344	0.725	0.726	0.801	2.82
Ascomorpha	Ascospp	0.196	0.222	0.239	0.302	1.38
Euchlanis dilatata	Euchdila	0.082	0.232	0.277	0.278	0.82

portance considering their lower densities or low variances after log-transformation.

The plot of monthly centroids of the zooplankton counts (Fig. 4b) and the correlations of the zooplankton counts with axes in the ordination space of the phytoplankton species (Fig. 5) confirm the absence of a clear trend over years in the zooplankton community. However, from the third axis of the ordination (not shown) and Fig. 5 we derive that *Daphnia* sp. and *Keratella cochlearis* were slightly decreasing. From the third zooplankton ordination axis alone we infer that also *Filinia* spp. *Trichocerca* spp. and *Anuraeopsis fissa* were decreasing over years. Some other species that seem to have decreased or increased were too low in numbers to infer a trend from this analysis. The year 1985 was conspicuous by having a secondary peak in *Daphnia* sp. in autumn.

In contrast to the phytoplankton densities which did not differ among the lakes (not presented in a figure) the zooplankton densities did slightly differ. The *Keratella* peaks were higher in

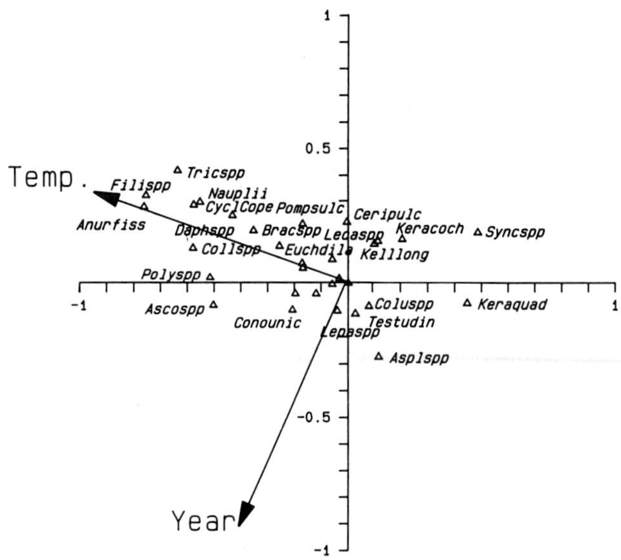

Fig. 5. Passive ordination (axis 1, abscissa: and axis 2, ordinate) of the zooplankton species in the same ordination space of temperature and year as in Fig. 3. Points are correlations of species with axes. The correlations of temperature and year are plotted as arrows.

Breukeleveen (P6) than at the other two stations during April and May and those of the rotifers were higher in lake Vuntus (P7) and the central lake Loosdrecht (P9) than in Lake Breukeleveen (P6) during July and August (Fig. 4c).

Discussion

The proportion of the total variance described by the hybrid analysis (2 constrained, 2 unconstrained axes) of the phytoplankton is low (33.7%) compared with the proportion of variance explained in other analyses either with the same material (*ca.* 50%), with data from other lakes (not published, max. *ca.* 60%) or with the zooplankton data analyzed in this paper (64%). This is due to several reasons. First, we chose to use only a few independent variables, thereby reducing the proportion of variance in the species data that could maximally be explained by the independent variables. Secondly, at the start of the WQL project less abundant species were considered to be of minor importance and, therefore, were lumped. Thus, adequate numbers have been counted only for the dominant species, which are a few in number. Densities for the other species have been derived from very low numbers of cells actually counted, their errors, therefore, are large. We, nevertheless, chose this analysis (using log-transformed data and the covariance matrix) because relatively more weight is given to the less abundant species, and less weight to species with less variation in abundance. Using untransformed counts generally results in information on a few important species only and in a high proportion of explained variance, because the variance in untransformed data of the other species is very low compared with the variance in the dominant species. However, in the present analysis this is less important, because few independent variables have been used and, therefore, the full correlation structure with the independent variables is displayed on the first axis for the zooplankton analysis and on the first two axes for the phytoplankton analysis. Retaining as much information as possible on the relative changes in abundance of all species in a low-dimensional ordination space, thus implicitly giving the abundant species a relatively lower weight, was considered to be more important than the statistical reasons for transformation, which would have led to the same transformation anyhow. If more independent variables were included in the model, the fit for species now of minor importance would have been better, although not significantly.

In the zooplankton community the abundances of species differ less than in the phytoplankton community. Therefore, more species have been counted with sufficient precision. Thus, the overall accuracy is much higher than that for most phytoplankton species. The proportion of the total variance that can be ascribed to temperature alone (the constrained first axis) is 31%; the second, unconstrained, axis (mainly describing the difference between spring and autumn) accounts for an additional 18% of the total variance in the species data. The next two, unconstrained, axes account for only 15% of the variance. The positive correlation (not shown) of the third axis with time (Year, $r = 0.249$) is confirmed by the results of the phytoplankton analysis (Fig. 5), but mainly concerns species having low numbers, which have been counted less accurately. Nevertheless, using supplementary information on the species that seem to have increased or decreased could provide more circumstantial evidence for changes in abundance.

Because, as expected, we observed considerable autocorrelation in space and time, a sound statistical test of the inferred relationships is hardly possible. Since the sampling units are not equally spaced in time, there is no simple permutation scheme for a Monte Carlo permutation test as provided by CANOCO. However, although it is possible to feed the program with a permutation scheme, we did not consider this necessary. Since most of the chemical and physical characteristics are related to each other in some way, deducing causal relationships would be incorrect.

The reduction in the external phosphorus load since 1984 most probably resulted in the decrease in total phosphorus observed in this study and by Van Liere *et al.* (1990b). However, the expected decrease in the abundance of filamentous cyanobacteria (Van Liere, 1986) did not occur. Several explanations can be given for this. Gons *et al.* (1986, 1992) report the effect of wind on this shallow lake system as an important cause for the sub-optimal light conditions for both green algae and macrophytes. However, this does not explain the increase in the filamentous cyanobacteria since Rijkeboer *et al.* (1990) indicate that from 1987 onwards the dominant algae in the Loosdrecht Lakes were phosphorus-limited rather than light-limited. The C/P ratios in the system have tended to increase since 1987 (Gulati *et al.*, 1992), also indicating that, with almost stable seston mass, phosphorus limitation is likely at this moment. Neither the resilience of the system (Van Liere *et al.*, 1990a), nor the existence of multiple stable states (Scheffer, 1990) are good explanations for the increase in filamentous cyanobacteria, since one can hardly expect an increase in numbers and biomass of phosphorus-limited cyanobacteria when total phosphorus is decreasing.

Although the reduction in external phosphorus load does appear to have resulted in a decrease in total P concentration of the lakes, the latter seem to have developed into an even more undesirable direction. Partly, however, variation in the climate may have caused the increase in filamentous cyanobacteria. All winters between 1987 and 1990 were mild and this may reduce the algal mortality rate, and, consequently, a high starting standing crop of algae in spring may result in extreme dominance of the filaments in summer. The raw data, however, do not clearly corroborate effects of climate on the observed trend. The observations of the more severe winter of 1990–1991 and thereafter may confirm the extent to which the climate was responsible for the observed trends in the Loosdrecht lakes. An alternative, even more speculative, explanation that probably cannot be checked is that during the study years *Prochlorothrix hollandica*, discovered in 1985 (Burger-Wiersma *et al.*, 1986, 1989) gradually superseded *Oscillatoria limnetica* or even some other *Oscillatoria* species. The two genera are difficult to distinguish in preserved samples; in fresh material, however, they can be distinguished using epi-fluorescence (Burger-Wiersma *et al.*, 1989; Van Liere *et al.*, 1989a), but this has been done only since summer 1988.

Strikingly, even though the phytoplankton community changed markedly, due mainly to a two-fold increase of filamentous cyanobacteria (including *Prochlorothrix hollandica*), the zooplankton community did not change much. Most species slightly decreased with time; only some species, e.g. *Asplanchna*, occurring in low numbers, seem to have increased. But, because their densities were low and, consequently, the error variances in the numbers high, the variance explained is low and their increase is, therefore, not significant. Apparently, most zooplankton species were influenced either by the stochastic changes in the climate, which is, however, not expected to result in a decrease, or by changes in the quantity and quality of their food. For example, green algae (e.g. *Scenedesmus* and *Ankistrodesmus*), considered to be good food for most zooplankters, decreased and filamentous cyanobacteria, which possibly hinder the uptake of food, especially at higher concentrations, as demonstrated by Dawidowicz (1990; Dawidowicz *et al.*, 1988), increased.

The ordination models presented here describe the main changes in the planktonic communities in Loosdrecht lakes using only a few independent variables. The absence of clear relationships with changes in nutrient concentrations of the system indicates that the range of nutrient concentrations encountered during the study years is too narrow, compared with the seasonal fluctuations, to draw conclusions. This may also be due to the considerable error variance in the observations due to sampling and counting methods, but, since the PCA of environmental data alone did not reveal strong trends in time, the former reason, a narrow range in nutrient concentrations, seems to be more important. Nevertheless, both the phytoplankton and the zooplankton communities indicate that the Loosdrecht lakes are hypertrophic, and we may expect that further reductions in phosphorus load finally will lead to restoration of the lakes. Without data from other comparable systems or, preferrably, experimental studies on the responses to nutrients of the now important species and the species expected to take over if the system becomes less eutrophic, we will not be able to predict the course of this restoration.

These ordination models describe the behaviour of small species groups or separate species in relation to temperature and in the course of time. The relationship with time had to be introduced because causal relationships with nutrients could not be deduced. Moreover, further lumping of the species can lead to a reduction of the error variance, but also to undesired central tendency effects: the decrease of one species and concurrent increase of an other within the same group will offset each other. Despite the apparent constancy in time or apparent lack of response to changes in a nutrient, the community may change considerably.

Besides the drawbacks of lumping based on rough taxonomic criteria, both descriptive models like the one in this study, as well as dynamic ecosystem models like PC-LOOS (Janse & Aldenberg, 1989, 1990; Janse *et al.*, 1992), should be based on a few, relatively homogeneous groups with respect to response of the species to relevant variables. If not, any model is likely to fail to describe and predict the changes in the ecosystem. Thus, it is unlikely that a dynamic model, calibrated on lumped data from this same system in the same period of time, will describe the changes that took place in the Loosdrecht lakes better than a descriptive model fitted on as many separate species as possible. However, the value of dynamic ecosystem models as a tool for understanding the mechanisms leading to changes cannot be denied.

Summarizing, no model based on the available data for the Loosdrecht Lakes only can be used for prediction purposes: collinearity and nonlinear relationships between temperature and nutrients do not allow reliable estimates of parameters for the species now present. It is in any case impossible to estimate parameters for species, now absent, that are likely to take over if the restoration measures will be successful. The present study indicates that the Loosdrecht lakes system is changing, but the relation between this change and the restoration measures is not yet clear.

Acknowledgements

The first author is especially thankful to Cajo ter Braak for stimulating discussions and Koos Vijverberg for his useful remarks on an earlier version of this typescript.

References

Burger-Wiersma, T., L. J. Stal & L. R. Mur, 1989. *Prochlorothrix hollandica* gen. nov., sp. nov.: a filamentous oxygenic photoautotrophic procaryote containing chlorophylls a and b: assignment to *Prochlorothrichaceae* fam. nov. and order *Prochlorales*, Florenzano, Balloni & Materassi 1986, with emendation of the original description. Int. J. Syst. Bact. 39: 250–257.

Burger-Wiersma, T., M. Veenhuis, H. J. Korthals, C. C. M. Van de Wiel & L. R. Mur, 1986. A new prokaryote containing chlorophylls *a* and *b*. Nature 320: 262–264.

Dawidowicz, P., 1990. The effect of *Daphnia* on filament length of blue-green algae. Hydrobiologia 191: 265–269.

Dawidowicz, P., Z. M. Gliwicz & R. D. Gulati, 1988. Can *Daphnia* prevent a blue-green algal bloom? A laboratory test. Limnologica 19: 21–26.

Gauch, H. G., 1982. Multivariate Analysis in Community Ecology. Cambridge University Press, Cambridge, 298 pp.

Gons, H. J., J. Otten & M. Rijkeboer, 1991. The significance of wind resuspension for the predominance of filamentous cyanobacteria in a shallow, eutrophic lake. Mem. Ist. ital. Idrobiol. 48: 233–249.

Gons, H. J., R. Van Keulen & R. Veeningen, 1986. Effects of wind on a shallow lake ecosystem: redistribution of particulates in the Loosdrecht Lakes. Hydrobiological Bulletin 20: 109–120.

Gons, H. J., T. Burger-Wiersma, J. H. Otten & M. Rijkeboer, 1992. Coupling of phytoplankton and detritus in a shallow, eutrophic lake (Lake Loosdrecht, The Netherlands). Hydrobiologia 233: 51–59.

Gulati, R. D., 1990. Zooplankton structure in the Loosdrecht lakes in relation to trophic status and recent restoration measures. Hydrobiologia 191: 173–188.

Gulati, R. D., K. Siewertsen & G. Postema, 1985. Zooplankton structure and grazing activities in relation to food quality and concentrations in Dutch lakes. Arch. Hydrobiol. Beih. Ergebn. Limnol. 21: 91–102.

Gulati, R. D., L. Van Liere & K. Siewertsen, 1991. The Loosdrecht lake system: Man's role in its creation, perturbation and rehabilitation. In: Terrestrial and Aquatic Ecosystems: Perturbation and Recovered. O. Ravera, Chichester Ellis Horwood. pp 593–596.

Gulati, R. D., A. L. Ooms-Wilms, O. F. R. Van Tongeren, G. Postema & K. Siewertsen, 1992. The dynamics and role of limnetic zooplankton in Loosdrecht lakes (The Netherlands) Hydrobiologia 233: 69–86.

Hofstra, J. J. & L. Van Liere, 1992. The state of the environment of the Loosdrecht Lakes. Hydrobiologia 233: 11–20.

Janse, J. H. & T. Aldenberg, 1989. Modelling the eutrophication of the shallow Loosdrecht Lakes. Verh. int. Ver. Limnol. 24: 751–757.

Janse, J. H. & T. Aldenberg, 1990. PCLOOS: eutrophication model of the Loosdrecht Lakes. WQL report 1990–1, RIVM, Bilthoven.

Janse, J. H., T. Aldenberg & P. R. G. Kramer, 1992. A mathematical model of the phosphorus cycle in the Loosdrecht Lakes and simulation of additional measures. Hydrobiologia 233: 119–136.

Kott, P., 1953. Modified whirling apparatus for the subsampling of zooplankton. Aust. J. mar. Freshwat. Res. 4: 387–393.

Lammens, E. H. R. R., P. J. Boesewinkel-de Bruyn, H. Hoogveld & E. Van Donk, 1992. Phosphorus-load, phytoplankton, zooplankton and fish stock in Loosdrecht Lake and Tjeukemeer: confounding effects of predation and food availability. Hydrobiologia 233: 87–94.

Otten, J. H., H. J. Gons & M. Rijkeboer, 1992. Dynamics of phytoplankton detritus in a shallow, eutrophic lake (Lake Loosdrecht, The Netherlands). Hydrobiologia 233: 61–68.

Pielou, E. C., 1977. Mathematical Ecology. 2nd edn. Wiley, New York, 385 pp.

Pielou, E. C., 1984. The interpretation of ecological data. A primer on classification and ordination. Wiley, New York, 263 pp.

Rao, C. R., 1964. The use and interpretation of principal component analysis in applied research. Sankhya A 26: 329–358.

Rao, C. R., 1973. Linear statistical inference and its applications. 2nd edn. Wiley, New York, 625 pp.

Rijkeboer, M., F. de Bles & H. J. Gons, 1990. Phosphorus, light and seston dynamics (in Dutch). Internal Report 1990–11, Limnological Institute, 85 pp.

Scheffer, M., 1990. Multiplicity of stable states in freshwater systems. Hydrobiologia 200/201: 475–486.

Ter Braak, C. J. F., 1987a. Ordination. In: Jongman, R. H. G., C. J. F. ter Braak & O. F. R. van Tongeren (eds), 1987. Data Analysis in Community and Landscape Ecology. PUDOC, Wageningen, 299 pp.

Ter Braak, C. J. F., 1987b. CANOCO-A FORTRAN program for canonical community ordination by [partial][detrended][canonical] correspondence analysis, principal component analysis and redundancy analysis (version 2.1). TNO Institute of Applied Computer Science, Wageningen, the Netherlands.

Ter Braak, C. J. F., 1989. CANOCO – an extension of DECORANA to analyze species-environment relationships. Hydrobiologia 184: 169–170.

Ter Braak, C. J. F., 1991. CANOCO update notes. GLW-DLO, Wageningen, the Netherlands (supplied with the program).

Van Liere, L., 1986. Loosdrecht lakes, origin, eutrophication, restoration and research programme. Hydrobiol. Bull. 20: 9–15.

Van Liere, L., S. Parma, L. R. Mur, P. Leentvaar & G. B. Engelen, 1984. Loosdrecht lakes restoration project, an introduction. Ver. int. Ver. Limnol. 22: 829–834.

Van Liere, L., L. Breebaart & Y. J. Dullemont, 1989a. Determining the relative number of prochlorophytes in lake phytoplankton using epifluorescence microscopy. Br. phycol. J. 24: 391–394.

Van Liere, L., L. Breebaart, W. Kats & J. J. Buyse, 1989b. The water quality of the Loosdrecht Lakes area. Commissie voor Hydrologisch Onderzoek TNO. Rapporten en Nota's 22: 265–279. (In Dutch)

Van Liere, L., R. D. Gulati, F. G. Wortelboer & E. H. R. R. Lammens, 1990a. Phosphorus dynamics following restoration measures in the Loosdrecht Lakes (The Netherlands). Hydrobiologia 191: 87–95.

Van Liere, L., O. F. R. Van Tongeren, L. Breebaart & W. Kats, 1990b. Trends in chlorophyll and total phosphorus in the Loosdrecht lakes, The Netherlands. Ver. int. Ver. Limnol. 24: 707–710.

Van Tongeren, 1991. CEDIT. Preliminary manual, Internal Report 1991–9. Limnological Institute, Nieuwersluis. 123 pp.

Hydrobiologia **233**: 119–136, 1992.
L. Van Liere & R.D. Gulati (eds), Restoration and Recovery of Shallow Eutrophic Lake Ecosystems in The Netherlands.

A mathematical model of the phosphorus cycle in Lake Loosdrecht and simulation of additional measures

J. H. Janse, T. Aldenberg & P. R. G. Kramer
National Institute of Public Health and Environmental Protection (RIVM), P.O. Box 1, 3720 BA Bilthoven, The Netherlands

Key words: model, simulation, eutrophication, phosphorus, P/C ratio, lake ecosystem

Abstract

The phosphorus cycle in the ecosystem of the shallow, hypertrophic Loosdrecht lakes (The Netherlands) was simulated by means of the dynamic eutrophication model PCLOOS. The model comprises three algal groups, zooplankton, fish, detritus, zoobenthos, sediment detritus and some inorganic phosphorus fractions. All organic compartments are modelled in two elements, carbon and phosphorus. Within the model system, the phosphorus cycle is considered as completely closed. Carbon and phosphorus are described independently, so that the dynamics of the P/C ratios can be modelled. The model has been partly calibrated by a method based on Bayesian statistics combined with a Range Check procedure.

Simulations were carried out for Lake Loosdrecht for the periods before and after the restoration measures in 1984, which reduced the external phosphorus loading to the lake from *ca.* 2 mgP m^{-2} d^{-1} to 1 mgP m^{-2} d^{-1}. The model outcome was largely comparable with the measured data. Total phosphorus has slowly decreased from an average 130 μgP l^{-1} to *ca.* 80 μgP l^{-1}, but chlorophyll-*a* (*ca.* 150 μg l^{-1}, summer-averaged) and seston concentrations (8–15 mgC l^{-1}) hardly changed since the restoration measures. About two-thirds of the seston consisted of detritus, while the phytoplankton remained dominated by filamentous cyanobacteria. The P/C ratio of the seston decreased from *ca.* 1.0% to 0.7%, while the P/C ratios of zooplankton, zoobenthos and fish have remained constant and are much higher. The system showed a delayed response to the decreased phosphorus loading until a new equilibrium was reached in *ca.* five years. Major reasons for the observed resilience of the lake in responding to the load reduction are the high phosphorus assimilation efficiency of the cyanobacteria and the high internal recycling of phosphorus. A further reduction of nutrient loading, perhaps in combination with additional measures like biomanipulation, will be the most fruitful additional restoration measure.

Introduction

Like many other shallow lakes in the Netherlands the Loosdrecht lakes have become highly eutrophic due to heavy nutrient loadings for the last forty years or so. This has led to almost permanent blooms of filamentous cyanobacteria with mean summer chlorophyll-*a* concentrations of 150–200 μg l^{-1}, high primary production and seston concentrations and loss of biological diversity (Hofstra & Van Liere, 1992). In order to reduce the algal growth, several measures have been taken to reduce the external phosphorus loading. From 1984 onwards, the source of summer inlet water, the polluted River Vecht, has been replaced by dephosphorized canal water.

This has resulted in a decrease in the average external phosphorus loading to the main lake from *ca.* 2 mg P m^{-2} d^{-1} in 1983 to 1 mg P m^{-2} d^{-1} from 1984 onwards (Engelen *et al.*, 1992; Buyse, 1988). This has, however, not led to a decrease of the seston concentration (Van Liere *et al.*, 1990).

The Water Quality research Loosdrecht lakes project (WQL) was carried out to assess and quantify the effects of reduction in phosphorus load on the lake ecosystem and to recommend additional water management measures (Van Liere *et al.*, 1984). A mathematical model, called PCLOOS, was developed at the National Institute of Public Health and Environmental Protection (RIVM), one of the participants in the WQL project. The aim of the model is to provide a better insight into the functioning of the system by integrating the results of the WQL project and to predict the effects of different restoration measures.

Several phosphorus models have been reported elsewhere, ranging from very simple to very complex ones. The most simple models, the so-called empirical or 'black box' models (OECD, 1982; Reckhow & Chapra, 1983; Sas, 1989), relate the in-lake total phosphorus concentration to the external loading and the water retention time and describe statistically the relationship between algal biomass and total P concentration. A second group are the dynamic models which describe phytoplankton growth as function of nutrient availability (see overview in Straskraba & Gnauck, 1985). Some of these models do, others do not include nutrient cycling in the sediment. The third group are detailed, multi-species models of the trophic web in aquatic ecosystems (*e.g.* Patten *et al.*, 1975).

The empirical models are, in fact, one-compartment models which cover the processes of inflow, dilution and net retention in the sediment. Such models are not sophisticated enough for the present study for two reasons. Firstly, in a shallow lake like Loosdrecht, with a very intense contact between water and sediment due to continuous mixing, the net retention is likely to decrease following the restoration measures (Sas, 1989). This implies that we need to include the upper sediment layer explicitly in the model. Secondly, the relationship between total phosphorus on one hand and algal biomass and total seston on the other hand is not unequivocal, so that a static description of this relationship is not sufficient. In the WQL project a dynamic model is needed, with a distinction between algal biomass and algal phosphorus and a variable ratio between them. Furthermore, detritus has to be considered as a separate compartment, also expressed in two units. Some dynamic phytoplankton models (*e.g.* Lehman *et al.*, 1975; Bierman & Dolan, 1981) use, in fact, a variable stoichiometry approach, but they either lack a sediment module or the description of the food web does not match the needs of the present study.

Apart from the algal growth in response to the phosphorus load we were also interested in the phosphorus exchange with higher trophic levels, especially zooplankton and fish. This required some modifications in the modelling of the phosphorus cycle. A basic feature of the PCLOOS model is that the phosphorus cycle is completely closed, except for external loading to and losses from the lake. This feature is important not only for describing algal growth or the exchange between water and sediment, but also for considering the phosphorus flow through the food web. It has been observed that at higher levels in the food chain the weight-specific phosphorus content of the organisms increases due to accumulation of phosphorus. The P/C ratio of the seston can be highly variable and is quite low in the Loosdrecht lakes (less than 1%; Gulati *et al.*, 1991), whereas this ratio is much higher for the zooplankton (2.0–2.5%; Gulati *et al.*, 1991) and zoobenthos (2.0–2.5%; Dr. J. Vijverberg, Pers. Comm.) and still higher for fish (*ca.* 5.5%; Dr. E. H. R. R. Lammens, Pers. Comm.). These differences can be maintained only if the animals have mechanisms to regulate their P/C ratio irrespective of the P/C ratio of their food. These mechanisms can involve a more efficient assimilation of phosphorus from the food and/or a reduced phosphorus excretion rate. In this respect, our model differs from several existing ones (*e.g.* Di Toro *et al.*, 1975; Di Toro & Matystik, 1980; Jørgensen,

1980). Aldenberg & Peters (1990) have stated that in a model in which the P/C ratios of the trophic levels differ, the P/C ratios of one or more compartments should be variable in order to maintain a closed P balance. In their model, only the detritus has a variable P/C ratio while the other compartments, phytoplankton and zooplankton, have a fixed one.

The PCLOOS model, version 2.4 (Janse & Aldenberg, 1990a, b, 1991) is based on the model by Aldenberg & Peters (1990), but extended with a sediment module, a more complete food web including zoobenthos and fish and variable P/C ratios of all the biotic and organic compartments. The model presented in this paper is PCLOOS version 2.5. It differs from the previous version in the incorporation of chemical adsorption, an improved modelling of some sediment processes and a number of minor changes in the phytoplankton, fish and zoobenthos growth equations. Moreover, the model was extended with a calibration routine based on Range-Check and Bayesian statistics.

Description of the lake

The model has been applied to the Loosdrecht lakes, a system of interconnected shallow lakes which originate from peat digging, located between Amsterdam and Utrecht, The Netherlands. A detailed description of the lakes system can be found in Gulati *et al.* (1990). The main lake, Lake Loosdrecht, has a water area of 989.1 ha, a mean depth of 1.91 m and an average water residence time of 0.7 years. Downward seepage averages 1.5 mm d^{-1} (Engelen *et al.*, 1992). The bottom consists of peat covered by a layer of loose organic matter of variable thickness. The lake is very much exposed to the wind and can be regarded as completely mixed.

Model structure

General features

The model comprises both the water column and the upper sediment layer (Fig. 1). The model is zero-dimensional in space: the lake is assumed to be completely mixed, a reasonable assumption regarding the high correlation of the data from the two sampling stations in Lake Loosdrecht, as can be derived from Breebaart *et al.* (1989). Carbon is used as the unit of biomass. The model is composed of eight organic compartments, modelled both in carbon and in phosphorus units, indicated as two partly overlapping boxes, and four inorganic phosphorus compartments, making twenty state variables altogether. Inorganic carbon is not included in the model: decomposition and respiration losses of CO_2 go to an 'outside' pool of dissolved inorganic carbon (DIC), from which CO_2 for algal growth is withdrawn. This assumption seems reasonable, because the phytoplankton growth in the lakes is not carbon-limited. For analogous reasons, other nutrients than phosphorus are not considered.

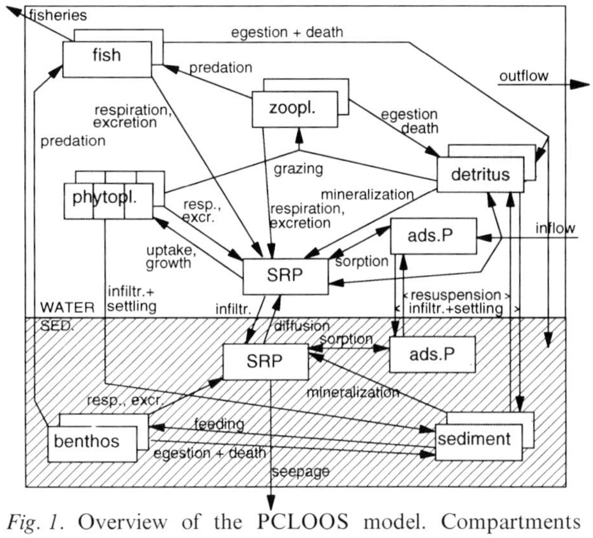

Fig. 1. Overview of the PCLOOS model. Compartments modelled in two units are indicated by two boxes behind each other. The algal compartment is composed of three functional groups: cyanobacteria, diatoms and green algae. The arrows in the figure denote both the carbon and the phosphorus fluxes. Some minor arrows have been lumped or left out for the sake of clarity.

Inorganic phosphorus is modelled both in the surface water and in the interstitial water. In both phases, a distinction is made between soluble reactive phosphorus (SRP), which is assumed to be biologically available, and adsorbed inorganic phosphorus. The organic matter in the lake, which

values on August 1 of the first year for comparison. If the simulation fell beyond one or more of these ranges, the parameter combination was rejected, the run was stopped and a new combination was chosen. If the simulation was within the ranges, the run was allowed to continue. This Range-Check procedure saved a considerable amount of computing time spent otherwise on unproductive runs. The successful runs were subjected to Bayesian statistics. For every measurement the difference between measured and simulated values was calculated and then used to compute the sum of squares for the variable:

$$\text{Sum Of Squares}_{\text{var}} = \sum_{t} (\text{obs}_{\text{var},\,t} - \text{model}_{\text{var},\,t})^2\,,$$

with t = time. The eight obtained sum of squares values, one for each measured variable, were used to calculate the marginal posterior density (Box, 1971):

$$\text{posterior density} = \prod_{\text{var}=1}^{8} (\text{Sum Of Squares}_{\text{var}})^{-\frac{1}{2}\cdot n_{\text{var}}}\,,$$

in which n = number of measurements, which can be different for different variables. The parameter combination with the maximum posterior density gives the best fit of the model to the measurements. This combination was used in the simulations depicted.

The model calculations were performed on a SUN Sparc station or on a 80386 PC, by means of the simulation program ACSL, level 10B, developed by Mitchell & Gauthier Associates (1991). The integration routine used was Gear's Stiff algorithm, with a variable order and variable step size, which were adapted according to preset error criteria. For our model, which includes processes with highly different time constants, this algorithm proved to be considerably faster than others. A ten-years simulation took 5–10 minutes per run. Initial values of the state variables were estimated from the data. Prior to a simulation, the model was allowed to equilibrate for a simulated period of three years.

Results

Simulations

Simulations have been carried out for the period 1982–1990 for three lakes of the Loosdrecht system, but here only the results for the largest lake, Lake Loosdrecht, are described. In the years up to 1984, the average total phosphorus concentration in the inflowing water (*viz.* the external P loading divided by the water inflow) was very high in the summer months, due to the inlet of highly polluted supply water from the River Vecht (Fig. 2). After restoration measures in 1984, these peak loadings have disappeared, but the loading from other sources remains high (Buyse, 1989; Engelen *et al.*, 1992). On annual basis, the external loading has been reduced by a factor 2 (Van Liere & Janse, 1992).

During the calibration of the model, about one run out of every six was successful, *i.e.* the results fell within the established ranges for the state variables. 200–300 runs were sufficient to find the maximum posterior density: the natural logarithm of it was −503. Simulations, together with the data, show that total phosphorus (Fig. 3) has tended to decrease steadily following the restoration measures, although the unexplained variation in the data is high. The solid line is the 'best

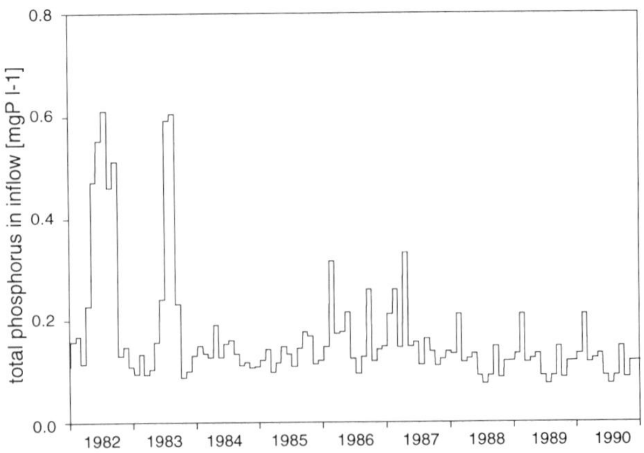

Fig. 2. Average total phosphorus concentration in the inflowing water to Lake Loosdrecht, 1982–1990. Based on data of Buyse (1988) and Engelen *et al.* (1992). The values for 1988–1990 are based on those of 1986.

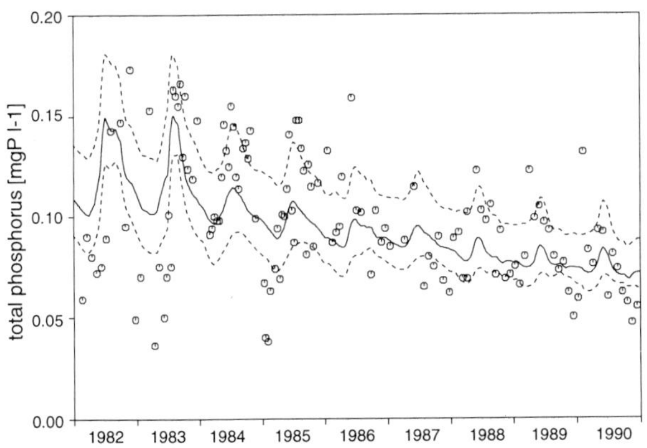

Fig. 3. Total phosphorus concentration in Lake Loosdrecht: simulation of nominal run (solid line), error bounds (dashed lines) and measurements (symbols), 1982–1990.

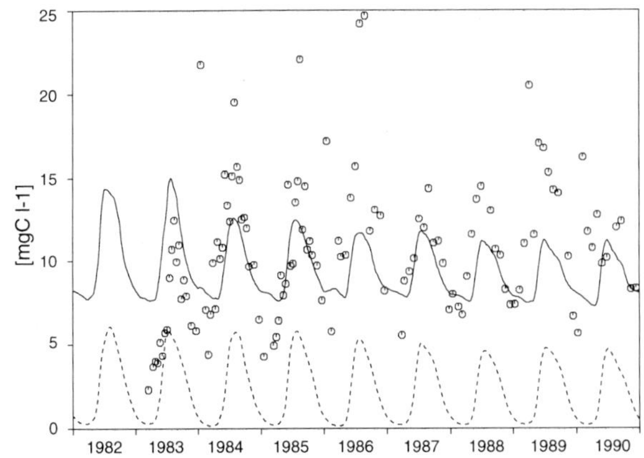

Fig. 5. Simulation of total seston (solid line) and cyanobacteria (dashed line) and measurements of total seston (symbols) in Lake Loosdrecht, in 1982–1990.

fit', the dashed lines indicate the ranges, due to variation of the 11 calibration parameters, of runs giving a natural logarithm of the posterior density higher than −520. A corresponding decrease in chlorophyll-*a* was, however, not observed (Fig. 4), nor in the total seston (Fig. 5) or Secchi depth, which still varies between 0.3 and 0.4 m. The cyanobacteria were the dominant group during most of the year, but in spring, also the diatoms developed a significant biomass, according to both the simulations and the field observations (Boesewinkel-De Bruyn *et al.*, 1988). Major part of the seston mass (*ca.* 60% in summer, > 90% in winter) consists of detritus; this is consistent with estimates by Gons & Van Keulen (1989) of an average detrital fraction in the seston mass of two-thirds. The zooplankton (Fig. 6), zoobenthos and fish stock are not much affected by the reduced nutrient loading. The simulated densities of fish and zoobenthos are in the range of field estimates: fish 275 kg ha^{-1} (corresponding to about 2.8 gC m^{-2}) (Lammens *et al.*, 1992) and zoobenthos 1 gC m^{-2} on the average (Dr. J. Vijverberg, Pers. Comm.). Strikingly, the seston concentration if calculated as phosphorus gives a stronger response to the decreased phosphorus input than the corresponding carbon concentration. The simulated P/C ratio of the seston grad-

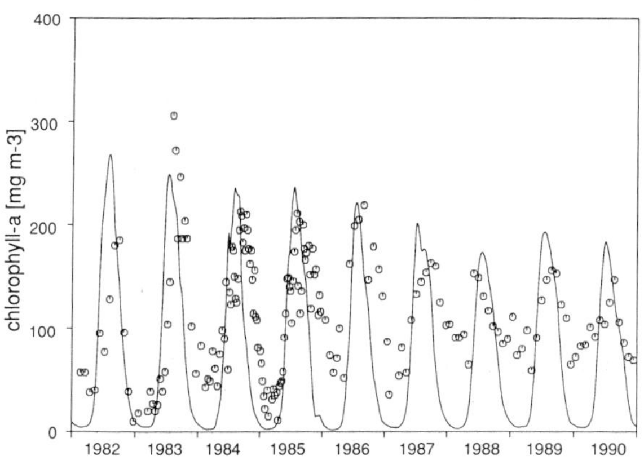

Fig. 4. Chlorophyll-*a* concentration in Lake Loosdrecht: simulation (line) and measurements (symbols), 1982–1990.

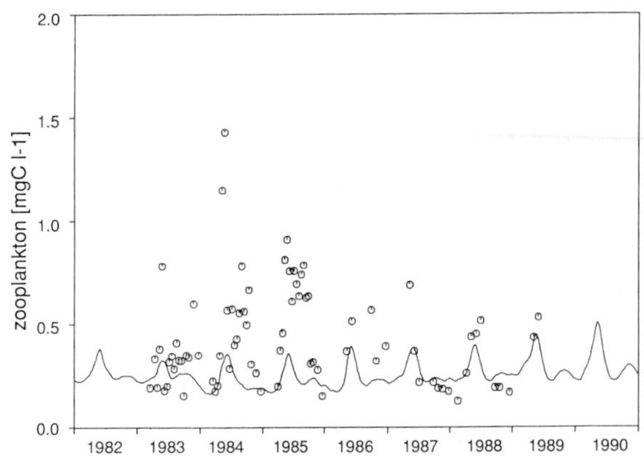

Fig. 6. Zooplankton concentration in Lake Loosdrecht: simulation (line) and measurements (symbols), 1982–1990.

ually decreases, both in the simulations and in the lake (Fig. 7), but those of the zooplankton, zoobenthos and fish do not change very much, because of the regulation mechanisms of the animals. The field zooplankton ratios also do not show a change (Gulati *et al.*, 1991). The phosphorus pool in fish, about 80 μg P l^{-1}, roughly equals the P in all the other compartments in the water phase together (Van Liere *et al.*, 1990), as found in many hypertrophic lakes. The SRP values (not shown) are very low during most of the year, according to both the model and the field data. As in the water phase, the simulated P concentration in the sediment (not shown) also decreases steadily because of the load reduction; although data are sparse, observations in the field seem to confirm this (Keizer & Sinke, 1992). The concentrations in the pore water vary considerably during the year, depending on the mineralization activity which is highly temperature dependent.

The characteristic response time of the model system to adapt itself to a decrease in nutrient loading is about five years. A new equilibrium, which has now been nearly reached, does not differ much from the one before the restoration measures, except for a decrease in phosphorus concentration, not accompanied by a corresponding decrease in chlorophyll-*a* and seston (Fig. 8). The shift to the left towards the maximum line based on a large number of Dutch lakes (CUWVO, 1987) clearly shows that the chlorophyll-*a*: total P ratio is among the highest recorded.

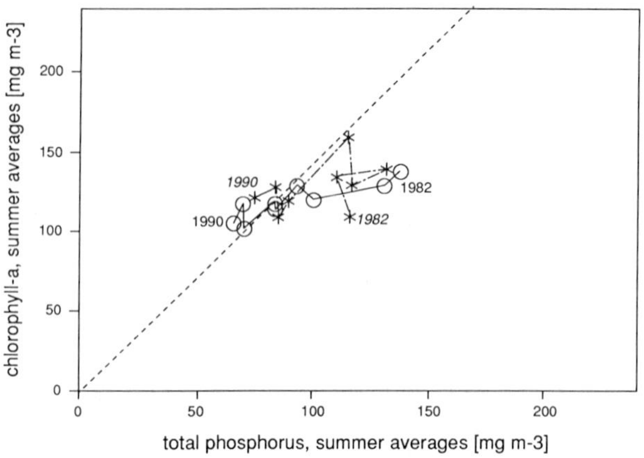

Fig. 8. Eutrophication diagram (summer-averaged chlorophyll-*a* against total phosphorus) in Lake Loosdrecht: simulation (circles) and measurements (asterisks), 1982–1990. The dashed line is the maximum ratio found in a database of Dutch lakes (Lijklema *et al.*, 1988).

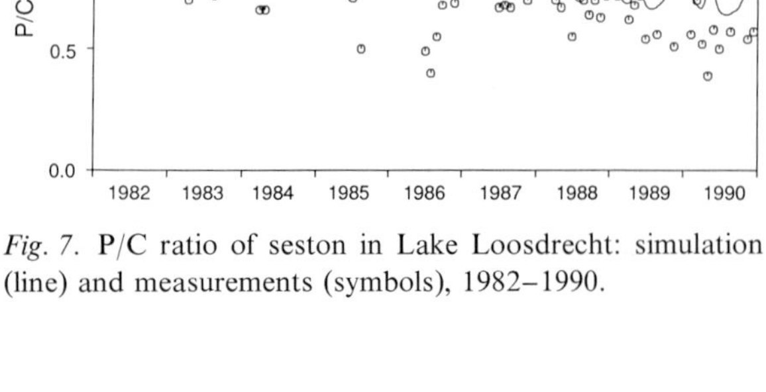

Fig. 7. P/C ratio of seston in Lake Loosdrecht: simulation (line) and measurements (symbols), 1982–1990.

The phosphorus turnover rates greatly differ for the different model compartments, as can be concluded from the simulated flow scheme, averaged for the summer of 1987 (Fig. 9). For the SRP compartment, which is very small in size, this rate is by far the highest (*ca.* 10 d^{-1}): SRP formed is taken up almost immediately by algae or, when these are saturated (for instance in winter), adsorbed onto iron-rich particles in the water. This pool of adsorbed phosphorus also has a high turnover rate (*ca.* 1 d^{-1}, on the average), due to rapid chemical exchange with the SRP pool. Of the organic compartments, the zooplankton has the fastest turnover, about 0.2 d^{-1}, the algae, detritus and zoobenthos have about 0.1 d^{-1} and the turnover rate of fish is only 0.02 d^{-1}. The compartment with the slowest turnover rate is the organic phosphorus in the sediment: about 0.01 d^{-1}. About 10% of the daily phosphorus needs of the phytoplankton is met, directly or indirectly, by the external loading, about 10% by desorption from the adsorbed P pool and only 3% by diffusion from the sediment. The larger part, *i.e. ca.* 75%, becomes available again by recycling via zooplankton, fish, detritus or the algae themselves. The phosphorus flow

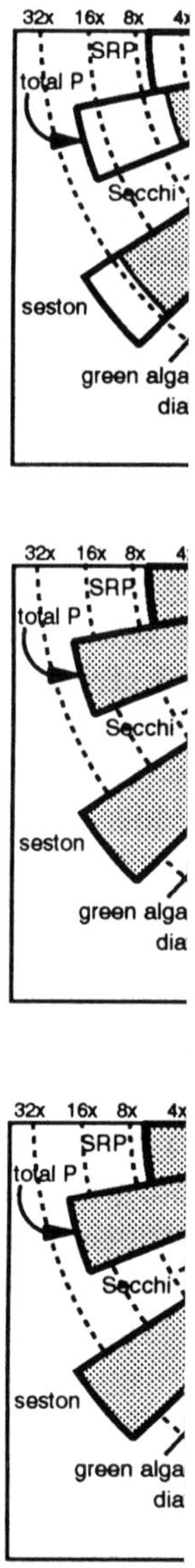

Fig. 12. Simu
Liere, 1992). I
pension. Meas
dark slices: si
averages.

Discussion

The lake ec

The overal
tem can b
schemes (F

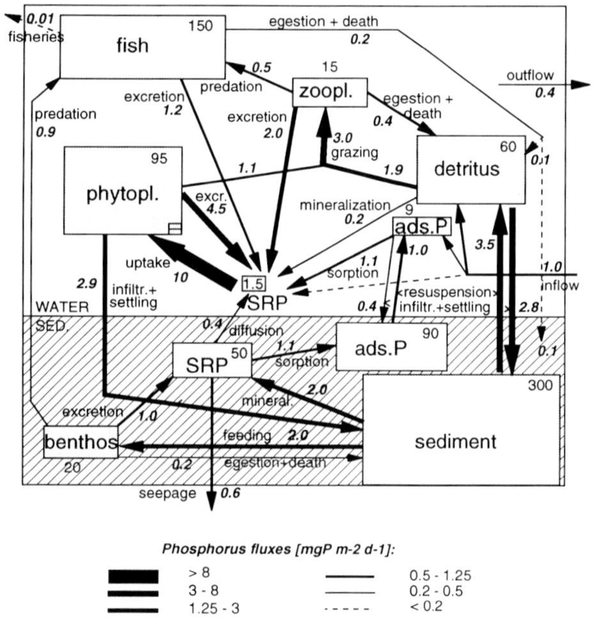

Fig. 9. Simulated phosphorus flow scheme of Lake Loosdrecht, summer average 1987. Concentrations in mgP m^{-2}, flows in mgP m^{-2} d^{-1}.

from the sediment to the water phase occurs mainly through resuspension; the flow due to fish feeding and the diffusive flux are much smaller.

The modelled diffusive flux approximately equals the flux measured in sediment cores (Boers & Van Hese, 1988; Keizer *et al.*, 1991); it has decreased a little since the restoration measures (Figs. 9 and 10).

Simulation of additional measures

Because the achieved reduction in phosphorus loading of the lakes did not lead to a discernible improvement in the water quality, additional measures were considered by the regional water quality authorities (see also Van Liere & Janse, 1992). The possible effects of the following measures have been modelled, with the same parameter settings as established during the calibration:

(a) No further change in the present phosphorus loading (1.0 mgP m^{-2} d^{-1}).
(b) A further reduction in the phosphorus loading to 0.35 mgP m^{-2} d^{-1} (= 0.1 gP m^{-2} yr^{-1}), *i.e.* a reduction which is the maximum technically possible.
(c) Extra flushing with $1 \cdot 10^6$ m^3 dephosphorised water per month. This is the amount technically possible given the present hydrological

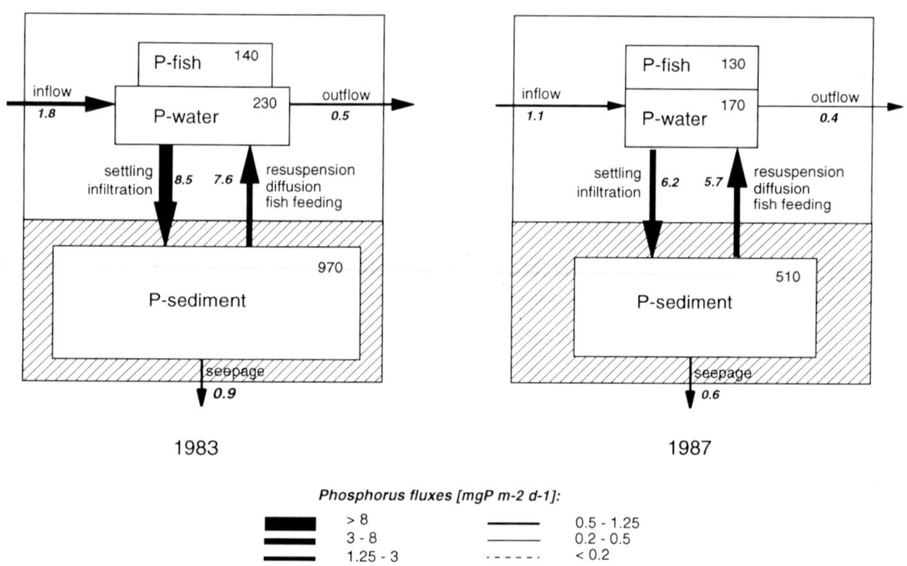

Fig. 10. Simplified phosphorus flow schemes of Lake Loosdrecht, yearly averages; concentrations in mgP m^{-2}, flows in mgP m^{-2} d^{-1}. (a) 1983. (b) 1987.

Spectrophotometric measurements

The best values for absorption and scattering spectra of pure water from 300 to 800 nm were estimated by Buiteveld & Donze (pers. comm.) based on data from: (1) Boivin (1986) for 400 to 436 nm, (2) Smith & Baker (1981) from 436 to 800 nm. Pure water absorbs light only very weakly in the blue and green regions of the spectrum (Fig. 2a). Absorption increases above 550 nm and is significant in the red region. Scattering by water is inversely proportional to wavelength, which combined with low absorption, causes the blue colour of pure water. In these lakes the molecular and Raman scattering effects are negligible compared with the scattering caused by particulate matter.

The spectral absorption properties of dissolved aquatic humus (Fig. 2b) were determined using the method described in Malthus and Dekker (1990). Absorption of light by aquatic humus declines exponentially with increasing wavelength according to Davies *et al.* (1987), although Buiteveld *et al.* (1987) found significant deviations from an exponential function. Data of Malthus (unpubl.) indicate that aquatic humus absorption in these lakes may be described by an exponential function provided a correction for residual scattering has been made. An increase in absorption covaries with an increase in lake trophic status. Aquatic humus removes blue light in the top few centimetres of the water column, imparting a yellow colour to the water. It is assumed that aquatic humus does not scatter light.

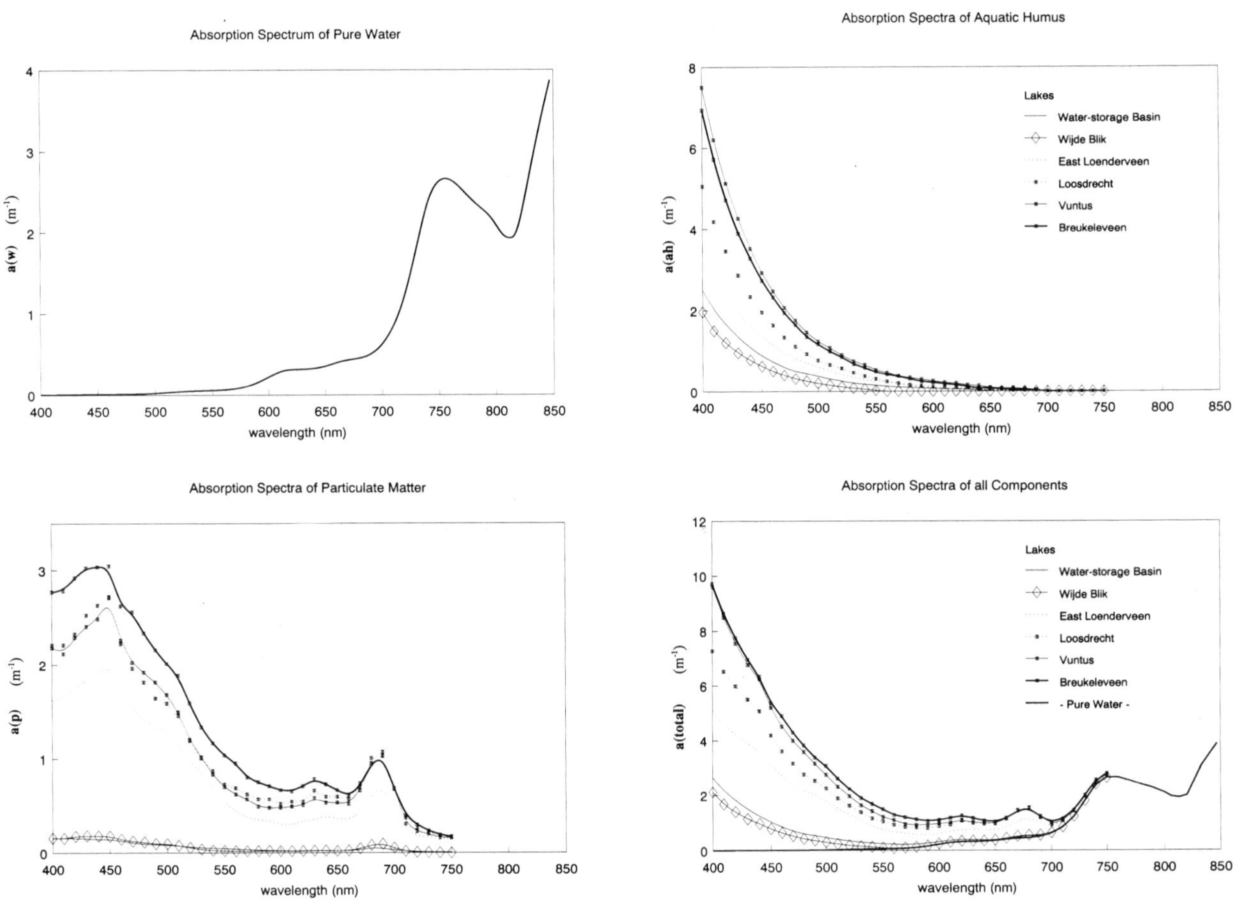

Fig. 2. Spectral absorption curves of (a) pure water (pers. comm. Buiteveld & Donze: see text); (b) aquatic humus; (c) suspended particulate matter; (d) all constituents. (From Dekker *et al.*, 1990c; note different scale Y-axis!)

The particles in the water consist of phytoplankton, dead organic matter and inorganic matter. Scattering by algae which are semitransparent and contain different materials within its cell structure is complex. Cyanobacteria contain gas vacuoles that have large breaking indices with the surrounding cytoplasm causing anomalous scattering at high angles (Donze *et al.*, 1987). Scattering by algae is roughly inversely related to wavelength. Scattering spectra of algae are given in Prieur and Sathyenadrath (1981) and Donze *et al.* (1987).

Figure 2c shows typical particulate absorption spectra for each lake from which several features are apparent: high light absorption at short wavelengths due to absorption by detritus and by algal pigments. A chlorophyll *a* absorption peak at 445 nm and an absorption shoulder at 485 nm, probably a β-carotene absorption feature, are evident. From 500 nm an exponential decrease in absorption occurs. A local maximum in absorption occurs at 630 to 640 nm. This is probably due to absorption by phycocyanin. At 675–680 nm the *in vivo* chlorophyll *a* absorption band is clear. From 680 nm onwards absorption by particulate matter decreases.

The absorption coefficient of the medium as a whole, at a given wavelength, is equal to the sum of the individual absorption coefficients of the components present (Kirk, 1983). Assuming a linear relationship between concentration and absorption, the absorption coefficient due to any one component is proportional to the concentration of that component (Beer's Law). Therefore

$$a(total) = a(w) + a(ah) + a(p),$$

where $a(w)$ is the absorption by pure water, $a(ah)$ by aquatic humus and $a(p)$ by suspended particulate matter. By summing the spectral absorption curves of water, aquatic humus and particulate matter, the total spectral absorption curve is derived (Fig. 2d). At short wavelengths absorption by aquatic humus in the eutrophic lakes is a factor 2 to 4 larger than by the particulate matter. The chlorophyll *a* absorption peak at 445 nm, clearly visible in Fig. 2c is almost undistinguishable in Fig. 2d due to the aquatic humus absorption. From 710 nm onwards the water absorption dominates the absorption spectrum. The pigment absorption peaks at 630–640 nm and 675–680 nm remain clearly visible in the graph.

Spectroradiometric measurements

The *in situ* downwelling and upwelling irradiance above and under the water surface was measured

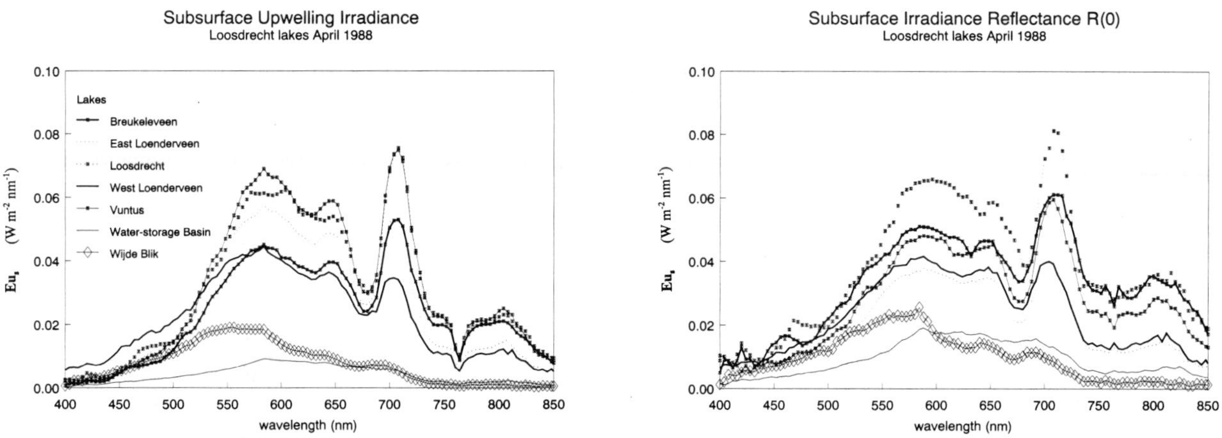

Fig. 3. a: The subsurface upwelling irradiance (Eu_s) measurements of 7 lakes (see also Tables 1 and 2). (From Dekker *et al.*, 1990c). b: The $R(0)$ (ratio of subsurface upwelling irradiance to subsurface downwelling irradiance) measurements of 7 lakes (see also Tables 1 and 2). (From Dekker *et al.*, 1991a).

using an underwater spectroradiometer at 25 sampling points in April 1988. These measurements reflect the combined spectral effects of absorption, scattering and solar-induced fluorescence from water, aquatic humus and the particulate matter. From these scans subsurface spectral irradiance reflectance $R(0)$ was calculated:

$$R(0) = Eu_s/Ed_s,$$

where Eu_s and Ed_s are subsurface spectral upwelling and downwelling irradiance. Representative $R(0)$ and Eu_s spectra are shown in Fig. 3. The following features are apparent in the spectra of the eutrophic lakes:

- Low $R(0)$ and Eu_s in the wavelength region of 400 to 500 nm due to absorption by aquatic humus and by algal pigments (e.g. chlorophyll *a*, β-carotene, zeaxanthin).
- Increased $R(0)$ and Eu_s up to 600 nm. Greater $R(0)$ and Eu_s in this wavelength region is the result of low absorption by photosynthetic pigments coupled with increased backscattering with increased particulate concentrations.
- A local minimum in $R(0)$ and Eu_s occurs at 630 to 640 nm. This is probably due to absorption by phycocyanin. This pigment occurs in cyanobacteria abundant in the eutrophic lakes of this system.
- At 675–680 nm the *in vivo* chlorophyll *a* absorption band is clear. Concentrations of chlorophyll *a* may reach 200 $\mu g\, l^{-1}$ in the most eutrophic of these lakes.
- At 700–720 nm a peak in $R(0)$ and Eu_s is visible. Vos *et al.* (1986) studied the nature of this reflectance peak and gave several explanations for its presence:
  (1) fluorescence of chlorophyll pigments;
  (2) anomalous scattering caused by the absorption peak at 675–680 nm;
  (3) a minimum in the combined absorption curves of algae and water.

Vos *et al.* (1986) concluded that option (3) is the most likely cause of this peak. This conclusion is supported by the local minimum in absorption, visible at 700–705 nm in Fig. 2d.

Quantum irradiance measurements

Quantum irradiance measurements (integrated over 400 to 700 nm) were made *in situ*. The model developed by Kirk (1983), based on radiative transfer theory and Monte Carlo simulations, was used to estimate the inherent optical properties for the *PAR* wavelength range. The model uses apparent optical properties such as net vertical irradiance attenuation coefficient and the underwater irradiance distribution. The apparent optical properties were determined from *in situ* measurements of upwelling and downwelling *PAR* (Malthus & Dekker, 1990; Dekker *et al.*, 1990c).

Table 2 shows typical values of absorption, *a* and scattering, *b* estimated from *in situ PAR* measurements in the Loosdrecht lake system. Scattering is the major controlling parameter of the underwater light field in the more eutrophic lakes. Absorption was also highest in the eutrophic lakes, resulting not only from higher pigment concentrations but from the increased scattering which increased the pathlength of photons, increasing their chance of absorption. *a* and *b* show strong covariance in these lakes ($r = 0.85$, $N = 111$, Malthus, unpubl. data).

The spectrophotometric and spectroradiometric measurements demonstrate that the spectral

Table 2. Some representative values of lake optical conditon parameters: vertical downwelling attenuation, *Kd*; vertical upwelling attenuation, *Ku*; absorption coëfficiënt, *a* and scattering coëfficiënt, *b* calculated from *PAR* measurements in the Loosdrecht lakes area

Lakes	*Kd* (m^{-1})	*Ku* (m^{-1})	*a* (m^{-1})	*b* (m^{-1})
Water-storage	0.7	0.6	0.52	1.32
Wijde Blik	0.8	0.7	0.54	1.35
East Loenderveen	3.4	2.9	2.02	12.20
West Loenderveen	3.6	3.2	2.27	11.35
Loosdrecht	4.4	4.0	2.54	16.13
Vuntus	4.2	3.8	2.36	17.68
Breukeleveen	5.3	4.8	3.07	19.52

area between 500 and 720 nm contains the relevant information for the determination of water quality parameters using remotely sensed data. Below 500 nm aquatic humus, together with pigments, absorbed almost all the light penetrating the water in these water bodies. From 710 nm onwards, absorption by water rises rapidly, thus reducing the amount of reflected light.

Information extraction from remote sensing data

Ideally the remotely-sensed upwelling radiance should be deconvoluted into each of the composing components of parameter specific absorption and scattering features. The exact relation between the desired information and the radiance component could then be established. Subsequently water quality feature extraction should take place, preferably as maps depicting the desired information. Such an ideal system is not yet operational. Intermediate information extraction methods are described in Dekker *et al.* (1990c). Indices for information extraction based on combinations of two or more bands are preferred because they utilise the spectral information in a reflectance signal from water analytically. Of these techniques, the rationing of spectral bands is the best known and most useful (Hilton, 1984). Ratios have the advantage of yielding information about the shape of the spectral signature curve and extraneous factors, acting equally on all bands, are annulled (Lillesand & Kiefer, 1987). A drawback in using band ratios is the enhancement of noise patterns that are uncorrelated in the component images (Lillesand & Kiefer, 1987). Therefore as few spectral bands as possible should be used in calculations. Furthermore, there is a normalisation problem, e.g. when two equally high or two equally low signals are rationed, they will present the same quotients.

To maximise the information content after spectral band rationing and minimise noise amplification, ratios of bands maximally 100 nm apart were used in this study. Philpot (1991) advises a maximum spectral distance of 50 nm when applying derivative ratio algorithms to avoid atmospheric effects in remote sensing. Spectral band ratios and water quality parameters were examined for correlations.

Remote sensing systems

The remote sensing systems discussed vary from high spectral resolution to low spectral resolution systems; i.e. an airborne imaging spectrometer to an airborne multispectral scanner to a space-based multispectral system (Table 3). Location of spectral bands of sensors used in the RSLL project between 400 and 850 nm are depicted in Fig. 4. Aerial photography is a spectrally low resolution analog imaging method. The ground-based spectrophotometric and spectroradiometric measurements and the air-borne spectrometry measurements were chosen as the reference data set for analysing the information extraction potential of the different remote sensing instruments.

The 17 spectra from the Programmable Multispectral Imager, PMI, in spectral mode taken from an altitude of 1000 m above lake level have been used to simulate the spectral bands of the PMI Chlorophyll #1 spatial mode, The CAESAR *Inland Water Mode* and the optical bands of the Thematic Mapper, TM. This enabled a focus on the comparison of relationships between optical water quality parameters and spectral bands or band combinations with improved spectral band positioning and decreased spectral band widths. Because the simulated spectral data originates from one data set, differences in measurements caused by radiometry (irradiance, atmospheric conditions, sensitivity), other sensor related differences and flight conditions, are excluded. The results of this comparison are given in Table 4.

The Programmable Multispectral Imager (PMI)

The PMI is an airborne imaging spectrometer. As a pushbroom imager it images a line of 1925 elements with five cameras. Both the modes of operation of this system were used: a spectral mode (288 spectral channels at 1.3 nm intervals, with 2.6 nm resolution from 430 to 805 nm) and a spatial mode (computer programmable selection of

Table 3. Main characteristics of the remote sensing instruments used in the RSLL research.

	Altitude (km)	Nominal pixel size (m)	Nr. of bands vis. & NIR	Spectral resolution (nm)	Dig. Number Range (max)	Swath (km)
Satellite Sensors:						
SPOT-HRV	832	20[1]	3	80–110	256	60
TM	705	30	4	60–120	256	180
Airborne Sensors[2]:	– spatial mode at 6 km flight altitude –					
PMI	0.5–10	4	8	2.6–300 +	4092/65536[3]	8
C-SM[4]	0.5–6	4	9	20–30	4000	2.6
C-IWM[5]	0.5–6	4	9	10–30	4000	2.6
	– spectral mode at 6 km flight altitude –					
PMI	0.5–10	20	288	2.6	4092/65536[3]	

[1] SPOT-HRV has a panchromatic resolution of 10 m

[2] The geometric performances are dependent on flight speed and/or integration time and on flight altitude; the spectral range of PMI is 430–805 nm; and CAESAR 400–1050 nm.

[3] 12 bit summation to 16 bit data

[4] C-SM = CAESAR *Sea Mode*

[5] C-IWM = CAESAR *inland water mode*

up to 8 wider spectral bands for high resolution spatial imaging). These modes are switchable in mid-flight. On 25 April 1988 a remote sensing mission was flown over the Loosdrecht lakes using the PMI. Several tracks were flown in the spectral and in the spatial mode (Dekker *et al.*, 1991).

Table 4. Best regression equations and correlations between the L_{air} simulated spectral bands (TM) and spectral band ratios and water quality parameters. (TM = Thematic Mapper; PMI Chl #1 and C-IWM = CAESAR *Inland Water Mode*; Max. Sig. = maximum level of significance at which correlation coëfficiënt $r > 0.80$; (6/7) = ratio of spectral *band 6* to spectral *band 7*); see Fig. 4 for spectral band location.

	TM Band		PMI *Chl #1* Band Ratio		*C-IWM* Band Ratio	
Max. Sign:	10%		5%		1%	
(N = 17)		r		r		r
SD	(–)	–	(6/7)	0.91	(2/5)	0.96
DW	(3)	0.92	(6/7)	– 0.94	(4/5)	0.95
Kd	(3)	0.87	(6/7)	– 0.91	(4/5)	0.96
Pigm	(2)	0.86	(6/7)	– 0.96	(4/5)	0.98
a	(3)	0.83	(6/7)	– 0.89	(4/5)	0.95
b	(–)	–	(6/7)	– 0.87	(3/5)	0.94

Results

Only two out of five cameras produced data without radiometric distortions. A post-flight uniformity correction was applied by the company operating the PMI, which had the effect of removing changing atmospheric pathlength radiance with changing angle of view across a swath. Despite these technical problems the results were promising.

Spectral mode data

Figure 5 shows upwelling radiance spectra, from the PMI from an altitude of 1000 m above lake level (For the same lakes as in Tables 1 and 2 and Figs 2 and 3). At wavelengths exceeding 500 nm

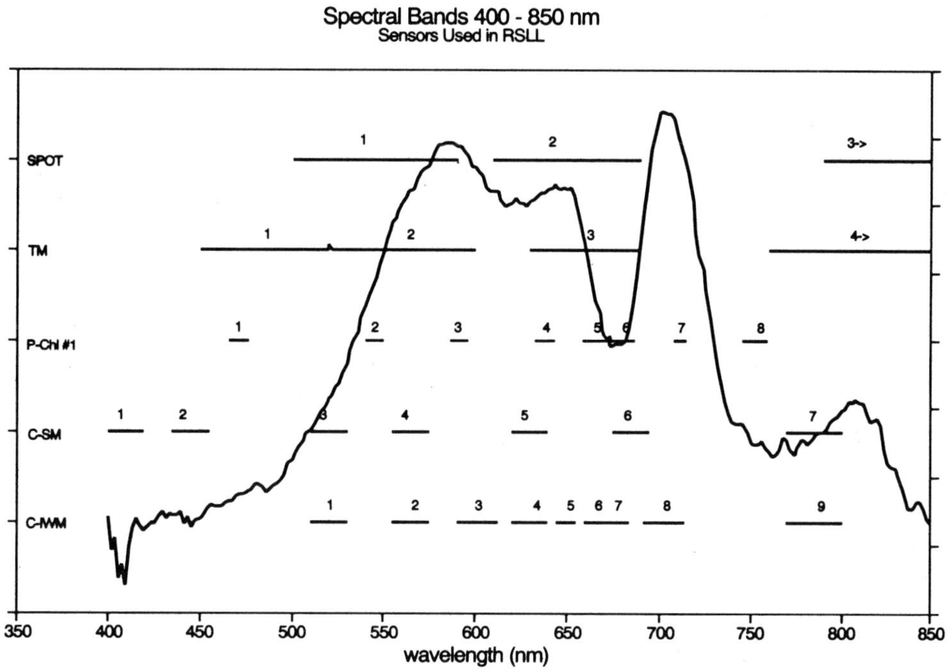

Fig. 4. Location of spectral bands of sensors used in the RSLL project between 400 and 850 nm. A subsurface reflectance spectrum (dimensionless) of Eastern Lake Loenderveen is superposed as an illustration of the location of spectral features. (Codes to spectral bandsets: SPOT = SPOT-HRV; TM = Landsat-TM; P-*Chl* #1 = PMI *Chlorophyll #1*; C-*SM* = CAESAR *Sea Mode*; C-*IWM* = CAESAR *Inland Water Mode*).

the same spectral characteristics were visible as in the subsurface upwelling irradiance spectra. Apart from differences in instrumentation, the difference between these and the subsurface spectra is surface reflection and atmospheric components in the airborne-measured spectra. The drop in the measured radiance values below 450 nm in the PMI data indicates a low radiometric accuracy in this region (Borstad, 1985).

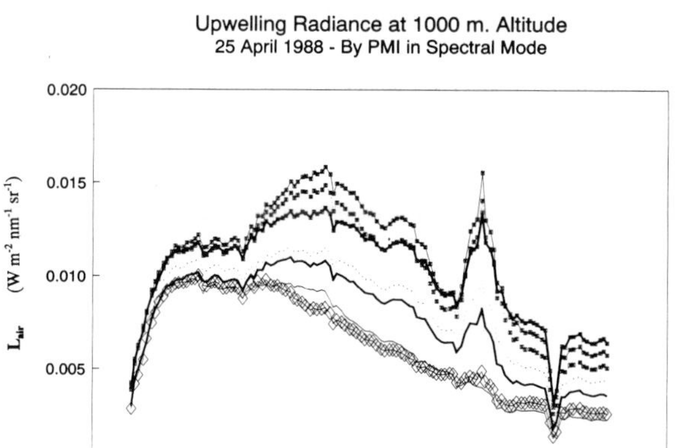

Fig. 5. Upwelling radiance measured at 1000 m above lake level (L_{air}) by the PMI in spectral mode. (From Dekker *et al.*, 1991)

Because the spatial mode overflights covered only 6 sample points (see further), 17 data points in the different lakes were sampled from the spectral mode PMI data. From these spectra integrated upwelling radiance values in the 8 spatial mode bandsettings were calculated (the spatial mode bands were simulated to obtain a significant data set). A regression analysis was carried out between ratios of these bands and the water quality parameters (only maximum correlation coefficients *r* are discussed here and given in Table 4). Seston dry weight, *DW*, sum of chlorophyll *a* and phaeopigments, *Pigm* and the vertical attenuation coefficient, *Kd*, were highly correlated with the ratio of the Chlorophyll #1 *band 6* to *band 7*. Secchi disc transparency, *SD*, was also highly correlated, but the independent *SD* measurements were distributed in two clusters: one cluster of three measurements at 200 to 225 cm and one at 30 to 50 cm transparency. No intermediate values were present. These results suggested that concentrations of *Pigm*, *Kd* and *DW* could be determined with relatively high accuracy in an image using the P(6/7) ratio.

Spatial mode image data

The Chlorophyll #1 bandset wavelengths are shown in Fig. 4. Colour Plate 1 is an image created by rationing the spectral *bands 6* and *7* of the Chlorophyll #1 spectral bandset. Spatial resolution was 2 m, but was reduced to 6 m for image processing purposes. No atmospheric corrections were made. The highest and lowest colour levels were assigned to the highest and lowest measurements of *Pigm*, *DW*, *Kd*, *a* and *b*. Intermediate values were linearly extrapolated between these colour levels, which appear to represent accurate values of water quality parameters as predicted by the band simulation modeling.

Due to deteriorating weather conditions simultaneously-measured water-based data were inadequate to validate the modelling statistically using the spatial mode data. Two sampling points in the northwestern part of Lake Loosdrecht, L4 and L7, were measured during the flight (Fig. 6). Points EL1, EL2 and EL3 were measured in the Eastern Lake Loenderveen and WL2 in the water-storage basin the following day. The measured values and the values obtained from the ratio image are compared in Table 5; the predicted data are in agreement with observed values.

Although Lake Wijde Blik is usually a clear lake (*SD* of 200 cm or more), a plume of eutrophic water (green, yellow and red colours) can be seen entering the lake from The Hilversums Canal, to the north of the lake, and curving to the southwest. The concentrations in the plume deduced from the image are: *DW*, 31.0–34.1 mg l^{-1} and *Pigm*: 83–92 $\mu g\,l^{-1}$. Wind direction during the remote sensing flight was from the northeast. It was assumed lake water flowed into this canal instead of from the canal into the lake. Water quality parameters of a sample taken in this canal (2 km to the east) on the flight date are: *DW*, 25 mg l^{-1}; *SD*, 40 cm and *Pigm*, 146 $\mu g\,l^{-1}$. These differences in estimation may be caused by the distance between the sampling point and the point of inflow, but indicate the influx of highly eutrophic water into Lake Wijde Blik.

The determination of this plume is an example

Table 5. Simultaneous *in situ* and remote sensing measurements. See Fig. 6. for location (sample = parameter value measured *in situ*; image = parameter value range deduced form the image; r^2 = coëfficiënt of determination between sample value and estimated image values; S_y = standard error of estimate for image values; from Dekker *et al.*, 1991a).

	DW (mgl^{-1})		*Pigm* (μgl^{-1})		*Kd* (m^{-1})		*b* (m^{-1})		*a* (m^{-1})	
	sample	image	sample	image	sample	image	sample	image	sample	image
L4	33.0	31.0–34.0	92	83–91	4.53	4.5–4.2	13.62	5.2–16.6	2.84	2.6–2.8
L7	37.5	24.1–37.0	98	92–100	4.43	4.6–4.8	14.37	13.8–15.1	2.66	2.8–3.1
EL1	22.7	24.7–27.7	68	66–73	3.41	3.4–37	13.26	12.3–13.7	2.19	2.1–2.4
EL2	23.5	24.7–27.7	66	66–73	3.51	3.4–3.7	12.47	12.3–13.7	2.07	2.1–2.4
EL3	22.8	21.5–24.6	59	57–65	3.43	3.1–3.3	12.20	10.8–12.2	2.02	1.9–2.1
WL2	3.7	2.6–5.6	6	4–12	0.93	0.8–1.1	1.68	2.1–3.4	0.65	0.5–0.7
r^2	0.97		0.99		0.98		0.95		0.96	
S_y	2.1		3		0.21		1.17		0.17	

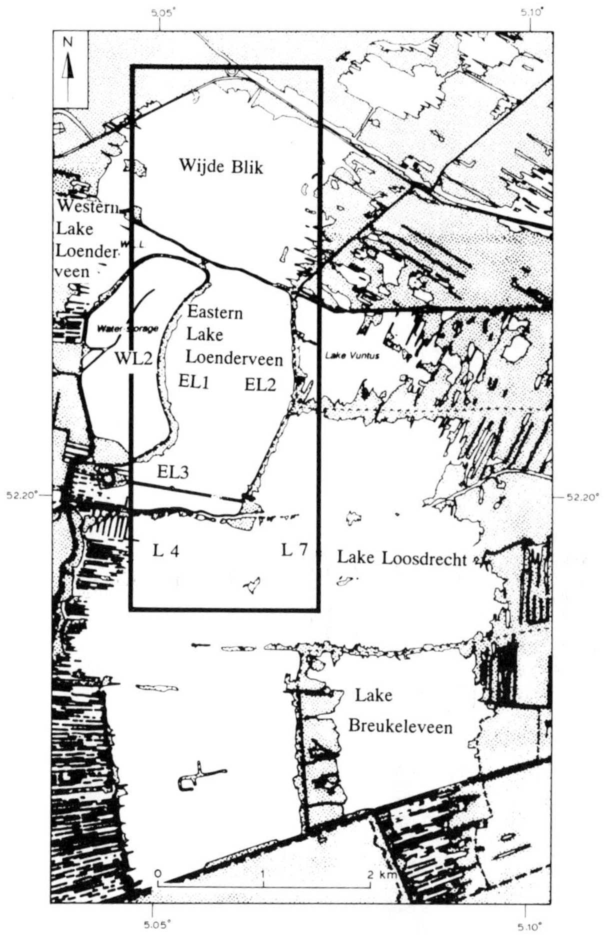

Fig. 6. Location of Colour Plate 1 and the simultaneously measured sampling points.

of the advantages of remote sensing for water quality research and monitoring. The synoptical view provided by remote sensing makes it possible to determine the presence and extent of such a phenomenon. More important, however, is that the analysis of the high spectral resolution data makes it possible to determine composition and concentration of the optical water quality parameters. The presence of the plume also demonstrates the disadvantages of point sampling data as a measure of lake water quality. Samples for routine water quality measurements are taken by the water management authorities in Lake Wijde Blik in the centre of the lake. Thus such sampling confined to one point in the lake is not representative of the entire lake (under these circumstances) and may lead to wrong assessments about water quality in the lake.

The other lakes that can be seen in the image, are Eastern Lake Loenderveen, Western lake Loenderveen, the water-storage basin and a part of Lake Loosdrecht. These lakes exhibit a more uniform (spatial) water quality. The water-storage basin shows gradients, however, which may be caused by the inlet of iron-rich water from the dephosphorisation plant in the southwestern corner of the lake (outside the limits of the image).

CAESAR

CAESAR is a multispectral pushbroom scanner, meant both for land and sea applications (Pouwels, 1987 and Looyen & Van Swol *et al.*, 1989).

For water applications there are two modes, each consisting of nine spectral bands: the *Sea Mode* and the *Inland Water Mode* (Table 3 and Fig. 4). The 'Sea Mode' is not suitable for detecting quality of inland waters, due to the spectral band selection, which is aimed at ocean type waters. In turbid, algae-rich and humus rich waters different spectral bands are necessary to adequately measure the desired spectral information (Dekker *et al.*, 1990a, b&c, 1991).

The CAESAR Inland Water Mode (C-IWM)

The *Inland Water Mode* was developed in the course of the RSLL Project (Dekker *et al.*, 1990b; and Table 3 and Fig. 4). This spectral bandset enables an accurate reconstruction of the spectral signature between 500 to 800 nm in these lakes (Fig. 7). A comparison of the spectral features in the section on the underwater light climate and the positioning of the C-IWM bandset illustrates the rationale in the selection of the bands. A detailed analysis of the choice of the C-IWM bandset can be found in Dekker *et al.* (1990a&b).

The spectral bands of the C-IWM were simulated using the PMI upwelling radiance spectra.

Plate 1. Spectral ratio image of (P 6/7). See legend for water quality parameter concentrations. See Fig. 6. for location and scale (reprinted from Dekker *et al.*)

The main results (Table 4) confirm that in eutrophic systems such as the Loosdrecht lakes estimation of in-lake parameters can best be achieved using band ratios based on wavelengths from 500 nm to 800 nm, with emphasis on the area between 600 nm and 720 nm. An advantage of using longer wavelengths is that atmospheric distortion of the remotely-sensed signal decreases with increasing wavelengths. The results show that an increase in correlation and significance features is strongly dependent on spectral band positioning and width.

Plate 1. (Continued)

Pigm (μgl^{-1})	DW (mgl^{-1})	kd (m^{-1})	b (m^{-1})	a (m^{-1})
0– 3	0 – 2.5	0.4–0.7	0.6–2.0	0.3–0.4
4– 12	2.6 – 5.6	0.8–1.1	2.1–3.4	0.5–0.6
13– 21	5.7 – 8.8	1.2–1.5	3.5–4.9	0.7–0.9
22– 29	8.9 – 11.9	1.6–1.8	5.0–6.3	1.0–1.1
30– 38	12.0 – 15.1	1.9–2.2	6.4–7.8	1.2–1.3
39– 47	15.2 – 18.2	2.3–2.6	7.9–9.3	1.4–1.6
48– 56	18.3 – 21.4	2.7–3.0	9.4–10.7	1.7–1.8
57– 65	21.5 – 24.6	3.1–3.3	10.8–12.2	1.9–2.0
66– 73	24.7 – 27.7	3.4–3.7	12.3–13.7	2.1–2.3
74– 82	27.8 – 30.9	3.8–4.1	13.8–15.1	2.4–2.5
83– 91	31.0 – 34.0	4.2–4.5	15.2–16.6	2.6–2.7
92–100	34.1 – 37.0	4.6–4.8	16.7–18.1	2.8–3.0
101–109	37.3 – 40.3	4.9–5.2	18.2–19.5	3.1–3.3
110 < <	40.4 < <	5.3 < <	19.6 < <	3.4 < <

Satellite sensor imagery

Although airborne remote sensing using very high spectral resolution data is most suited for quantitative assessment of water quality, relatively low spectral and spatial resolution satellite imagery may also be useful (Dekker & Peters, 1992). Due to their sufficient spatial resolution, the Thematic Mapper (TM) on board the Landsat satellite and SPOT-HRV sensors on board the SPOT satellite are suited for imaging medium sized or smaller inland waters, such as the Loosdrecht lakes. The data are easily available at a moderate acquisition cost (see Table 3 for main characteristics of these systems). Furthermore, continuity in data acquisition is guaranteed: Landsat 6 will be launched soon and Landsat 7 and SPOT 3 and 4 are in the planning phase. Here only Landsat-TM data are

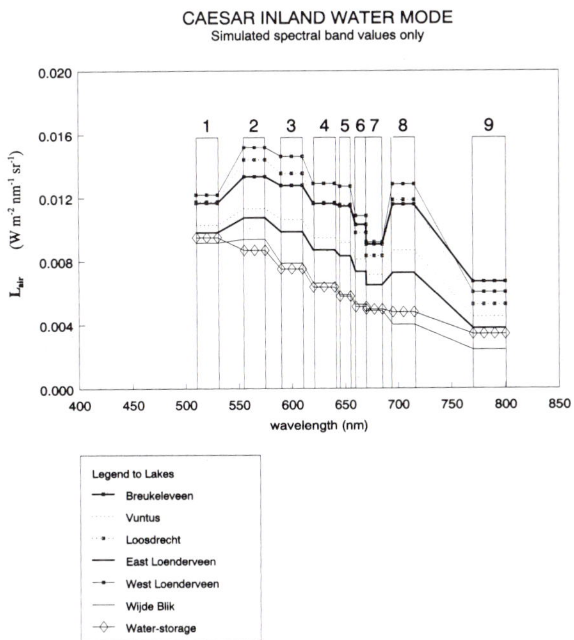

Fig. 7. L_{air} spectra of 7 lakes in the Loosdrecht lakes system as reconstructed from simulated CAESAR *Inland Water Mode* L_{air} spectral bandset values. (From Dekker *et al.*, 1990b).

Table 6b. Regression equations using the log-transformed data for July 5, 1987. (Regression formula: $Y = a * X^b$; coëfficiënts are determined using the formula: *LN* Y = LN a + b LN X)(from Dekker *et al.*, 1990c)

Regression equation	r	$(r^2)\%$
Ln *SD* = + 42.75 − 11.95*Ln TM3	− 0.82	68
Ln *DW* = − 49.39 + 16.07*Ln TM2	0.85	73
Ln *DW* = − 40.87 + 14.56*Ln TM3	0.93	86
Ln *Pigm* = − 55.37 + 18.39*Ln TM2	0.87	77
Ln *Pigm* = − 43.54 + 15.97*Ln TM3	0.91	83

the ($s_{y'}$). Thus, in many cases it is the low radiometric sensitivity of the TM which is the limiting factor in an accurate assessment of water quality parameters (Dekker & Peters, 1992).

There was a drawback attached to using linear regression analysis on untransformed data: when applying the regression equations to the full range of digital data present in TM images, negative values were frequently calculated. An exponential curve of the form $Y = a * X^b$ fitted the data well. There is physical basis for this: (1) negative values for the parameters do not occur; (2) most of the correlation between radiance values and the water quality parameters is based on the scattering properties of the water constituents. An increase in scattering will also increase the multiple scattering thus giving the increase an exponential character. Thirdly, the data from highly eutrophic lakes varies more in time than data from oligotrophic or mesotrophic lakes. Thus, a log-transformation normalises the variance across the range of water qualities encountered.

In contrast to the results from the non-transformed data, the log-transformed data had a higher correlation of *SD*, *DW* and *Pigm* with *band 3* than with *band 2* (Table 6b). The log-transformation caused calculated dimensions of the dependent variables based on the bivariate linear regression results to be more realistic, especially at low digital numbers for the *DW* and *Pigm* concentrations and at high *DN*s for *SD* transparencies. At the other end of the scale, however, the values become unrealistically high for *DW* and *Pigm*. Therefore the lowest and highest *DN* values in Table 7 which lead to the calculation of unrealistic values are marked with a lower than (<) or greater than (>) sign instead of an estimated parameter value. In Table 7 for *DN* 28 & 29, therefore, only a minimum value at *DN* 28 is given for *DW* and *Pigm*.

Interpretation of water quality map based on a colour coded TM band 2 image of 5 July 1987

The lakes Nieuwe Polder, Wijde Gat and Hollands Ankeveen appear to be similar in water quality (Plate 2b). Lake Vuntus differs (up to 2 *DN*s higher) from lakes Loosdrecht and Breukeleveen. Lake Vuntus had a concentration of cyanobacteria of 210.000 ml^{-1} and a *Pigm* of 189 $\mu g\, l^{-1}$ compared with 140.000 ml^{-1} and a *Pigm* of 143 $\mu g\, l^{-1}$ in Lake Loosdrecht and 140.000 per ml with a *Pigm* of 162 $\mu g\, l^{-1}$ in Lake Breukeleveen. Lake Vuntus seems to be more yellowish because of higher green and high red re-

Table 7. Concentration intervals of water quality parameters on 5 July 1987 calculated for TM *band 2* DN values using the log-transformed regression equations from Table 6b. (Ref. Colour Plate 2b.)

TM2	*DW*(mg/l)	*Pigm*(µg/l)	*SD*(cm)	colour code
DN22 –	1 – 2	3 – 7	439 –255	blue
DN22 –	2 – 4	7 – 15	255 –151	light blue
DN24 –	4 – 7	15 – 32	151 – 92	dark green
DN25 –	7 –14	32 – 66	92 – 57	light green
DN26 –	14 –26	66 –133	57 – 36	green-yellow
DN27 –	26 –47	> 133	36 – 23	yellow
DN28 –	> 47	> 133	23 – 15	orange
DN29 –	> 47	> 133	< 15	salmon

Plate 2a. TM pseudonatural colour composite Thematic Mapper image on July 5, 1987; *bands 1, 2,* and *3* are represented in blue, green and red respectively. (From Dekker & Peters, 1992)

Plate 2b. Water quality map based on a colour coded TM *band 2* image of the 5 July 1987. Concentration of water quality parameters can be determined by comparing the *DN* values in the image with the calculated values in Table 7. (From Dekker & Peters, 1992.)

flectance, possibly due to the high amount of cyanobacteria. In *band 3* Lake Vuntus is similar to these lakes. Other exceptions are Lake Naardermeer and the southeastern Lake Naardermeer-Bovenste Blik, which differ by two *DN*s. The lakes Stichts Ankeveen and Hollands Ankeveen are almost invisible because their reflectance values approach those used to mask out the land reflectance. This is probably due to emerse macrophyte coverage. The results of June 16 1986 are similar.

The algorithms developed through statistical

analyses are based on the assumption that a 17 by 17 pixel area is a representative sample for the average water quality measured there (Dekker *et al.*, 1990c). If each lake were homogeneous (i.e. well mixed) the concentration maps of basis of these algorithms would show no variation at all within the lakes. In reality gradients and patterns are observed (Plates 1 and 2a–b). There is enough variability among the lakes for the calculated regression line to be valid. So, the hypothesis that the optimized spectral sample is representative for the average measured parameters is not rejected.

Except for the ($s_{y'}$) as a limiting factor for accuracy of determination, the absolute range of *DN* levels is also limiting. The amount of concentration levels maximally discernible expressed as *DN*s, with a minimum threshold of at least 10 pixels per level, is 6 for *band 2* and 9 for *band 3* on both dates. If a threshold of 100 pixels per *DN* level is applied the maximum range would decrease to 5 or 6 levels. The two images discussed here were taken around maximum solar elevations (*ca.* 58 degrees) on cloud-free days and represent optimal images that can be achieved over water at this geographic location. It is concluded that Landsat TM images are of limited value for the determination of surface water quality. Suboptimal images will not be applicable for these purposes.

The use of remote sensing in water quality management and research

A simultaneous image of water quality over a large area can be obtained in minutes (by satellite sensor) or hours (airborne remote sensing). Using airborne systems a particular water body can be imaged consecutively on one day or, weather permitting, on consecutive days. Surface streamflow dynamics may be visualised and calculated. In lake restoration projects remote sensing can help to establish changes during the course of the restoration phase. The technique can be applied to measure the dispersion of suspended material during dredging operations of polluted water bottoms. In general, spatial variability of optical water quality parameters can be determined. This may lead to optimalisation of sampling networks. Insight is obtained into how representative point samples are for an entire water body. Remote sensing enables non-intrusive monitoring (e.g. nature preservation areas). Remote sensing images may have a demonstrative function, since such information generally has a higher impact factor in public awareness than figures or tables. Remote sensing data may have an archive function: unlike water samples digital remote sensing data can be stored for long term and will always be available for further analysis, e.g. new information extraction methods can be used on existing data. This latter was the case with the Thematic Mapper images; only after acquisition of the images, was the relevant water quality data assembled.

Developments in remote sensing of inland water quality: general conclusions

Any research involving remote sensing of water quality in the optical and nearby infrared wavelengths involves analysis of the spectral signature of the water under observation. The spectral signature is a result of the effects that spectral scattering, spectral absorption and, possibly, fluorescence have on the water penetrating irradiance. These are in turn the result of the effects of the water itself and its constituents in dissolved, colloidal or suspended form on incoming light. The upwelling radiance from the water column at the surface is a convolution of the spectral effects of these constituents on the incoming irradiance. Bottom reflectance is assumed to be negligible in the following discussion. Although it is an important topic in remote sensing of water, the techniques of removal of atmospheric influences are not discussed.

A thorough research should incorporate the following aspects to obtain reproduceable results.

- Measurement of surface and subsurface upwelling (ir)radiance) and of optical properties, both inherent and apparent.
- Determination of the relationship between

the optical properties and the optically active water constituents. This should preferably be carried out analytically or else statistical relationships may be used, provided an analytical explanation of the results is possible.
- investigation of the covariance of chemical substances with the optically active water parameters. The statistical relationships between remote sensing data and chemical parameters is based upon a covariance with one of the optically active parameters; most such relationships will remain site specific and dependent on time of day and season.

With this knowledge the performance of a remote sensing instrument can be analysed. The spectral bands should selectively measure spectral features. Information extraction will inevitably involve combining spectral band information. As few spectral bands as possible must be used, to minimise both the required amount of data and the increase of the noise relative to the signal, and to maximise the physical interpretability of the data.

For the measurement of specific spectral features, such as photosynthetic pigment absorption, narrow bandwidths are required. Narrow bands cause a decrease in radiance levels. To obtain adequate signal to noise ratios the radiometric resolution must be high at the typically low radiance levels of water bodies. A trade-off must, therefore, be made with regards to spectral band width.

For spectral signature analysis imaging spectrometers are very useful. They are the only instruments that can measure the spatial variability of the spectral signature of water bodies. Thus they can also determine the relevance of surface-based spectral measurements (which are point measurements) to the entire water body. However, the amount of data generated by imaging spectrometers is extremely high, is of an abstract nature, and needs expert analysis methods to extract the necessary information.

Multispectral scanners for water quality assessment should make use of specific spectral bands, positioned to measure the expected spectral signatures efficiently. They are relatively simple compared with imaging spectrometers and are presently at a higher operational level. The amount of data will depend on the number of spectral bands and the radiometric and spatial resolution features, but will generally be a factor 10 to 20 lower than for imaging spectrometers. If the spectral bands are positioned correctly, the information extraction is relatively easy provided the appropriate spectral signatures are known.

Satellite sensor systems, such as the Thematic Mapper and SPOT-HRV, lack sufficient spectral and radiometric resolution to accurately measure water quality features. By virtue of their operational status and comparatively low costs per square km imaged, they are feasible options for the purposes of large-scale monitoring at low water quality feature resolution. The near-infrared channels of both satellite systems have been used to monitor algal surface layers (not encountered in this research) (Buiteveld *et al.*, 1987).

The range and resolution in wavelengths on the future Medium Resolution Imaging Spectrometer (MERIS) planned on the European Space Agency Polar Platform to be launched in 1997, the Moderate Resolution Imaging Spectrometer (MODIS) planned on the Earth Observation System (EOS-A) platform to be launched in 1998 and the High Resolution Imaging Spectrometer (HIRIS) planned on the second EOS-A platform to be launched after 1998, instruments will make it possible to monitor water quality accurately from space in both a qualitative and quantitative manner. The CAESAR *Inland Water Mode* illustrates that nine spectral bands are sufficient to determine several water quality related optical parameters to a high degree of accuracy. This illustrates the possibility of data reduction when systems such as MERIS, MODIS and HIRIS become operational. In this case a reduction of the amount of spectral bands measured over the range of 400 to 850 nm from 45 (in case of complete coverage at 10 nm resolution) down to 9 spectral bands without losing relevant information seems possible.

Until the next generation of earth observation satellites is operational towards the end of the nineties, airborne remote sensing is the main area of research for water quality assessment of inland waters. The airborne systems will pave the way to optimise the use of these satellite based imaging spectrometers as soon as they are operational. The positive aspects of satellite remote sensing will then be combined with the advantages of the current airborne systems.

Specific recommendations to facilitate this transition are:

(1) to perform spectral signature analysis of all water bodies under consideration for surveillance, monitoring, or analysis from the air or from space;
(2) to develop reliable methods for information extraction from the measured spectral signatures;
(3) to continue research making use of imaging spectrometers for thorough fundamental analysis of the spectral signature as it is measured by remote sensing;
(4) to develop, test and operationalise multispectral scanners for water quality analysis;
(5) to make use of the results of airborne campaigns to prepare the water quality community for this novel manner of studying, monitoring and understanding water quality, especially in the spectral and spatial domain.

Acknowledgements

Many institutions contributed to the RSLL project. In alphabetical order they are: Laboratory for Microbiology at the University of Amsterdam; The Limnological Institute; Moniteq Ltd. of Canada; Municipal Waterworks of Amsterdam (GWA); National Aerospace Laboratories (NLR); Programme Bureau of the BCRS; Free University of Amsterdam; Water Pollution Control Authority Amstel- en Gooiland (ZAG).

The author is especially indebted to the following persons for being allowed to disrupt their own research or other activities, whenever it was thought to be necessary: Loes Breebaart, drs. Tineke Burger-Wiersma, drs. Jan Jaap Buysse, dr. Herman Gons, drs. Jasper Griffioen, drs. Ben Kal, Andre Lameris, dr. Louis van Liere, Ir. Wim J. Looyen, dr. Frank Müller Karger, Ir. Hein Noorbergen, Marchien A. van Oostende, drs. Steef Peters and drs. Pierre Verstraelen.

Many more people contributed to this research, especially the personnel of the Institute of Earth Sciences at the Vrije Universiteit and of the Limnological Institute who carried out the field-based measurements and laboratory analyses.

References

Boivin, L. P., W. F. Davidson, R. S. Storey, D. Sinclair & E. D. Earle, 1986. Determination of the attenuation coefficient of visible and ultraviolet radiation in heavy water. Appl. Optics 25: 877–882.

Borstad, G. A., H. R. Edel, J. F. R. Gower & A. B. Hollinger, 1985. Analysis of test and flight data from the Fluorescence Line Imager. Can. Spec. Publ. J. Fish. aquat. Sci. 83: 38 pp.

Buiteveld, H., F. De Jong, R. Spanhoff & M. Donze, 1987. Shape and variability of the absorption spectrum of aquatic humus. Symp. on Rem. Sens. Resource Developm. & Env. Managem., Enschede, the Netherlands, August 1986, IAH/Netherl. Soc. for Rem. Sens.: 703–705.

Davies-Colley, R. J., W. N. Vant & R. J. Wilcock, 1988. Lake water colour: comparison of direct observations with underwater spectral irradiance. Wat. Res. Bull. 24: 11–18.

Dekker, A. G. & S. W. M. Peters, 1992. The use of the Thematic Mapper for the analysis of eutrophic lakes: A case study in The Netherlands. Int. J. Rem. Sens.

Dekker, A. G. & E. Seyhan, 1988. The Remote Sensing Loosdrecht Lakes Project. Int. J. Rem. Sens. 10 & 11: 1761–1773.

Dekker, A. G., T. J. Malthus & E. Seyhan, 1990a. Improving quantitative analysis of inland water quality using high spectral resolution imaging and non-imaging data. Proc. IGARSS'90 Symp., Washington, May 20–24, 1990: 117–120.

Dekker, A. G., T. J. Malthus & E. Seyhan, 1990b. An inland water quality bandset for the CAESAR system based on spectral signature analysis. Proc. International Symposium Remote Sensing and Water, Enschede, August 1990, IAH/Neth. Soc. for Rem. Sens., The Netherlands: 597–606.

Dekker, A. G., T. J. Malthus & E. Seyhan, 1991. Quantitative modelling of inland water quality for high resolution MSS-systems. IEEE Trans. on Geosc. and Rem. Sens. 29: 89–95.

Dekker, A. G., T. J. Malthus, S. W. M. Peters & E. Seyhan, 1990c. The Remote Sensing Loosdrecht Lakes Project. BCRS Report 90-28. Dutch Remote Sensing Board, Delft, The Netherlands: pp. 99.

Donze, M., G. B. J. Dubelaar & J. W. M. Visser, 1987. Anomalous behaviour of forward and perpendicular light-scattering of a cyanobacteria due to intracellular gas vacuoles. BCRS report 87-08, Dutch Remote Sensing Board, Delft, The Netherlands: pp. 43.

Gitel'son, A. A. & A. M. Nikanorov, 1988. Remote monitoring of ecological condition of aquatic ecosystems. Proc. USA-USSR Symposium, Athens, Georgia, Oct 19–21, 1987. on: Fate and Effects of Pollutants on Aquatic Organisms and Ecosystems: 166–182.

Gitel'son, A. A., A. M. Nikanorov, G. Y. Szabo & F. Szilagyi, 1986. Etude de la qualité des eaux de surface par teledetection. Proc. Budapest Symp., July 1986, on: Monitoring to Detect Changes in Water Quality Series: IAHS Publ. no. 157: 111–121.

Gulati, R. D., L. Van Liere & K. Siewertsen, 1991. The Loosdrecht lake system: Man's role in its creation, perturbation and rehabilitation. Terrestrial and Aquatic ecosystems: Perturbation and recovery. Ellis Horwood Ltd.: 593–606.

Hilton, J., 1984. Airborne remote sensing for freshwater and estuarine monitoring. Wat. Res. 18: 1195–1223.

Jerlov, N. G., 1976. Marine optics. Elsevier, Amsterdam, The Netherlands.

Kirk, J. T. O., 1981a. Monte Carlo Study of the nature of the underwater light field in, and the relationships between optical properties of, turbid yellow waters. Aust. J. Mar. Freshwat. Res. 32: 517–532.

Kirk, J. T. O., 1981b. Estimation of the scattering coefficient of natural waters using underwater irradiance measurements. Aust. J. Mar. Freshw. Res. 32: 533–539.

Kirk, J. T. O., 1983. Light and photosynthesis in aquatic ecosystems. CSIRO, Canberra, Australia. Cambridge University Press: pp. 401.

Lillesand, T. M. & R. W. Kiefer, 1987. Remote sensing and image interpretation. John Wiley & Sons, New York: pp. 721.

Looyen, W. J., 1988. Geometric aspects of digital CAESAR images. Nederl. Geod. Tijdschrift Geodesia, 30: 159–162.

Looyen, W. J. & R. W. Van Swol (Eds), *et al.*, 1989. CAESAR: Performance and first evaluation of application possibilities. Report BCRS-89-06, Final Report Project TE-1.3, June 1989. Netherlands Remote Sensing Board, Delft, The Netherlands: pp. 101.

Malthus, T. J. & A. G. Dekker, 1988. Spectral light attenuation in a hypertrophic lake system (Loosdrecht lakes, the Netherlands). Verh. int. Ver. Limnol. 24: 711–714.

Miyazaki, T., H. Shimizu & Y. Yasuoka, 1987. High-speed spectroradiometer for remote sensing. Appl. Optics 26: 4761–4766.

Noordzij, W. J., F. B. van der Laan & W. Verhoef, 1989. National Point of Contact, Products and Services, detailed guide to imagery available. NPOC Annex to Dutch Remote Sensing Newsletter 1989-no 38, (no page numbering).

Palmer, K. F. & D. Williams, 1974. Optical properties of water in the near infrared. J. Opt. Soc. Am. 64: 1107–1110.

Philpot, W. D., 1991. The derivative ratio algorithm: avoiding atmospheric effects in remote sensing. IEEE Trans. on Geosc. and Rem. Sens. 29: 350–357.

Prieur, L. & S. Sathyendranath, 1981. An optical classification of coastal and oceanic waters based on the spectral absorption curves of phytoplankton pigments, dissolved organic matter and other particulate materials. Limnol. Oceanogr. 26: 671–689.

Seyhan, E., N. J. J. Bunnik, W. Verhoef & J. van Kuilenburg, 1974. Measurements of spectral signatures for water quality monitoring. NIWARS Publication NO. 24, Delft, The Netherlands. Paper presented at the First General Conference of the Remote Sensing Society, Birmingham, Great Britain, Sept. 1974: pp. 29.

Smith, R. C. & K. Baker, 1981. Optical properties of the clearest natural water (200–800 nm). Appl. Optics 20: 177–184.

Vos, W. L., M. Donze & H. Buiteveld, 1986. On the reflectance spectrum of algae in water: the nature of the peak at 700 nm and its shift with varying concentration. Commun. on San. Eng. and Water Management, nr. 7, ISSN-0169-6246, TU Delft, The Netherlands 86-22: pp. 29.

Hydrobiologia **233**: 161–163, 1992.
L. Van Liere & R.D. Gulati (eds), *Restoration and Recovery of Shallow Eutrophic Lake Ecosystems in The Netherlands.*

Water quality research in the Loosdrecht lakes: proposals for the follow-up restoration measures

Marien A. De Ruiter
Province of Utrecht, P.O. Box 80300, 3506 TH Utrecht, The Netherlands

Key words: eutrophication, departmental policy, Loosdrecht lakes, recovery, water management

Abstract

Reduction of the external phosphorus load in 1984 did not lead to recovery of Loosdrecht lakes. Therefore the Provincial Water Authorities of the Province of Utrecht proposed, after consulting the research group Water Quality Research Loosdrecht lakes (WQL), measures to reduce the present external phosphorus loading from 0.35 to 0.1 g $Pm^{-2} y^{-1}$. Mathematical modelling suggested that such a reduction would benefit the water quality of the lakes. The results of the WQL investigation did not lead to additional measures (dredging, biomanipulation, chemomanipulation).

Introduction

The provincial bylaw entrusts the Provincial Water Authorities of Utrecht with the responsibility of maintaining water quality within their provincial jurisdiction. This has already resulted in initiatives of the Province of Utrecht to restore the water quality by providing sewerage systems for the community inhabiting the Loosdrecht lakes area, to divert domestical and industrial loads. Besides, in close cooperation with Amsterdam Municipal Water Works, supply of water to the lakes from River Vecht has been replaced by water from Amsterdam-Rhine Canal (Gulati *et al.*, 1991; Hofstra & Van Liere, 1992). Although the concentration of total-phosphorus in the lakes has showed a slight decrease, the concentration of chlorophyll *a* has not decreased significantly (Van Liere *et al.*, 1990). Also the required water clarity for swimming water purposes, a Secchi-disc transparency of 1 m was not achieved. Therefore, Provincial Water Authorities directed their attention to additional measures, for example biomanipulation. This initial effort in Lake Breukeleveen, one of the lakes in the Loosdrecht lakes system, to stimulate large-bodied zooplankton by reducing the planktivorous fish standing crop did not produce the desired results (Van Donk *et al.*, 1990).

This paper briefly discuss future proposals of the Provincial Water Authorities for restoration of the Loosdrecht lakes.

The proposed water management measures

The recommendations of the workgroup WQL (Water Quality Research Loosdrecht lakes) have shown clearly that further reduction of external phosphorus loading to the lakes is still the crucial measure in restoration of the lakes in this area (see in Van Liere & Gulati, 1992). At present the total external phosphorus sources are (Engelen *et al.*, 1992):

1. The Polder Bethune water is supplied to the lakes to prevent chloride concentration from rising, because the chloride concentration in the water from Amsterdam-Rhine Canal en-

Hydrobiologia **233**: 165–170, 1992.
L. Van Liere & R.D. Gulati (eds), Restoration and Recovery of Shallow Eutrophic Lake Ecosystems in The Netherlands.

The significance of the Loosdrecht lakes research project for eutrophication policy in The Netherlands

H. W. Kroes
Ministry VROM (Housing, Physical Planning and Environment), Directorate of International Environmental Affairs, P.O. Box 450, 2260 MB Leidschendam, The Netherlands

Key words: eutrophication, policy, environment

Abstract

The Project on the Water Quality Research in the Loosdrecht Lakes (WQL) has come up to the expectations of the Netherlands Environment Ministry. The results reaffirm the main lines of national eutrophication policy drawn up in 1979 (Policy Document on Phosphates) and further developed in the eighties. Interesting new insights have been gained, for example into the role of sediment and seston as well as into the relative importance of trophic levels. It is not possible, however, to definitely establish the effect of WQL on eutrophication policy.

The Loosdrecht project is an example of genuine ecological research, incorporating several disciplines, placing the object of research into its surroundings, emphasising the relation nature-culture and committing itself to certain value judgements and policy choices. As a consequence, there is a striking resemblance between the evolution of the project itself and that of ecosystems.

All in all the Loosdrecht project should be regarded as a paradigm for future ecosystem studies.

Introduction

Some ten years ago two motives led to the conception of the Loosdrecht Lakes research project: a scientific one and a political one. From a scientific viewpoint there was a need to gain insight into the functioning of an aquatic ecosystem as a whole, in relation to its environment. From the viewpoint of policy and water management there was a need to obtain information on the beneficial effects of reducing the phosphate load of the Loosdrecht lakes. The ministry of Housing, Physical Planning and Environment (VROM) made a financial contribution to the project because of the second motive. Of course, this decision was influenced by the hope that enlarging the knowledge on the Loosdrecht lakes would be useful for extrapolation to other surface waters in the Netherlands: VROM is a national policy-making body, not a regional water board.

In this paper I will try to assess the success of the project (indicated in this paper as WQL, Water Quality research Loosdrecht lakes) by answering the following four questions:

a. Has WQL come up to the expectations of VROM in a general sense?
b. Has WQL affected policy on eutrophication abatement or, in a broader sense, Dutch government policy with respect to surface water quality? Is it possible to indicate a pattern of cause and effect and the extent of the influence?
c. Has the government, especially VROM, benefited from WQL in other respects?

d. Should WQL be an example for a similar project in the future?

In trying to answer these questions I will also touch on some general aspects of the relation between research and environmental policy.

WQL and VROM's expectations

Each year the Directorate-General for Environment of the Ministry of VROM contracts with a great number of research institutes, universities, technical consultant bureaus etc. for carrying out research. The objective is to update basic information for environmental policy both for the short and the long term, rather than to promote scientific research as such.

An essential difference exists between studies expected to provide specific information for the short term and those intended to enlarge the general basis of knowledge on environmental problems. VROM finances both categories as well as the intermediate forms, but expectations as to the outcome are not identical for the various types of studies.

In case of a short term study immediate applicability of the results in policy development is expected. For example, after the accident in the Swiss firm Sandoz in 1986 the Inspectorate for Environmental Hygiene (a VROM agency) asked a technical consultancy bureau to study the risks of water polluting accidents in Dutch industrial plants, and to draw up a list of such plants. The results were used soon thereafter in the international negotiations on the abatement of Rhine pollution and for formulating a national policy to prevent such accidents. Results of a strategic long term study, on the other hand, at best may be used for incorporation in long-term policy, but sometimes only serve to develop general lines of thought in policy-making departments.

The WQL research project is an example of a strategic long-term study. Therefore, VROM's expectations were modest as to the direct applicability of the results. The type of information collected is important for the long term and can be one of the many contributions to an intricate and generally slow process of policy making (as will be shown in the next section). In this sense, WQL has come up to our expectations.

Eutrophication: research and policy

The second and most important question that has to be dealt with, is whether WQL so far has had an impact on Dutch national eutrophication policy. I cannot give an unambiguous answer to that question, in spite of the fact that the project has lasted for ten years. Answering that question in my view is as complex a task as unraveling the

Table 1. Research on eutrophication 1960–1991. If only the year of start is given, the research is still continuing.

Period	Name of surface water and research organisation
1960–	Plas Vechten ('Lake Vechten'), Limnological Institute.
1967–	Friese meren ('Frisian lakes'), Limnological Institute and Fosfri-project.
1967–	IJsselmeer, Randmeren ('Lake IJssel' and 'Bordering lakes'), Limnological Institute, National Agency for the IJsselmeerpolders, National Institute for Wastewater Treatment.
1972–1982	Eems-Dollard, Project Biological Research Ems-Dollart Estuary.
1975–1981	Grote Rug (Reservoir for drinking water supply, Dordrecht), National Institute for Drinking Water Supply.
1975–	Reservoirs for drinking water supply (e.g. Biesbosch), Drinking water supply companies.
1975–	Nieuwkoopse plassen, Reeuwijkse plassen etc. (lakes in the Rijnland area), Water Board 'Rijnland', National Institute of Public Health and Environmental protection.
1979–	Loosdrechtse plassen ('Loosdrecht lakes'), some 10 research institutes: WQL.
1986–	Kortenhoefse/Ankeveense plassen (lakes north of the Loosdrecht area), Zuiveringsschap Amstel- en Gooiland and Agricultural University Wageningen.
1987–	Rhine, National Institute of Public Health and Environmental Protection.

processes in the eutrophication of the Loosdrecht Lakes.

To assess the possible impact of eutrophication research in general and WQL in particular on policy making, I have summarised (Tables 1 and 2) the major research projects since the sixties and the highlights in policy making since the early seventies. The information shows that, although much has been done and is still in progress both in research and in policy making, it is almost impossible to establish a direct cause-and-effect relationship between the two fields. Generally speaking, the only certainty is that research projects started much earlier than the policy making, and that individual research projects last longer than policy making projects.

Finding the causal relations would require an extensive and detailed study. On the basis of my own experience I fear, however, that the essential relations would, in all probability, remain hidden. The process of policy making is complicated; crucial moments can only be pointed at by insiders, and that too only in a subjective way, difficult to verify by outsiders. The role of individual persons can be important, but the role of collective decision making is important too, and, finally, progress in policy making sometimes strongly depends on accidents. The Sandoz accident for example, resulting in the Rhine Action Programme (ICPR, 1987), triggered a marked acceleration in international and national eutrophication abatement policy.

I can illustrate the above with an example taken from policy developments in the seventies. The decision to produce a policy statement on phosphates under the term of office of Mrs. Vorrink, Minister of V&M ('Public Health and Environmental Hygiene', the predecessor of VROM), was not taken after a linear process with consecutive phases of research, policy preparation and policy decisions, but more or less by coincidence. A report of scientists (KNCV, 1976) had been published, which in its turn was the result of a mixture of scientific considerations and ideas of individual persons (Golterman & others) on a desirable policy. There were action groups on the subject of detergent phosphates and the Dutch Parliament had questioned the Minister about

Table 2. Policy making on eutrophication 1972–1991.

Period	Document/event
1972	'Urgentie nota milieuhyiëne'. Government statement on urgent issues in environmental hygiene, including a rough sketch of the eutrophication problem.
1978	'KNCV-rapport'. Royal Netherlands Chemical Society: recommendations on phosphate policy.
1979	'Fosfatennota'. Government policy statement on measures from 1980 to 1985 to reduce eutrophication: phosphate removal in sewage treatment plants, banning phosphates from detergents, international actions, regional actions.
1980	Agreement with detergent manufacturers in the Netherlands to reduce phosphate content of textile detergents to below 7,5% (as P) in 1983.
1985	Objectives of Fosfatennota not realised.
1986	Sandoz accident (Rhine, Basel).
1987	Rhine Action Programme. One of the objectives: a 50% reduction of phosphate load of the Rhine between 1985 and 1995.
1987	Third International Conference on the Protection of the North Sea. One of the objectives: a 50% reduction of phosphate load of the North Sea between 1985 and 1995.
1987	Decree on the application of animal manure (Soil Protection Act). Restricts the amounts of phosphate to be applied per ha of grassland or arable land.
1989	'Nationaal Milieubeleidsplan'. National Environmental Policy Plan with objectives on the reduction of eutrophication of water and soil.
1990	Decree on limit values for phosphate in effluents of sewage treatment plants (Surface Water Pollution Act).
1991	Draft Decree on limit values for nitrogen in effluents of sewage treatment plants (Surface Water Pollution Act).

phosphates in detergents. Within the ministry a discussion had been going for some time on whether or not a policy statement on phosphates should be published; the minister's opinion had not yet been asked.

Mrs. Vorrink was less patient than some of her officials and, adding a remark to a draft answer to parliamentary questions, asked when, at last, the 'Fosfatennota' would be ready for publication. This tipped the balance in the departmental offices in favour of those who wanted to publish. In this way, the policy statement eventually appeared. It was not Mrs. Vorrink, however, but her successor, Dr. Ginjaar, who signed the 'Fosfatennota' (Minister V&M & Minister V&W, 1979).

What would have happened without this coincidence? Would the statement have been published anyway? And would it have been published if the research results up to that moment had not been available?

Such questions are difficult to answer. It is possible, however, to try and find a pattern from a subjective viewpoint, i.e. from the impressions of individual scientists and policy makers. I did not conduct an enquiry, but on the basis of my experience, I conclude that the impact of research on policy:

- depends on the question whether a link between the two has been established in advance in a particular case, and on the closeness of that link,
- depends on the complexity of the problem,
- is often difficult to isolate from the multiplicity of factors that may influence the decision making by a minister and his/her advisers directly, but also indirectly through effects on public opinion, parliament, regional authorities, etc.

For the WQL project *links* between research findings and specific policy decisions were not stated at the start of the project. Before making a financial commitment VROM did not state whether or not policy would be adjusted according to the outcome of the investigations. Consequently, there was no reason to expect that on the completion of the project important decisions would be made on this particular basis. In general the results of the investigations seem to support the policy developed in the past fifteen years or so. Important among these are: (1) external phosphate flows to surface waters have to be reduced drastically and (2) if the external load remains high, the in-lake restoration measures, such as dredging and biological management will be vain. External phosphate loading can only be reduced sufficiently by tackling all sources, i.e. direct discharges from sewage treatment plants and (phosphate fertilizer) industry, pollution from agricultural areas and transboundary pollution, for example via the River Rhine. In addition, as the 'Fosfatennota' indicates, sufficient attention should be paid to regional and local circumstances by drawing up regional management plans to complement the national and international policy.

Without any doubt, the impact of WQL on government policy would have been more spectacular if the results had been more surprising, for example if a hitherto neglected key factor in eutrophication had been discovered. Nevertheless, in addition to this confirmation of the basic elements of eutrophication policy, WQL has provided policy makers with insight into many important aspects, for example the role of the sediment, the seston and the higher trophic levels. Many examples are given in the other papers in this volume.

In this connection it is interesting to quote from the opening address at a symposium on eutrophication in 1979 by the then Environment Minister (Ginjaar, 1980):

> 'The role of the scientist nowadays concerned with the problems of eutrophication is not longer that of a pioneer, explorer or prophet, as it was ten years ago, but that of a patient filler-in and finisher, with an eye for fine detail. This last is no less important than the first; but it is different. The great discovery has been made: it bears a capital P.'

For the management of the Loosdrecht Lakes themselves WQL has evidently produced a wealth of information. Details on this subject may be found in the paper by De Ruiter (1992).

The eutrophication problem is 'not a simple

one'. Therefore, one cannot expect eutrophication research to produce simple recipes for policy making. The relations between research and policy in a given field will therefore be diffuse and iterative, and extend over a long period of time. For the scientist as well as the civil servant and the politician this can be frustrating. The scientists may not recognise his results being put into practice. On the other hand, politicians and civil servants may be impatient and tempted to think that such studies do not contribute to decisions in policy making.

This takes us to the third factor affecting the impact of research on policy making: the 'paths' along which influence is exerted. The main paths of influence in water management policy in the Netherlands roughly run from scientists to policy-making officials in the ministries. Rarely is there a direct impact of research results on the opinion of politicians. One should realise that there are close contacts between policy making groups in the ministries and aquatic research institutes. The National Water Management Agency of the Ministry of Transport and Public Works ('Rijkswaterstaat') has policy making branches as well as research divisions. The Environment Ministry (VROM) has close connections with the National Institute of Public Health and Environmental Protection (RIVM) in the field of environmental research. Provinces and Water Boards often employ their own research scientists. In these cases, therefore, the relationship between research and policy making is 'built-in'.

Some other institutes, however, do not have such a close connection with policy making. This is true for the major participants in the WQL project, viz. the Limnological Institute, the Free University, Amsterdam, and the University of Amsterdam. They do have relations with policy making authorities, albeit mostly on a temporary basis, through incidental research contracts. I have the impression that scientists in these institutes sometimes think their research does not get due attention from the authorities. This may make them refrain from attempts to get involved in policy making. In stead of trying to influence policy more directly, for example by approaching politicians and officials in the ministries, they concentrate exclusively on research.

Other benefits for policy from WQL

Apart from the often elusive impact of research on the material content of policy, there are various effects on the *way* policy is made.

Policy makers may, for example, want to justify decisions already made in their own minds with research projects still to be carried out. They thus use the scientist as a notary, i.e. to provide a document signed by an expert in the field. In Germany this practice has even been institutionalised in the so-called 'Gutachten' produced by scientists both for companies and authorities.

Research projects also are used sometimes to draw attention to specific environmental problems when it is impossible to reach a solution by negotiations. Authorities as well as citizens use this instrument. A good example is the study that was started by the municipality of Rotterdam a few years ago on the discharges of industries responsible for the pollution of Rhine sediment in Rotterdam harbour (Rotterdam, 1986). The municipality wanted to force a reduction of the discharges of heavy metals and other toxic substances to a level not yet attained in the Rhine negotiations between national governments. In such a case research projects are being used as a crowbar. A third motive for having research carried out for other reasons than the enlargement of knowledge can be a wish to postpone decisions. Research projects may be used as an excuse for not having to decide immediately.

I think that if WQL has acted as notary, crowbar or excuse at all, it has done so only to a limited extent. In any case, the Environment Ministry, in deciding to finance parts of the programme, mainly sought the enhancement of knowledge on eutrophication in a general sense; it did not attempt to justify its policies or to force decisions and certainly did not try to find an excuse for refraining from taking decisions. What I would like to emphasise is that scientists should know when they are being used as a notary, as a crowbar or as an excuse. Of course, scientists can

terms of quantities of the water involved and qualitative differences between these, were important in understanding the seasonal nutrient dynamics (Engelen *et al.*, 1992). Moreover, interactions between the various systems in the area, both in the past and recent, have had been strongly influenced by human activities in the past as well as at present. The main among these latter being the recent increased need for public water supply and the regional transfer of water. Unique features of the hydrological research were: (1) the development of two- and three dimensional models for ground water flow, based on monthly and annual water balances: and (2) field experiments on turbulence and diffusion using tracer techniques and laboratory studies on wind driven currents. As regards surface water movements, the remote sensing studies appear to simultaneously provide the much-needed complementary information. The hydrological studies strongly plead for conserving the ground water in the catchment area (the Gooi Region). Though the studies of Engelen *et al.* (1992) suggested flooding the deeper polder areas to reduce seepage losses, the repercussions of such a measure, if taken, need to be seriously considered (see later in this paper).

The studies on phosphorus recycling in the sediments (Keizer & Sinke, 1992) indicate that regeneration *via* mineralization of phosphorus of (*ca.* 3 mg P m^{-2} d^{-1}) is an order of magnitude higher than the diffusive release from sediments (0.3 mg P m^{-2} d^{-1}) under aerobic conditions at 20 °C. Interestingly, this diffusive release decreased from about 1 mg P m^{-2} d^{-1} in 1984 to 0.3 mg P m^{-2} d^{-1} in 1990. Only a very small part of the phosphorus appears to be adsorbed by Fe(III) in the top sediment layer. This is attributed to a marked downward seepage loss of inorganic phosphorus (0.8 mg P m^{-2} d^{-1}), as well as to uptake of this phosphorus during resuspension. Only the upper 2-cm sediment which contains *ca.* 20% potentially bioassimilable phosphorus is important. Thus, seepage, sediment burial, and its diagenesis and mineralization are important factors in the loss of phosphorus from the cycle. The model predictions suggest that any further reduction of external phosphorus load may cause a gradual decrease in total phosphorus concentration (Keizer & Sinke, 1992).

The simulation of the phosphorus cycle using the dynamic mathematical model PCLOOS (Janse *et al.*, 1992) involved mass and rate processes at all the trophic levels, as well as the inorganic phosphorus fractions. The input of the model were the hydrological water- and matter-balance data (Engelen *et al.*, 1992). Also carbon was included in modelling of the organic compartments, independent of phosphorus, so that the P/C ratio's could be modelled too. However, the separation of seston into phytoplankton and detritus gave discrepancies in the flow of phosphorus in the open water, as well as for growth rates compared with values based on primary production (Van Liere & Janse, 1992), although these discrepancies were not large. The outcome for phosphorus was largely comparable with the empirical data that indicate a decrease in total-phosphorus concentration from *ca.* 130 μg P l^{-1} to 80 μg P l^{-1}', but no such decrease was predicted for concentrations of seston and chlorophyll *a*. A delayed response to the reduced phosphorus loading and a new equilibrium in about five years was predicted. Both internal recycling and high assimilation efficiency of cyanobacteria are given as the causes for the delay in response to rehabilitation measures. Additional restoration measures such as flushing, chemomanipulation, or biomanipulation are considered as ineffective complementary measures at the present level of phosphorus loading (Van Liere & Janse, 1992). Therefore, research on phosphorus dynamics and modelling presents a strong plea for further reduction in the present rate of external loading, *viz.* 0.35 g P m^{-2} y^{-1}.

The discovery of *Prochlorothrix hollandica* (Burger-Wiersma *et al.*, 1986, 1989), the first free-living oxychlorobacterium isolated from a freshwater lake, was an impressive scientific achievement. This filamentous form, which measures *ca.* 150 μm in length and about 2 μm in width, resembles *Oscillatoria limnetica* in its superficial appearance, but, unlike cyanobacteria, it contains chlorophyll *b*, and lacks phycobilins, as well as has other biochemical characteristics that distinguish it from cyanobacteria. In Loosdrecht lakes *P. hollandica* forms an important proportion of the total densities of filamentous prokaryotes, including cyanobacteria. With actual growth rates generally less than *ca.* 0.1 d^{-1}, its maximal densities (*ca.* 120.000 ml^{-1}) occur in mid summer when the prevailing temperatures (18–22 °C) result in growth rates exceeding 0.1 d^{-1}. *Prochlorothrix*, which has been considered as a missing link between Cyanophyta and Chlorophyta, has aroused renewed interest in molecular and evolutionary biology of prokaryotes.

Studies on seston dynamics in shallow lakes are rare and information on the effects of changes in wind velocity and direction on the spatial and temporal distribution is very scanty. The present studies on Loosdrecht lakes (Gons *et al.*, 1992) have demonstrated that wind action is the main driving force in the stirring up of sediments and resuspension and translocation of seston. The extent to which seston is translocated depends greatly on fetch and wind direction. The action of wind in these lakes, thus, appears to be more important than turbidity caused by bream browsing in the sediments in search of food and the 'normal' water-based recreational activities of man. Apparently frequent resuspension of seston, coupled with the dominance of filamentous prokaryotes and brown colour of the water, due to dissolved humic substances, are the main factors for the poor light climate in these shallow, peaty lakes (Gons & Otten, 1991).

The relationships between the dominant phytoplankton species and detritus were tested in the laboratory in the specially developed Laboratory Scale Enclosures (LSE's) (Rijkeboer *et al.*, 1991). These systems allowed studying the effect of different loads of phosphorus on growth rates of *Oscillatoria* spp. and *Prochlorothrix*. The extension of studies using the LSE's led to the development of the Epipelon Trap (ET) to collect the overflow (among which the detritus produced), and to follow its fate further. It may be mentioned that detritus, which forms two-thirds of total seston in the lakes, was found to be an important phosphorus buffer apparently due to the affinity of soluble phosphorus to adsorb onto this detritus (Rijkeboer *et al.*, 1991). On the basis of a steady-state model of the dynamics of phytoplankton detritus, the areal organic dry weight concentration of detritus in the lake is *ca.* 60 g DW m^{-2} distributed equally in the seston and epipelon (Otten *et al.*, 1992). The mineralization rate of the refractory detrital material, based on laboratory studies at 20 °C is only 0.005 d^{-1}, but in the longer term will be even lower. Thus, refractory detritus tends to accumulate, and may influence the water quality for a long time to come.

The zooplankton composition in the lakes is typical of highly eutrophic lakes: 1) dominance in the Crustacea of small-bodied cladocerans, mainly *Bosmina* spp. and cyclopoid copepods; and 2) a very high diversity (species composition) of rotifers, with at least 20 taxa and 40 species though only a few are dominant. (Gulati, 1990). A complete absence of large-bodied daphnids (*Daphnia galeata*) common in many other eutrophic lakes was quite striking; the cause of the virtual disappearance in the more recent years of the small-bodied daphnids (*D. cucullata*) is not known. These changes in the daphnid population, and lack of larger-sized species, appears to be due to intensive predation by planktivorous fish, mainly bream (*Abramis brama*).The food quality and quantities are apparently influenced by the very high concentrations of filamentous prokaryotes ($> 10^8$ filaments l^{-1}), with their likely adverse effects on the grazing activities of crustaceans. In contrast to the *Daphnia* species, the bosminids are well adapted to the abundant food in the lakes and are less vulnerable to predation by planktivores. About two-thirds of daily mean grazing by crustaceans of 4–10% was attributable to the populations of *Bosmina* alone. On the other hand, rotifers, especially the smallest among them, appeared to be important grazers of food particles $< 10\ \mu m$, dominated mainly by flagellates, green algae and bacterioplankton (Wilms *et al.*, 1990; Ooms-Wilms, 1991).

The phosphorus excretion by crustaceans, though low compared with values from the literature, covered about one-third of the daily phytoplankton phosphorus demand (Den Oude

& Gulati, 1988). In addition, the rotifers are assumed to regenerate relatively much more phosphorus daily (Gulati *et al.*, 1992). The P/C ratios of the particulate matter exhibit an increase with trophic level, with the more constant zooplankton ratio of *ca.* 0.02 (Gulati *et al.*, 1991), being intermediate between those of seston (0.009) and fish (*ca.* 0.03).

The role of fish in phosphorus recycling through its excretion and egestion as well as leaching from dead fish could be crucial but remains unquantified. In view of the fish standing crop of *ca.* 320 kg ha^{-1} in these lakes the phosphorus in the fish was estimated at about 160 mg P m^{-2} (Van Liere & Janse, 1992), *i.e.* about 50% of the total phosphorus in the water column. Even though fish may be considered as a phosphorus sink, considering their phosphorus-stock, they form a continuous source of phosphorus release into the system through their excretion, egestion and mortality.

Besides the achievements and lacunae of the WQL research there are several areas in which our knowledge of the Loosdrecht lakes' ecosystems is very sketchy. More information is needed on composition and structure of fish and their food preferences. Bottom fauna, dominated by *Chironomus pulmosus*, was found to be rich but its importance has not been quantified. Besides, the role of ciliates, which were found abundant in the benthic collections (unpublished data) in these eutrophic systems has remained virtually unknown. Both ciliates and flagellates may form important pathways through which bacterial energy is channelled and grazer and decomposer food chains are linked. In short, though a great deal has been achieved by the WQL, much still remains to be done to advance our knowledge about these shallow, eutrophic and 'resilient' lake ecosystems.

Spin-off and social relevance

A large coordinated and integrated long-term research programme, as the one in Loosdrecht lakes, offers an infrastructure which has a great scope to attract other research subjects. Only when a programme of multidisciplinary research has been established can the ecosystem modellers start their work. However, WQL chose to start from the onset with the mathematical modelling in order to have an integration between modelling, and field research and laboratory studies. The logistics and infrastructure of WQL made it possible to collect at an extremely short notice a large number of samples over a vast area for groundtruth verification of data collected by remote sensing research. Together with the proper remote sensing research this resulted in a water quality model, which can be extended easily to other lakes (Dekker *et al.*, 1992).

Finally; the integration of various disciplines and cooperation among scientists with different interests and expertise, and working in different institutes was among the WQL's major achievements. The contacts and exchange of scientific information through discussons and joint publications led to greater understanding of the functioning of the ecosystem, but also increased the understanding of the scientific problems of one another's discipline. It led also to a platform for discussions with policy makers, water managers and money givers, culminating in the biannual plenary workshops, with all participants involved.

The 'mission oriented' research programme of WQL resulted in an efficient infrastructure for consultancy to the water quality authorities. The WQL scientists have been active participants in various committees to advise the Water Quality Authorities of the Province of Utrecht, but also in committees dealing with eutrophication problems in other Dutch lakes, but also European lakes in general. This resulted in several reports on hydrology (see Engelen *et al.*, 1992), flushing (Van Liere, 1986), dredging (Boers *et al.*, 1985; Higler *et al.*, 1987), and on the reactions of suspended matter (Gons, 1987; Gons & Van Keulen, 1989). A workshop on recovery of ecosystems from eutrophication was organized as part of a cooperative international research programme of the European Community (Giussani *et al.*, 1991).

However, the most important 'mission' in this context was the advice on additional measures (for a review see Van Liere & Janse, 1992).

Summarizing: all the WQL-authors of this volume agreed that the most important measure was a further reduction in external phosphorus load. This measure was preferred to measures to reduce the'internal' phosphorus load, and to ecosystem management by chemomanipulation or biomanipulation.

To achieve improvement in the water quality choices have to be made, and targets to be assessed. Hofstra & Van Liere (1992) assessed the target values for the ecosystem as those in the period 1930-1940, when the lakes were mesotrophic, macrophyte dominated, and rich in floral and faunal species. Choice of such a 'reference' is arbitrary but has some realism. The oligotrophic situation, as it existed at the beginning of the century, is presumably impossible ever to accomplish. And the mesotrophic situation (1930-1950) was stable for many years at a higher external phosphorus load than the present one (Van Liere *et al.* 1991).

Engelen *et al.* (1992) suggest a return to the 'original' hydrological situation, the return of the groundwater from the Pleistocene sandridge of the region The Gooi. However, this choice is as arbitrary as the one above. The groundwater quality also has changed; it contains a fairly high amount of soluble phosphorus (Hettling, 1985), it is not clear what will be its fate on its way to the lakes. Keizer & Sinke (1992) and Janse *et al.* (1992), as mentioned, calculated the present downward seepage as an important loss factor for phosphorus. Now the external load and the loss factor are nearly in equilibrium (Van Liere & Janse, 1992). If the upwelling ground water flow were to be restored, by means of inundation of the Polder Bethune, the downward seepage from the Loosdrecht lakes will be reduced almost completely (Engelen *et al.*, 1992). Although the lakes then need less supply of water, the phosphorus entering the lakes will have a longer retention time. In short, the incoming phosphorus will remain in the lake ecosystem, thus entering the foodchain to be stored in lake sediment. If no phosphorus were to be lost from the system by seepage recovery would be counteracted.

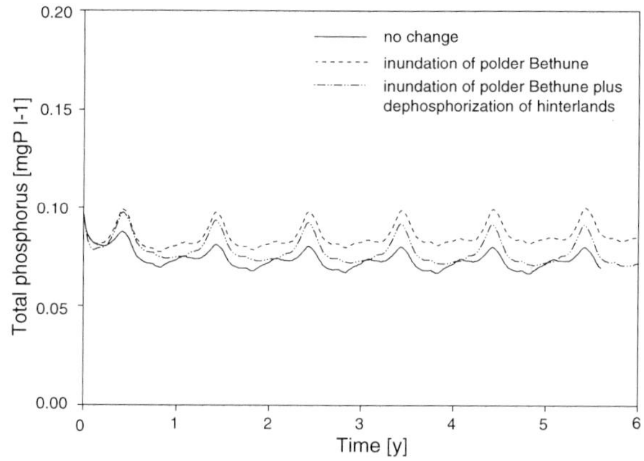

Fig. 1. Simulation of total-phosphorus concentration in the Lake Loosdrecht using three scenarios. No change means external phosphorus loading as at present (0.35 g P m^{-2} y^{-1}); see Janse *et al.*, 1992). Hydrological situation as in 1986.

Janse (pers. comm.) simulated the inundation of the Polder Bethune using the hydrological water- and matter-balance model (Engelen *et al.*, 1992), and its output as input for the ecosystem model PCLOOS (Janse *et al.*, 1992) to evaluate the influence of this inundation. The effect of the formation of 'Lake Bethune' on the phosphorus concentration in (Fig. 1) was compared, with the '*status quo*' (present situation: 0.35 g P m^{-2} y^{-1}). Clearly the inundation of Polder Bethune has a negative effect on Loosdrecht lakes' water quality. The water quality of the lakes will deteriorate, and even an additional dephosphorization of water from the eastern catchment area will not counteract this negative effect on water quality sufficiently.

On achieving reduction of phosphorus, with an uninundated Polder Bethune, model calculations revealed that, only after reduction to 50% of the present load, will the concentrations of total-phosphorus and chlorophyll *a* decrease (Janse *et al.*, 1992).

However, Loosdrecht lakes still will have its detritus both suspended, sedimentated and often resuspended (Gons, 1987). The rate of recovery of these lakes will be related directly to the ratio between loss of this detritus and input of organic matter *via* primary production (Gons *et al.*, 1992; Otten *et al.*, 1992). Thus, a further reduction in the external P-load will play a crucial role.

Acknowledgements

We gratefully acknowledge Jan Janse's model calculations, and his permission to use them, which resulted in the presentation of Fig. 1.

References

Boers, P. C. M., Th. E. Cappenberg & P. Van lersel, 1985. Baggerexperiment in de Loosdrechtse Plassen, voorbereidend onderzoek (A dredging experiment in L. Loosdrecht, spade-research, in Dutch). WQL-report 1985–10.

Burger-Wiersma, T., M. Veenhuis, H. J. Korthals, C. C. M. Van Der Wiel & L. R. Mur, 1986. A new prokaryote containing chlorophylls *a* and *b*. Nature 320: 262–264.

Burger-Wiersma, T., L. J. Stal & L. R. Mur, 1989. *Prochlorothrix hollandica* gen. nov., sp. nov.: a filamentous oxygenic photoautotrophic prokaryote containing chlorophylls *a* and *b*: assignment to *Prochlorotrichacae* fam. nov. and order *Prochlorales* Florenzano, Balloni and Materassi 1986, with emendation of the ordial description. Int. J. Syst. Bact. 39: 250–257.

Dekker, A. G., T. J. Malthus, M. M. Wijnen & E. Seyhan, 1992. Remote sensing as a tool for assessing water quality in Loosdrecht lakes. Hydrobiologia 233: 137–159.

Den Oude, P. J. & R. D. Gulati, 1988. Phosphorus and nitrogen excretion rates of zooplankton from the eutrophic Loosdrecht lakes, with notes on other P sources for phytoplankton requirements. Hydrobiologia 169: 379–390.

De Ruiter, M. A., 1992. Water quality research in Loosdrecht lakes: proposals for the follow-up restoration measures, Hydrobiologia 233: 161–163.

Engelen, G. B., B. F. M. Kal, J. J. Buyse & F. G. M. Van Pruissen, 1992. The hydrology of the Loosdrecht lakes area. Hydrobiologia 233: 21–38.

Giussani, G., L. Van Liere & B. Moss, 1991. Ecosystem Research In Freshwater Environment Recovery. Proceedings of a Workshop, Pallanza, Italy, Sept. 26–29, 1990. Mem. Ist. ital. Idrobiol. 48: 1–362.

Gons, H. J., 1987. Relationship between transparency and suspended matter in L. Loosdrecht, in regards to the standards of swimming-water (in Dutch). WQL-report 1987–3.

Gons, H. J. & R. Van Keulen, 1989. Relationship between transparency and suspended matter in L. Loosdrecht, final report (in Dutch). WQL-report 1989-4.

Gons, H. J., J. H. Otten & M. Rijkeboer, 1991. The significance of wind resuspension for the predominance of filamentous cyanobacteria in a shallow, eutrophic lake. Mem. Ist. ital. Idrobiol. 48: 233–249.

Gons, H. J., T. Burger-Wiersma, J. H. Otten & M. Rijkeboer, 1992. Dynamics of phytoplankton and detritus in a shallow, eutrophic lake (Lake Loosdrecht, The Netherlands). Hydrobiologia 233: 51–59.

Gulati, R. D., 1990. Zooplankton structure in Loosdrecht lakes in relation to the trophic status and the recent restoration measures. Hydrobiologia 191: 173–188.

Gulati, R. D., K. Siewertsen & L. Van Liere, 1991. Carbon and phosphorus relationships of zooplankton and its seston food in Loosdrecht lakes. Mem. Ist. ital. Idrobiol. 48: 279–298.

Gulati, R. D., A. Ooms-Wilms, O. F. R. Van Tongeren, G. Postema & K. Siewertsen, 1992. The dynamics and role of limnetic zooplankton in Loosdrecht lakes (The Netherlands): Hydrobiologia 233: 69–86.

Hettling, H., 1985. A hydrological inventarization of the Loosdrecht lakes area (in Dutch). WQL-report 1985–2.

Higler, L. W. G., S. Parma & L. Van Liere, 1987. Dredging in Loosdrecht lakes, report of a workshop, (in Dutch). WQL report 1987–5.

Hofstra J. J. & L. Van Liere, 1992. The state of the environment of the Loosdrecht lakes. Hydrobiologia 233: 11–20.

Janse, J. H., T. Aldenberg & P. R. G. Kramer, 1992. A mathematical model of the phosphorus cycle in Lake Loosdrecht and simulation of additional measures. Hydrobiologia 233: 119–136.

Keizer, P. & A. J. C. Sinke, 1992. Phosphorus in the sediment of the Loosdrecht lakes and its implications for lake restoration perspectives. Hydrobiologia 233: 39–50.

Kroes, H. W., 1992. The significance of the Loosdrecht lakes research project for eutrophication policy in The Netherlands. Hydrobiologia 233: 165–170.

Ooms-Wilms, A. L., 1991. Ingestion of fluorescently labelled bacteria by rotifers and cladocerans in lake Loosdrecht as measures of bacterivary: preliminary results. Mem. Ist. ital. Idrobiol. 48: 269–278.

Otten, J. H., H. J. Gons & M. Rijkeboer, 1992. Dynamics of phytoplanktonic detritus in a shallow, eutrophic lake (Lake Loosdrecht, The Netherlands). Hydrobiologia 233: 61–68.

Rijkeboer, M. & H. J. Gons, 1988. The relationship between oxygen exchange and changes in seston in laboratory scale enclosures. Verh. int. Ver. Limnol. 23: 756–761.

Rijkeboer, M., F. De Bles & H. J. Gons, 1991. Role of sestonic detritus as a P- buffer. Mem. Ist. ital. Idrobiol. 48: 251–260.

Van Liere, L. (ed), 1986. Flushing as a tool to combat eutrophication in L. Vuntus (in Dutch). WQL Report 1986–3.

Van Liere, L., 1992. Publications of the workgroup WQL (Water Quality Research Loosdrecht Lakes) 1983–present. Hydrobiologia 233: 179–186

Van Liere, L. & R. D. Gulati, 1992. Restoration and recovery of shallow eutrophic lake ecosystems in The Netherlands: epilogue. Hydrobiologia 233: 283–287.

Van Liere, L. & J. H. Janse, 1992. Restoration and resilience to recovery of the Lake Loosdrecht ecosystem in relation to its phosphorus flow. Hydrobiologia 233: 95–104.

Van Liere, L., J. Ebert, W. Kats & J. J. Buyse, 1991. The water quality of Loosdrecht lakes reviewed. Mem. Ist. ital. Idrobiol. 48: 219–232.

Van Liere, L., S. Parma & R. D. Gulati, 1992. Working group Water Quality Research Loosdrecht Lakes: its history, structure, research programme, and some results. Hydrobiologia 233: 1–9.

Wilms, A. L., R. D. Gulati & G. Postema, 1990. First attempt to measure the clearance rate of *Anuraeopsis fissa*. Verh. int. Ver. Limnol. 24: 742–744.

concentration in the Veluwemeer decreased dramatically up to March 1980, when the concentration approximated the concentration in the polder water (Fig. 3). Soluble reactive phosphorus (SRP) decreased concomitantly to values around the detection limit. The chlorophyll *a* concentration also dropped dramatically, but already in January 1980 stabilization set in at a level of about 60 mg m^{-3}, which is above the average concentration in the water used for flushing. The particulate-N concentration in Veluwemeer dropped from 4.9 gN m^{-3} when flushing was started to less than 2.0 gN m^{-3} in May 1980 (Fig. 4).

The main cause of improvement of Veluwemeer water is the dilution with supply water. The concentrations of total-P, Kjeldahl-N and chlorophyll *a* in Veluwemeer were very much related to the fluctuations in the quality of the supply water ($r^2 > 0.95$). If the dilution of the water is the driving force for quality improvement, the nutrient concentration in Veluwemeer will start resembling the nutrient concentration in the supplied water more and more, as more polder water is supplied. Although this is the case for the total-P concentration in Veluwemeer, the situation for chlorophyll *a* and nitrogen is more complex.

The stabilization of the chlorophyll *a* concentration on a higher level than the polder water implies that flushing is counteracted by algal growth. Figure 5 shows growth curves for a number of algal groups derived from the algal growth model BLOOM (Los, 1991) and an average flushing rate. The curves are a function of the water temperature and have been calculated at optimum light- and nutrient situation. At water temperatures below 2 °C, the growth rate of algae will not offset the effect of lake's flushing, so that the algae will continue to be washed out. However the net wash out effect will be 20 to 30% per month. At water temperature above 2 °C algae can increase in numbers effectively, irrespective

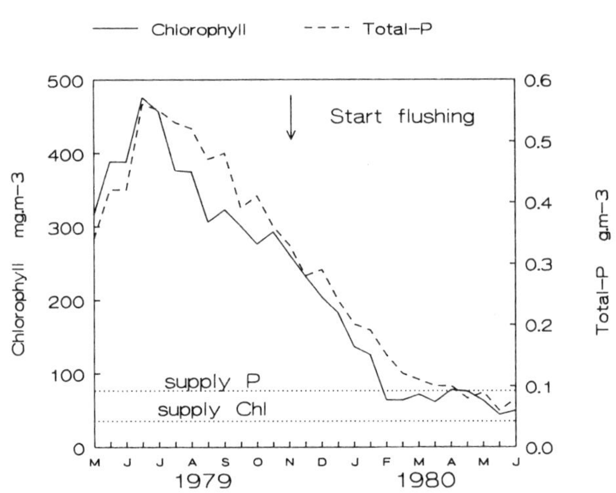

Fig. 3. The effects of the first winter flushing (1979/1980) on chlorophyll *a* and total-phosphorus concentrations (the horizontal dotted lines represent the concentrations in the supply water).

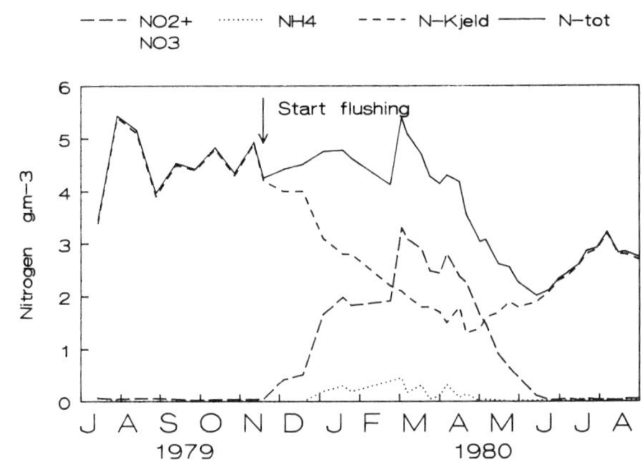

Fig. 4. Nitrogen concentrations in Veluwemeer, before and after the winter flushing of 1979/1980.

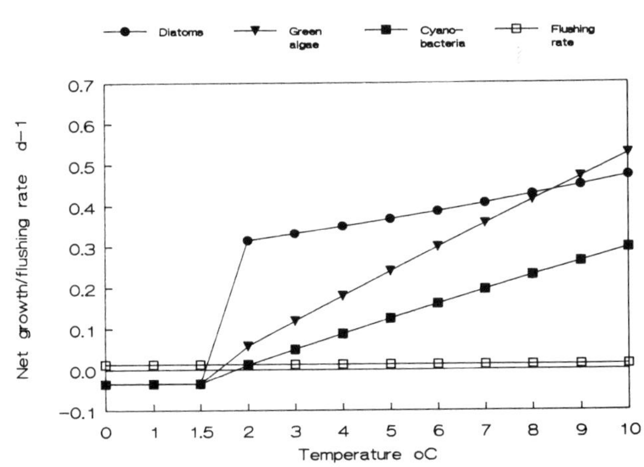

Fig. 5. Growth rates of the various algae species groups as function of the water temperature (from: Los, 1991) and the flushing rate in Veluwemeer.

of flushing. Thus, it is reasonable to assume that from early spring onwards (February/March) algal growth could prevent the chlorophyll *a* concentration from decreasing to the level of the supply water.

Comparing the concentrations of nitrogen compounds in Veluwemeer and the supply water, it appears that at the start of flushing the nitrogen in the lake water predominantly contains particulate-N and concentrations of nitrate and ammonium are low. The supply water, however, is rich in ammonium ($1\ gN\ m^{-3}$) and nitrate ($3\ gN\ m^{-3}$). When the two water types are mixed, the chlorophyll *a* concentration and the Kjeldahl-N concentration decrease, but concentrations of nitrate an ammonium rise rapidly to $2.5\ gN\ m^{-3}$ and $0.5\ gN\ m^{-3}$ respectively, in February and March (Fig. 4). In summer a decrease in nitrate and ammonium sets in, probably because of denitrification at rising temperatures. Algal growth exhausts the dissolved forms of nitrogen, with the result that in May 1980 all nitrogen in Veluwemeer occurred again in particulate form (Kjeldahl-N).

The dilution of Veluwemeer, started in 1979, gives rise to a permanent improvement of the water quality (Fig. 6). The flushing in winter 1980/1981 had, therefore, less impact than the flushing in 1979/1980 (Fig. 7). The differences in nutrient concentrations between supply water and lake water had already been reduced considerably. Figure 7 shows an increase in total-P concentration again in March 1981, presumably because of sediment release, but from 1981 onwards the phosphorus release from the sediment decreased and stabilized at a very low level from 1985 onwards (Fig. 8). The sediment release, calculated by the model SED (Van der Molen, 1991) and measured by continuous flow experiments according to Boers & Van Hese (1988), decreased proportionally more than the external load because of a number of processes, related to flushing:

– The reduced organic matter production and

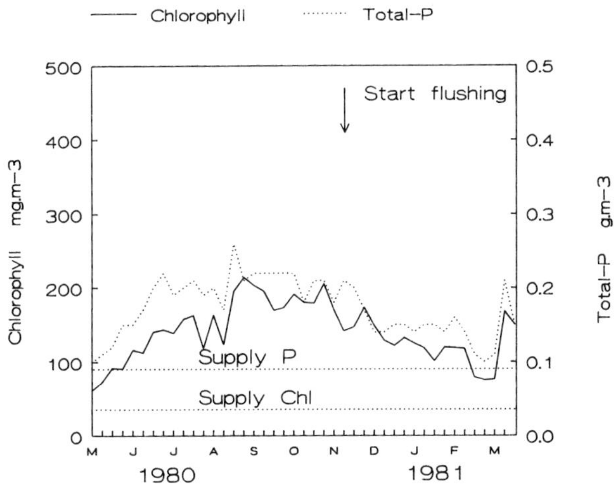

Fig. 7. The effects of the second winter flushing (1980/1981) on concentrations of chlorophyll *a* and total-P (the horizontal dotted lines represent the concentrations in the supply water).

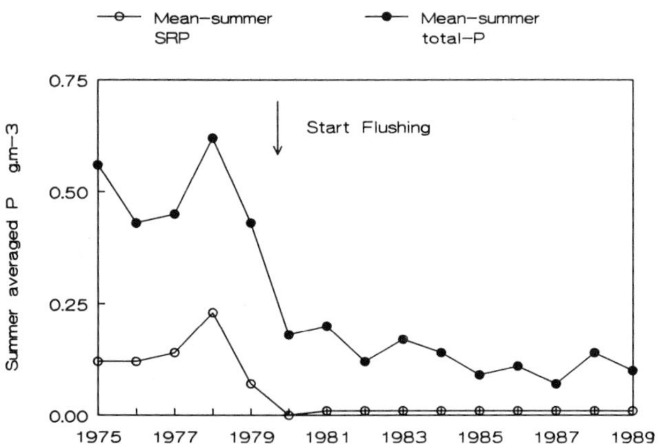

Fig. 6. Summer average total- and soluble reactive phosphorus during 1975–1989.

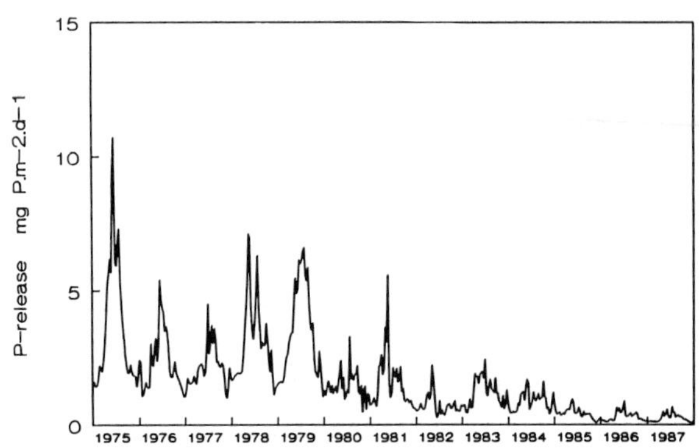

Fig. 8. Release of phosphorus in Veluwemeer based on model SED (from: Van der Molen, 1991).

the considerable supply of nitrate in the polder water have caused improvement of the sediment oxidation.

- Moreover, the supply of calcium and to a lesser extent the supply of iron with the polder water (Brinkman & Van Raaphorst, 1986; Van der Molen, 1990) have possibly led to an increased fixation of phosphorus in the bottom and a decreased release rate.

Further, flushing has caused a decrease in the biovolume of the cyanobacteria (Fig. 9), while the number of algae remained unchanged high (Fig. 10). There was no distinct shift in the composition of the algae species up to 1985. This means that flushing did not initially lead to the realization of the objective to prevent the cyanobacteria dominance. However, the decrease in biovolume resulted in an increased transparency (Fig. 11).

Situation since 1985

The cyanobacteria disappeared from 1985 onwards, although the phosphorus concentrations had already decreased considerable. In Fig. 12 the number of the cyanobacteria, as a percentage of the total number of algae, have been plotted, as they occurred in the growing seasons (April–October) of 1980 to 1989 in Veluwemeer and Wolderwijd respectively. During the period 1985 to 1987 three successive cold winters occurred. The number reduction was probably influenced by the reduced productivity in winter (low temperatures), poor light conditions under the ice cover (Adams *et al.*, 1984) as well as by increased sedimentation under the ice cover. During a similar cold winter in 1978/1979 the cyanobacteria dominance was not broken, but only temporarily reduced under prolonged ice cover (Fig. 13).

The following qualitative model can be devel-

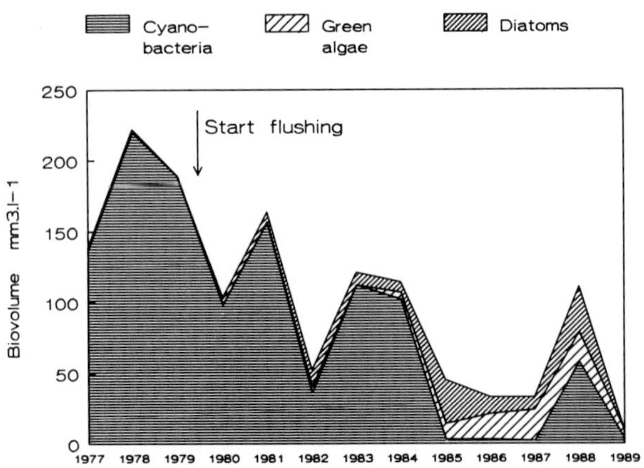

Fig. 9. Stacked presentation of the summer average of the biovolume of algae in Veluwemeer per species group over the period 1977–1989.

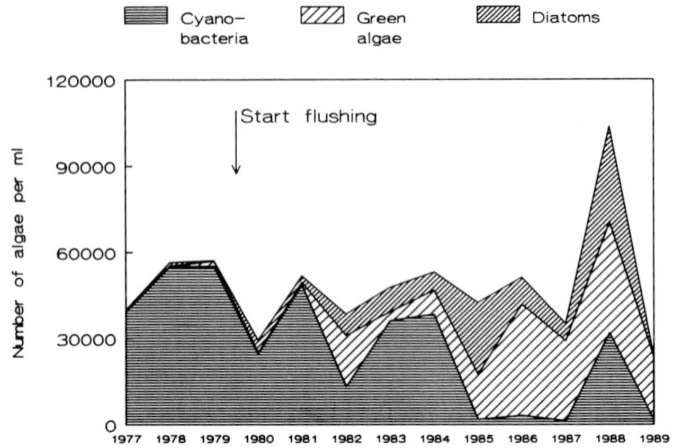

Fig. 10. Stacked presentation of the summer average of the densities of algae in Veluwemeer per species group over the period 1979–1989.

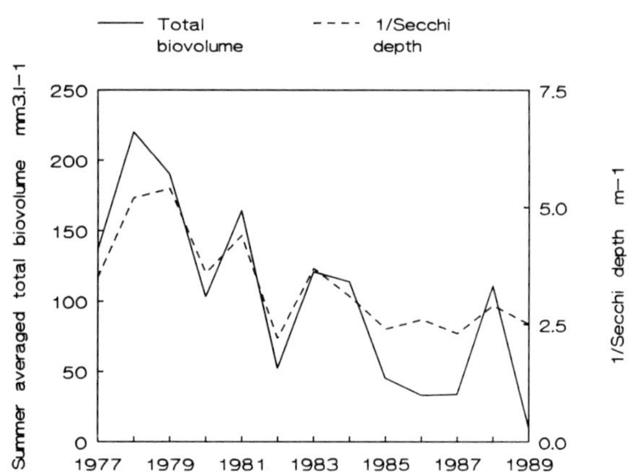

Fig. 11. Summer averages of the biovolume of the algae and the reciprocal Secchi depth values in Veluwemeer for 1977–1989.

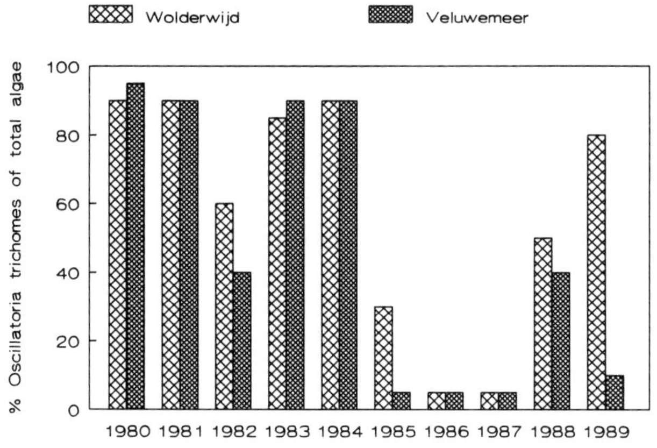

Fig. 12. Percentages of the *Oscillatoria agardhii* in the algae population in Veluwemeer and Wolderwijd during the period 1980–1989.

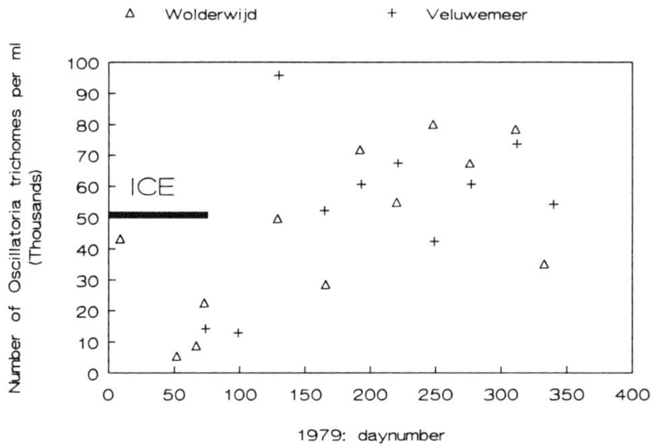

Fig. 13. Influence of a period of prolonged ice coverage on the numbers of cyanobacteria in Veluwemeer and Wolderwijd during the winter of 1978/1979.

oped to explain the shift in algae composition during the period 1985–1987 in Veluwemeer and Wolderwijd. *Oscillatoria agardhii* has a relatively low growth rate, particularly at high light intensities (Loogman, 1982), but it can maintain its position because the loss rates are low also. This species is hardly eaten by zooplankton and its sedimentation rate is low (Reynolds & Walsby, 1975). Therefore, the number of cells at the beginning of a growing season are important for the dominance in the growing season. The combination of a severe winter and scarcity of phosphorus has the following effect: the densities of *Oscillatoria agardhii* will decrease dramatically during ice coverage and as soon as the water temperature exceeds the critical value of 2 °C diatoms and green algae will take up all the available phosphorus. This causes a further deterioration in the competitive position of the cyanobacteria. Table 2 shows this process after the winter of 1984/1985. The direct contribution of flushing to the reduction of cyanobacteria in Veluwemeer in the winter of 1984/1985 is limited.

Although phosphorus levels still were relatively low, the number of cyanobacteria started to increase again after the mild winter of 1987/1988. Low numbers of cyanobacteria in 1989 in Veluwemeer, compared to Wolderwijd, are caused by low availability of phosphorus.

During the summer another mechanism may play a role in the disappearance of *Oscillatoria agardhii.* The photosynthesis capacity of *Oscillatoria agardhii* gradually decreases and may even be reduced to zero at higher light intensities (Loogman, 1982). The increased transparency during summer from 1985 onwards therefore contributes to unfavourable conditions for *Oscillatoria agardhii.*

Effects on lake-transparency

The transparency in Veluwemeer increased, but less than expected. Figure 11 showed a continuing decrease of the total biovolume since 1985, whilst stagnation occurs in the (reciprocal) Secchi-depth values. The improvement of trans-

Table 2. Number of algae per ml in Veluwemeer during the winter period 1984/1985

Month	Cyano-bacteria	Green algae	Diatoms
Oct 1984	52775	4238	1038
Nov. 1984	33038	2138	600
Dec 1984	29700	3075	375
Jan 1985	10650	1050	0
Feb 1985	54575	5138	75
Mar 1985	2425	2160	1250
Apr 1985	2871	16019	50699
May 1985	4156	17265	32840

10^6 m^3 may occur due to infiltration and evapotranspiration. These losses must be compensated, for maintaining a certain water level. In order to maintain the original total-P concentration in Veluwemeer, *ca* 50 10^6 m^3 will be additionally required during the summer period to compensate for the 'concentration effect'.

In view of the really acceptable (phosphorus) quality of Veluwemeer in the present situation, it seems that the efficiency of flushing is low at present. The quality of water in the bordering lake Wolderwijd offers more prospects for improvement. Therefore, we recommend to shift the flushing operations from Veluwemeer towards Wolderwijd in the coming years.

References

Adams, W. A., W. P. Adams, P. A. Flavelle & N. T. Roulet, 1984. Variability of light beneath a modified portion of the snow and ice cover of a lake. Verh. int. Ver. Limnol. 22: 65–71.

Afghan, B. K. & J. F. Ryan, 1975. A modified procedure for the determination of nitrate in sediments and some natural waters. Envir. Letters 9: 59–73.

Berger, C. & A. Bij de Vaate, 1983. Limnological studies on the eutrophication of Lake Wolderwijd, a shallow hypertrophic *Oscillatoria*-dominated lake in the Netherlands. Schweizerische Zeitschrift für Hydrologie 45: 458–479.

Boers, P. C. M. & O. Van Hese, 1988. Phosphorus release from the peaty sediments of the Loosdrecht Lakes. Wat. Res. 22: 355–363.

Brinkman, A. G. & W. Van Raaphorst, 1986. De fosfaathuishouding in het Veluwemeer. Ph.D. thesis, Technical University Twente, The Netherlands, 481 pp.

Cooke, G. D., E. B. Welch, S. A. Peterson & P. R. Newroth (eds), 1986. Lake and reservoir restoration. Ann Arbor Science, Butterworth Publishers, 392 pp.

Dankers, N. & R. Laane, 1983. A comparison of wet oxidation and loss on ignition of organic material in suspended matter. Envir. Techn. Letters 4: 283–290.

Gunnison, D. & M. Alexander, 1975. Resistance and susceptibility of algae to decomposition by natural microbiol communities. Limnol. Oceanogr. 20: 64–70.

Hosper, S. H. & M-L. Meijer, 1986. Control of phosphorus loading and flushing as restoration methods for Lake Veluwe, the Netherlands. Hydrobiol. Bull. 20: 183–194.

Jewell, W. J. & P. L. McCarthy, 1971. Aerobic decomposition of algae. Envir. Sci. Technol. 5: 1023–1031.

Loogman, J. G., 1982. Influence of photoperiodicity on algal growth kinetics. Ph.D. thesis, University of Amsterdam, The Netherlands, 112 pp.

Los, F. J., 1991. Mathematical simulation of algae blooms by the model BLOOM II. Delft Hydraulics, Delft, The Netherlands, 113 pp.

Moed, J. R., 1973. Effects of combined action of light and silicon depletion on *Astrionella formosa*. Verh. int. Ver. Limnol. 18: 1367–1374.

Murphy, J. & J. P. Riley, 1962. A modified single solution method for the determination of phosphate in natural waters. Analyt. Chim. Acta 27: 31–36.

PER, 1982. Projektgroep Eutrofiëringsbestrijding Randmeren. Eutrofiëringsonderzoek Wolderwijd-Nuldernauw. RWS-nota, Lelystad, The Netherlands, 166 pp.

PER, 1986. Projektgroep Eutrofiëringsbestrijding Randmeren. Bestrijding van de eutrofiëring van het Veluwemeer-Drontermeer. RWS-nota, Lelystad, The Netherlands, 296 pp.

Reynolds, C. S. & A. E. Walsby, 1975. Water-Blooms. Biol. Rev., 50: 437–481.

Van der Molen, D. T., P. C. M. Boers & L. Lijklema, 1989. Gegevens-analyse nutriënten en zwevend materiaal Veluwemeer. H_2O 22: 760–762.

Van der Molen, D. T., 1990. Inventarisatie gegevens m.b.t. de waterhoeveelheden en makro-ionensamenstelling van de beken, kwel, neerslag en de polderuitslag op het Veluwemeer. RIZA werkdokument 90.153X, Lelystad, The Netherlands, 29 pp.

Van der Molen, D. T., 1991. Simple, dynamic model for the simulation of the release of phosphorus from sediments in shallow, eutrophic systems. Wat. Res. 25: 737–744.

Hydrobiologia **233**: 197–210, 1992.
L. Van Liere & R.D. Gulati (eds), Restoration and Recovery of Shallow Eutrophic Lake Ecosystems in The Netherlands.

Lake restoration with and without dredging of phosphorus-enriched upper sediment layers

Joop Van der Does,[1] Pierre Verstraelen,[2] Paul Boers,[3] Jan Van Roestel,[4] Rudi Roijackers[5] & Gerard Moser[1]
[1] *Water Board of Rijnland, P.O. Box 156, 2300 AD Leiden, The Netherlands (correspondence address);* [2] *Water Pollution Control Authority Amstel & Gooiland, P.O. Box 1061, 1200 BB Hilversum, The Netherlands;* [3] *Institute for Inland Water Management and Waste Water Treatment, P.O. Box 17, 8200 AA Lelystad, The Netherlands;* [4] *Oranjewoud Ltd, P.O. Box 8590, 3009 AN Rotterdam, The Netherlands;* [5] *Agricultural University Wageningen Department of Nature Conservation, P.O. Box 8080, 6700 DD Wageningen, The Netherlands*

Key words: lake restoration, eutrophication, sediment, dredging, water quality management

Abstract

Human activity has been the cause of continuing decline of water quality in most Dutch lakes. Development of lake restoration programmes must take into account the lake functions. Major reduction of the nutrient and pollutant loading is the primary step in lake restoration. Still, the recovery of eutrophic lakes is retarded frequently because of internal phosphorus loading by the lakes' sediments. Sediment dredging, as an additional tool for water quality management to accelerate accomplishing the desired water quality, is studied. In this paper we evaluate the preliminary results of eight lake restoration projects in the Netherlands. The lakes are compared in order to estimate the magnitude of the internal phosphorus loading. Dredging as an additional measure was carried out twice in the peatlake Geerplas. In the Nieuwkoop Lakes only the external phosphorus loading was substantially reduced from 0.9 to 0.2 g P m^{-2} y^{-1}. Provisional results of these two shallow peatlake restoration projects focussed on eutrophication abatement with and without dredging, are presented. Both show a decrease in phosphorus concentration in the lakes. The necessity to dredge the lakes is discussed.

Introduction

Excessive loading with nutrients, particularly phosphorus, often leads to increased algal growth and poor water quality in most Dutch lakes. Consequently many human and nature related functions are hampered. To combat this process of deterioration, restoration programmes that have been drawn up relate to major reduction of loading with nutrients in surface water and groundwater. Despite restoration measures the success of the programmes focused on reduction of the external phosphorus loading is often limited (Van der Does & Klapwijk, 1985, 1987; Van Liere *et al.*, 1990).

Apparently very large reductions of the phosphorus loading are needed to force a lake into another trophic category (Cullen & Forsberg, 1988). There might be a resistance of the foodweb to changes (Gulati *et al.*, 1990; Van der Vlugt *et al.*, 1992). In this respect foodweb manipulation techniques have proven to be successful in small lakes, but their use in large lakes is still a matter of research. Another cause for a delayed or only partial response on external loading reduction may be internal loading by the sediments. Phosphorus release from enriched upper sediment layers retards the decrease of phosphorus concentration in shallow lakes. The reduction of the internal loading can be met by dredging the phosphorus rich upper sediment layers (Björk, 1988; Cooke *et al.*, 1986) or by phosphorus inactivation techniques (Cooke *et al.*, 1986; Boers, 1991; Prepas *et al.*, 1990). Such a selective dredging of phosphorus is both costly and time consuming. The latter method, being much more economic and faster, still needs to be proven (Boers *et al.*, 1992). For these reasons, lake restoration programmes must consider the need for additional restoration techniques that might be necessary besides external loading reduction. The necessity also depends on the lakes objectives, which are related to assigned lake functions (e.g. recreation, drinking water production, nature conservation) and type of watersystem. An example is given in Table 1.

This paper focusses on the response of lake phosphorus concentrations to reduction of external phosphorus loading by diminishing the phosphorus sources. The net internal phosphorus loading of the lake is estimated in different ways. Two lake restorations with and without dredging of phosphorus enriched upper sediment layers are compared. Lake restoration measures that were taken, are also related to water quality objectives and hydrological circumstances.

Table 1. Examples of objectives for phosphorous concentrations, which differ according to the type of water system and its function (Ministry of Agriculture, Nature Management and Fisheries, 1990; Ministry for Transport and Public Work, 1989; Province of North Holland, 1990; Province of South Holland, 1991):

Type of water system	Function	Objective for total phosphorus concentration	
Shallow lake	AMK*	Summer average	<0.08 mg P l^{-1}
Shallow lake	Nature conservation	Summer average	<0.05 mg P l^{-1}
Deep lake	AMK	Summer average	<0.15 mg P l^{-1}
Ditch	AMK	Summer trimester	≤ 0.30 mg P l^{-1}

* AMK is the General Environmental Quality in The Netherlands.

Material and methods

Research area

The shallow peat lakes studied are situated in the low-lying centre of The Netherlands. The area borders on the dunes in the west, the North Sea Canal, Lake IJ and Lake Gooi in the north, a Pleistocene ice-pushed ridge in the east and the river Oude Rijn in the south (Fig. 1). The lake area is surrounded by cities i.e. Amsterdam, Leiden, The Hague, Utrecht and Hilversum and is called 'the green heart'.

Originally the green heart was part of a peatland belt lying just above sea level and reaching from the north-west of The Netherlands to the south-west. It consisted of swamps and marshes, slow flowing rivers and few shallow lakes. Water economy was characterized by groundwater flow from the adjacent sand ridges and dunes and an upward groundwater seepage causing a constant water surplus. Dense vegetation retained the water and allowed only decelerated water discharge into the rivers. Since the early Middle Ages, agricultural development, peat mining, canalization and organized drainage caused a lowering of the groundwater table, peat mineralization and subsidence, and, consequently, land inversion. Finally, in the last three centuries, land reclamation and drinking water extraction from the sand ridges and dunes caused a further drop of the groundwater table and increased groundwater flow towards the deep polders. The year around surplus changed into a winter surplus that is pumped into the rivers and canals and summer shortages that are neutralized by water supplies from the heavily polluted rivers and canals. Nowadays the green heart lies some 2 m below sea level and comprises many shallow lakes (1–3 m deep) created by peat mining, several deep lakes (5–50 m deep) created by sand mining, slow flowing rivers, canals, drained wetlands that are used as pasture lands, and some undrained wetlands.

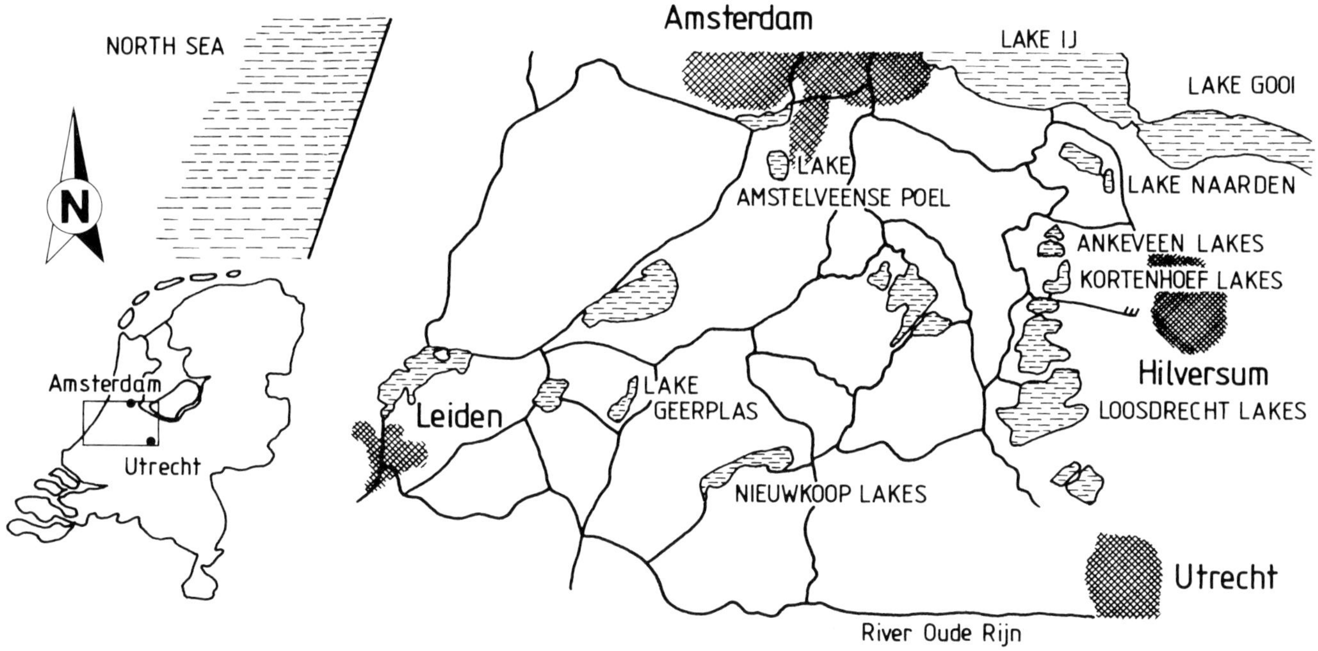

Fig. 1. Research area with the situation of the investigated lakes.

Hydrological and morphological data of the investigated lakes are presented in Table 3 and discussed with the results.

Monitoring of water and sediment quality

Water samples were taken monthly and sediments were sampled occasionally. Concentrations of nutrients, chlorophyll *a*, Secchi disk depth, dry weight, Fe, and LOI (Loss on ignition) were determined according to Dutch standard methods. The used symbols are explained in Table 2. The availability of phosphorus in sediments and its impact on the phosphorus concentration in lakes is estimated as: P – release and bioavailable – P from the sediment. Phosphorus release rates from intact sediment cores were sampled and measured according to the continuous flow technique (Boers & Van Hese, 1988). Bio-available phosphorus of the top 5 cm of the sediment was quantified as the yield of algal biomass in batch cultures containing algae using sediment as the sole phosphorus source (Bruning & Klapwijk, 1984; Klapwijk & Bruning, 1986; Boers *et al.*, 1984). Maximal algal biomass was expressed as summer average chlorophyll-*a* and calculated from summer averages of total P and total N according to Lijklema *et al.* (1988):

- $\text{Chl-}a = 1428 * \text{P-lake}$,
- $\text{Chl-}a = 89.6 + 89.6 * \text{N-lake}$.

Net internal loading (NIL) expected after a reduction of the external loading was calculated from TP, TP/2 μm, and Fe concentrations in sediments according to Boers (1991):

- $\log(\text{NIL} + 1) = -0.11 + 0.82 * \log(\text{TP})$
- $\log(\text{NIL} + 1) = 1.87 + 1.48 * \log(\text{TP/Fe})$,
- $\log(\text{NIL} + 1) = -0.77 + 1.08 * \log(\text{TP}/{<}2\ \mu\text{m})$.

Results and discussion

Hydrological considerations of lake restoration

The lakes investigated have a surface area ranging from 0.3 to 19.6 km^2 (Table 3). Most of the water bodies in the lakes are surrounded by marshlands and pastures with ditches. The lakes are part of polder systems that need to be main-

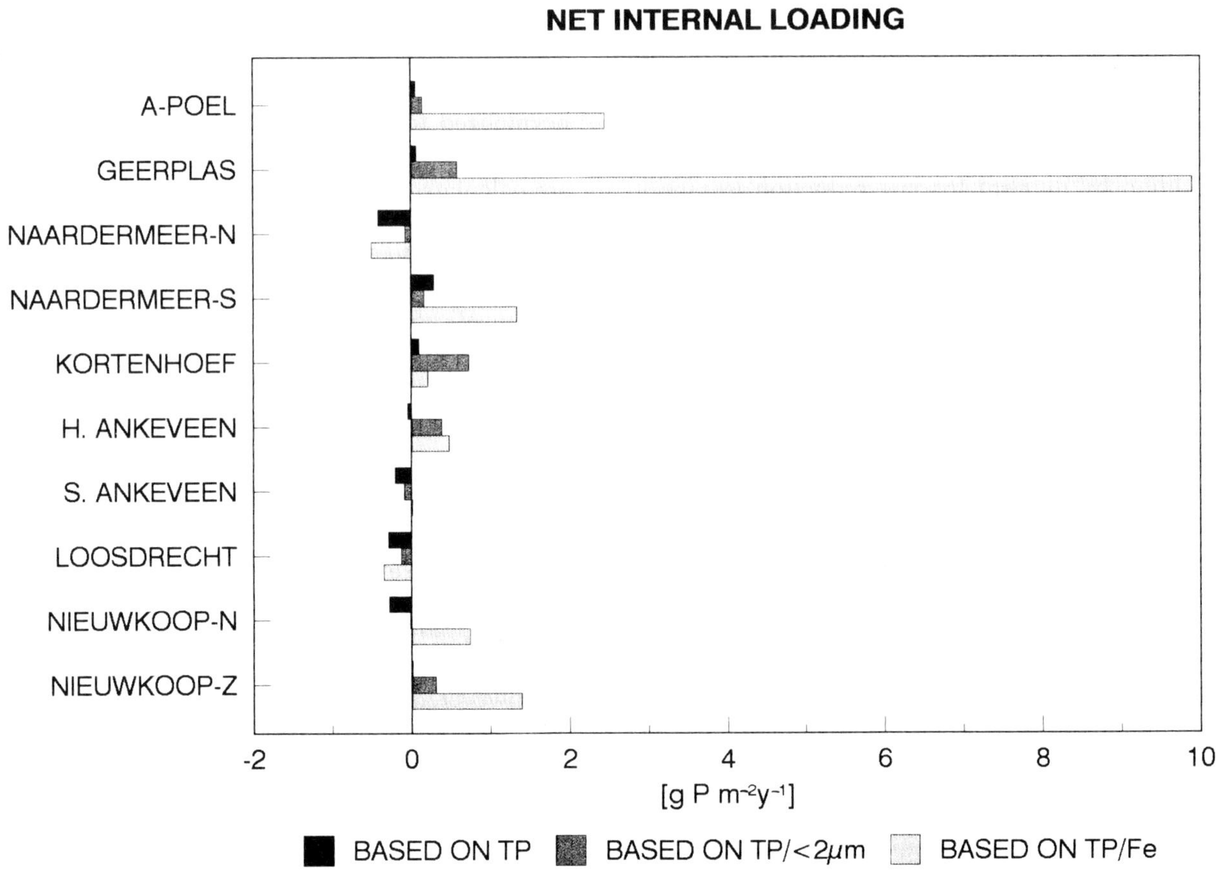

Fig. 3. Calculated net internal loading (NIL) of the investigated lakes; based on total phosphorus (TP) in sediments, based on TP and Fe in sediments and based on TP in sediments normalized to the clay fraction, according to Boers, 1991.

ments overestimate the *in situ* sediment release if SRP-concentrations in the continuous flow reactor are much smaller than under field conditions; differences in redox at the sediment/water interface between incubated cores and *in situ* sediment may affect the adsorption capacity of the sediment considerably. Moreover hydrological factors such as seepage, infiltration and resuspension, are excluded in laboratory experiments.

Algal available phosphorus of the sediment top layer (0–5 cm) shows the potential available phosphorus at a certain moment (Table 4), but cannot be linked to a annual net internal phosphorus loading.

To assess possible retardation of lake restoration by reduction of the external phosphorus loading, water quality managers are especially interested in NIL during summer. Net internal loading based on measured release rates (NIL_{rel} in the different lakes exceeds L_{in} in some lakes: Amstelveense Poel, Geerplas, Naarden-South and Ankeveen (Fig. 4). Since the net internal loading is difficult to quantify the lakes are ranked relatively (Table 6), according to the following criteria, that may show the magnitude of the internal loading; calculated NIL (based on TP, TP/Fe and TP/$<2\ \mu$ m), phosphorus release rates, % algal-availability P, ratio NIL_{rel}/L_{in} and P-concentration (summer average). This results in a sequence of the lakes based on an average score shown in Table 6.

According to this enumeration internal phosphorus loading is relatively highest in lake Geerplas, followed by the lakes Nieuwkoop – Zuideinderplas, Naarden-South, Amstelveense Poel and Kortenhoef. The relatively lowest internal

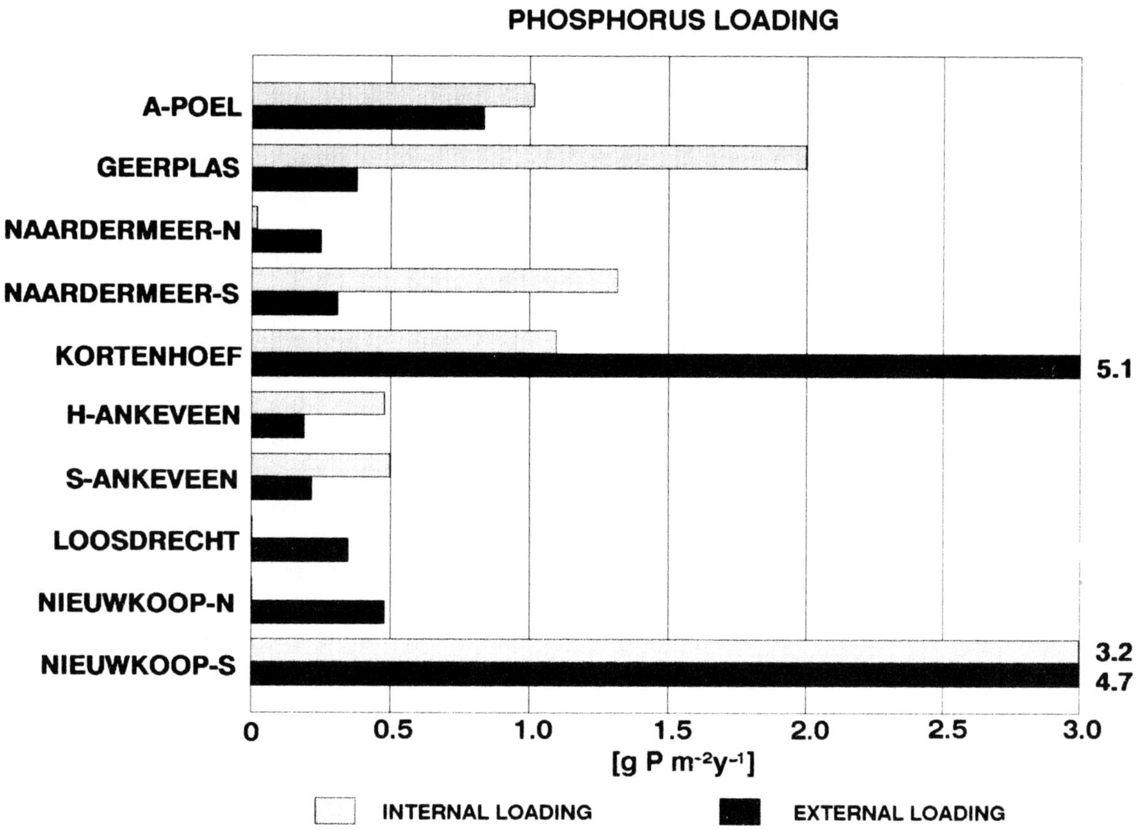

Fig. 4. External phosphorus loading (L_{in}) compared to the estimated net internal phosphorus loading based on phosphorus release rates (NIL_{rel}) of the different lakes.

Table 6. Lakes ranked on the average score of each lake according to the criteria that may express the magnitude of the net internal loading (NIL); calculated NIL by TP, TP/Fe and TP/<2 μm; P_{rel}; % Algal available P; NIL_{rel}/L_{in} and P-lake. A low score indicates a high net internal loading.

Lake	NIL TP	NIL TP/Fe	NIL TP/<2 μm	P_{rel}	ALGAL %	NIL/ L_{in}	P-lake	Average score
Geerplas	3	2	1	2	3	1	2	2.0
Nieuwkoop-Zuideind.	5	4	3	1	5	6	1	3.4
Naarden-South	1	5	4	3	5	2	4	3.4
Amstelveense Poel	4	6	2	5	1	5	3	3.7
Kortenhoef	2	1	7	4	2	7	5	3.9
Hollands Ankeveen	6	3	6	7	4	3	6	4.9
Stichts Ankeveen	7	9	8	6	5	4	10	5.9
Nieuwkoop-Noordein.	8	7	5	10	8	10	7	7.6
Naarden-North	10	8	10	8	9	8	9	8.6
Loosdrecht	9	10	9	9	10	9	8	8.9

phosphorus loading will be in the Loosdrecht Lakes followed by Naarden-North and Nieuwkoop Noordeinderplas. Additional measures to reduce internal loading seem urgently in the lakes with average scores below 4.

Reduction of the internal loading is considered if the external loading has been far reduced in comparison with the internal loading for example in lakes Geerplas and Naarden (Fig. 4). In the lakes Nieuwkoop – Zuideinderplas, Kortenhoef and Amstelveense Poel reduction of the external loading is mostly needed. The response of the lake after reduction of the external phosphorus loading may indicate if the internal loading should also be reduced (Fig. 2).

Dredging criteria for peaty lakes

Dredging of the nutrient-rich upper sediment layers, i.e. surface sediment skimming is applied to reduce the internal phosphorus loading. While preparing lake restoration projects the following criteria were recommend – derived from the ranked lakes in Table 6 – to select lakes for additional dredging measures:

Characteristics	Criteria
– external P-loading	far reduced in relation with the residence time
– P-lake	>0.08 mg P l^{-1}
– TP-upper layer	>0.08 mg P g^{-1}
– bio-available P in the upper layer	$>25\%$
– P-release rate in upper layer	>2.5 mg P m^{-2} d^{-1}
– TP lower layer	<0.70 mg P g – 1
– bio-available P in lower layer	$<20\%$

In two lakes, with highest net internal loading (Table 6), experimental restoration projects with a different approach. were started: lakes Geerplas and the Nieuwkoop lakes.

The external P-loading in both lakes is far reduced in 1989 and 1990. Lake Geerplas has a long residence time (>1 year), therefore the high net internal loading is also reduced by dredging. In the Nieuwkoop-Zuideinderplas the short residence time in combination with a nutrient-poor water supply is used to deplete the phosphorus load in the sediment in order to decrease the net internal loading without dredging.

Restoration of lakes Geerplas: dredging

Since lake Geerplas is characterized by a relatively high internal loading and reduction of the external loading alone may not lead to its recovery, sediment dredging was considered as an additional measure. Dredging was focussed on removing the nutrient-rich fine-grained top mud layer using a two sided suction head. The thickness of the mud layer removed varied in thickness from 1 to 100 cm, without causing high turbidity.

The first attempt to remove the lake's sediment was hardly successful (period April–October 1989), as the fluid mud layer was transported from the undredged area to the area from where the sediment had been removed by dredging. The upper sediment layer proved to be very mobile. Within a month wave-induced changes in mud thickness up to 30 cm were measured (Fig. 5).

The measurements of the changes in fluid mud level in lake Geerplas are large. A theoretical explanation is found in the behaviour of the fluid mud layer in which three sub-layers can be distinguished: a lower consolidated mud layer at the bottom, a middle layer of fluid mud suspension and a upper layer of diluted mud. Fluidization of the mud layer is initiated by wave induced pressure variation of the bed which lead to increasing water pressure in the pores and hence to a reduction of the internal soil shear strength (Van Rijn, 1989). Transport of the upper dilute mud suspension layer is caused by advection and diffusion of the water flow. The wind induced circulation flow is an important transport mechanism and related to the water depth of a lake. The bathometric features of lake Geerplas show that in the deeper middle part of the lake, the wind induced horizontal flow is contrary to the wind, while at less deeper borders the water flow corresponds to the wind direction, according to Sünderman (1979). Transport of the middle fluid mud suspension layer may take place by gravity forces, for instance by sloping bottom. Wind causes a surface slope in a lake. In that case the interface slope between the water and the fluid mud layer is opposite to the surface slope and an order of magnitude larger, due to the density differences between the water column and the fluid mud layer (Schönfeld & Kranenburg, 1981). The mobility of mud can be minimised by increasing the lake's depth, by reducing the transport of the fluid mud

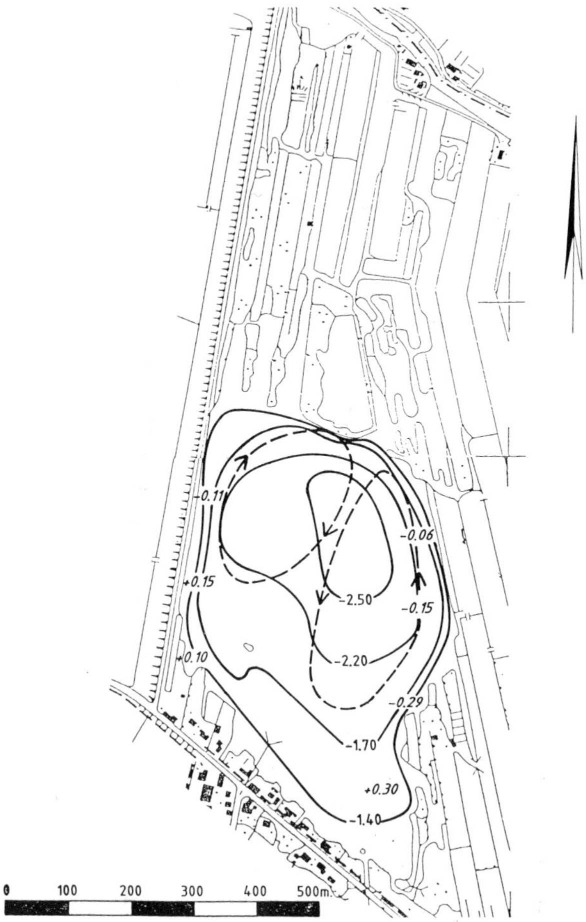

Fig. 5. Bathometric features of Lake Geerplas with depth profiles (solid lines) and circulation by prevailing winds (broken lines) and results of mud transport. For example *+15* is the localy increased thickness of the mud layer between October 6th and October 27th, 1989.

layer with screens or by reducing the fetch length of the wind. Reducing horizontal mud transport by screens was tested in lake Geerplas.

The screens jutted about 1 m above the sediment and stretched up to about 1 m below the water surface. They could withstand wave action and stormy conditions. Differences in sediment thickness of up to 30 cm were observed between both sides of the screens during a stormy 4 week period (January–February 1990). In this period the sediment thickness increased about 2 cm within the protected area of 5.000 m^2, whereas on the outside of the screens sediment thickness increased up to 30 cm. Thus the horizontal fluid mud transport can be suppressed. Apparently fluid mud transport with higher concentrations fine grained material seems to occur in 1 m above the original sediment upper layer.

The restoration programme in lake Geerplas was completed in 1990 by further reducing the external phosphorus load from 0.38 to 0.13 g P $m^{-2} y^{-1}$. Waste water discharges into the lake were stopped and the lake was hydrologically isolated. The inlet water was dephosphorized before it entered the lake since 1989.

During a second attempt from October 1990 till May 1991 surface sediment was skimmed with preventing horizontal mud transport. Screens were used to separate the three compartments: 1) dredged compartments; 2) compartments which were dredged and 3) untreated compartments. Phosphorus concentrations in the lake decreased to 0.1 mg P l^{-1} at the end of the second dredging operation (Fig. 6).

Restoration of the Nieuwkoop Lakes: no dredging

The Nieuwkoop Lakes consist of two main basins; lake Zuideinderplas and lake Noordeinderplas (Fig. 1). The restoration programme was executed in 1990. Point sources of phosphorus discharge (households and horticulture) were reduced. A watershed was created between areas for nature management and agriculture. Transit water flow and hydraulic loading decreased. Phosphorus removal was done by treating the inlet water with iron chloride. The external phosphorus loading in 1990 decreased from 0.92 to 0.22 g P $m^{-2} y^{-1}$ for the whole Nieuwkoop Lakes area (Van der Does & de Jong, 1992). The lakes Noordeinderplas and Zuideinderplas differ greatly in residence time, because water supply and outlet of excess of precipitation from the Noordeinderplas passes through the Zuideinderplas. Therefore these lakes differ in their nutrient loading, water quality and sediment loading. Phosphorus release rate in the Noordeinderplas was almost nil, but extremely high in the Zuideinderplas (Table 4).

Notwithstanding the high internal phosphorus

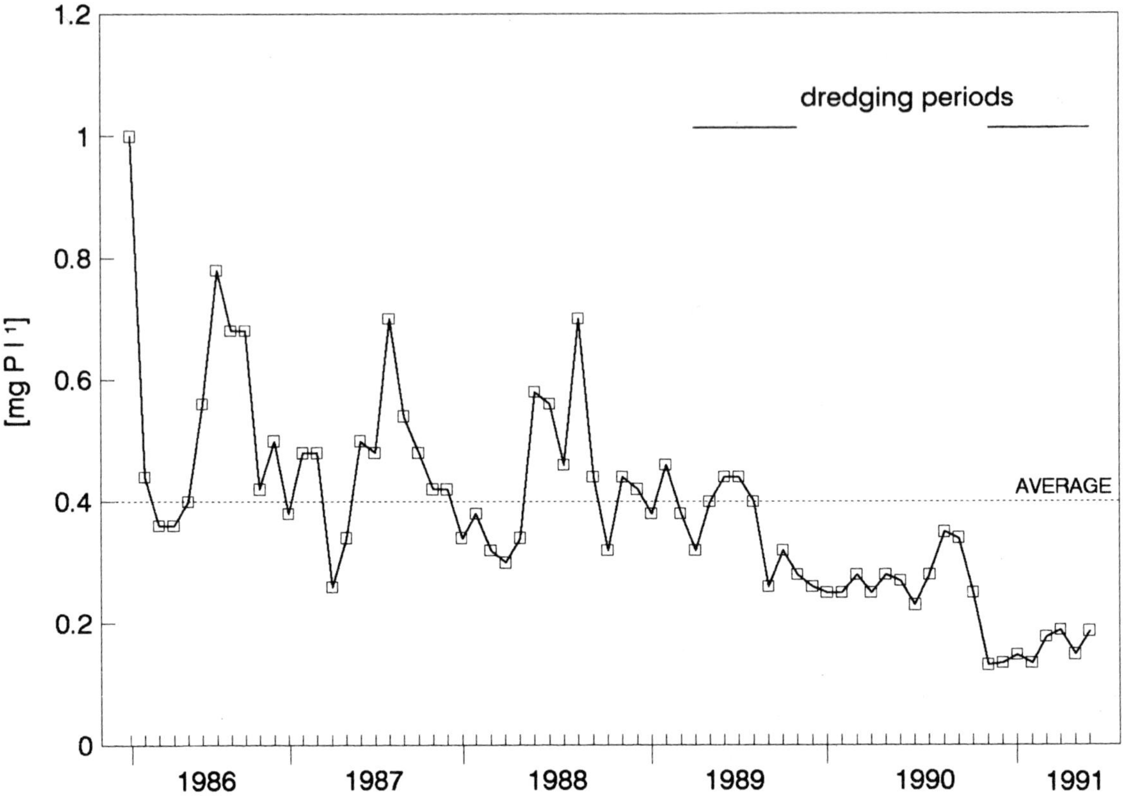

Fig. 6. Changes in the total phosphorus concentration in lake Geerplas from Jan. 1986–May 1991. The dredging periods: 1989 and 1990/1991.

loading in the Zuideinderplas, dredging is not yet carried out pending an evaluation of the lake's response to the reduction in external load. The Zuideinderplas is chosen as pilot location to study the effect of its low residence time (4–5 months) in the removal of phosphorus released by the sediments. Development of the total phosphorus concentration in lake Zuideinderplas is presented in Fig. 7. Reduction of the external phosphorus load in the Zuideinderplas (about 90%) resulted in a decrease of 41% of the lake's total phosphorus concentration (summer average) in 1990. This means a decrease of 46% in P-lake of the percentage of reduction of the external phosphorus loading in the first year after the measures were implemented. In the Noordeinderplas no significant changes were observed so far.

Sas (1989) indicated that phosphorus concentrations in treated lakes will stabilize after about 3–5 times the residence time. Therefore the experiment in the Nieuwkoop lakes will continue for several years.

Conclusions and recommendations

- Assignment of different lake functions which lead to contradictory measures should be avoided.
- Hydraulic loading of peaty lakes has to be controlled and can be used as a tool for lake restoration by reducing the external phosphorus loading or removal of phosphorus released by the sediments.
- More accurate nutrient- and water balances are still of most importance for evaluation and prediction of effects of lake restoration measures.
- Decrease of phosphorus concentrations can be achieved with and without dredging of nutrient enriched upper sediment layers. The choice be-

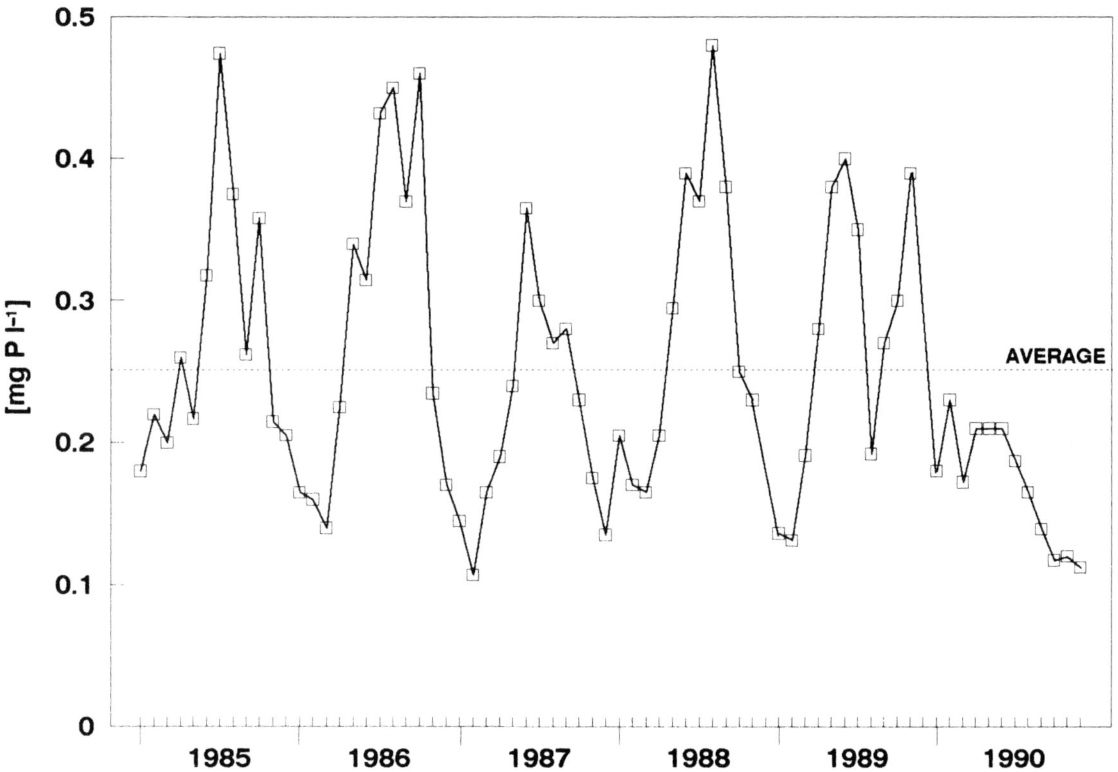

Fig. 7. Concentration of the total phosphorus in lake Nieuwkoop Zuideinderplas in the period 1985–1990. External phosphorus loading was reduced by *ca.* 90% in 1990.

tween dredging or not depends on specific hydrologic circumstances (e.g. depth and residence time) and sediment characteristics (e.g. depth profile and net internal loading).

- Nitrogen reduction without phosphorus reduction is not recommended for the investigated shallow lakes.
- Mobile sediments must be controlled if dredging is applied for reduction of internal nutrient loading.
- Horizontal sediment transport can be controlled by screens. This new technique should be used to minimize the area of sediment removal and the costs of restoration of shallow lakes.

References

Björk, S., 1988. Redevelopment of lake ecosystems. A case-study approach. Ambio 17: 90–98.

Boers, P. C. M., J. W. T. Bongers, A. G. Wisselo & Th. E. Cappenberg, 1984. Loosdrecht Lakes restoration project: Sediment phosphorus distribution and release from the sediments. Verh. int. Ver. Limnol. 22: 842–847.

Boers, P. C. M. & O. Van Hese, 1988. Phosphorus release from the peaty sediments of the Loosdrecht Lakes (The Netherlands). Wat. Res. 22: 355–363.

Boers, P. C. M. (1991). The release of dissolved phosphorus from lake sediments. Ph.D. Thesis. Agricultural University of Wageningen. 139 pp.

Boers, P. C. M., J. Van der Does, M. P. Quaak, J. C. Van der Vlugt & P. A. Walker, 1992. Fixation of phosphorus in lake sediments using iron (III) chloride: experiences, expectations. Hydrobiologia 233: 211–212.

Bruning, C. & S. P. Klapwijk, 1984. Application of derative spectroscopy in bioassays estimating algal available phosphates in lake sediments. Verh. int. Ver. Limnol. 22: 172–178.

Burger-Wiersma, T., 1991. Orienterend onderzoek naar stikstof fixatie in Nederlandse meren. Inventarisatie van stikstof fixatie in Nederlandse meren gedurende het groeiseizoen van 1990. Report Nr. 1991-1. Laboratorium voor Microbiologie, Universiteit van Amsterdam (in Dutch).

Cooke, G. D., E. B. Welch, S. A. Peterson & P. R. Newroth, 1986. Lake and reservoir restoration. Buttersworth Publishers, Boston.

Cullen, P. & C. Forsberg, 1988. Experiences with reducing

point sources of phosphorus to lakes. Hydrobiologia 170: 321–336.

Gulati, R. D., E. H. R. R. Lammens, M.-L. Meyer & E. Van Donk (eds), 1990. Biomanipulation – Tool for Water Management. Developments in Hydrobiology 61. Kluwer, Dordrecht. Reprinted from Hydrobiologia 200/201.

Klapwijk, S. P. & C. Bruning, 1986. Available phosphorus in the sediments of eight lakes in the Netherlands. In: P. G. Sly, (ed.), Sediment and Water Interactions, pp. 391–398. Springer Verlag, New York.

Lijklema, L., J. H. Janse, R. M. M. Roijackers & M.-L. Meijer, 1988. Eutrophication in the Netherlands. H_2O 21: 462–467 (in Dutch with an English summary).

Lijklema, L., 1991. Response of lakes to the reduction of phosphorus load. Hydrobiol. Bull. 24: 165–170.

Ministry of Agriculture, Nature Management and Fisheries, 1990. Nationaal Natuurbeleidsplan (In Dutch). Staatsuitgeverij, 's-Gravenhage.

Ministry for Transport and Public Works, 1989. Derde nota op de waterhuishouding. (In Dutch with an English summary). Staatsuitgeverij, 's-Gravenhage.

Prepas, E. E., T. P. Murphy, J. M. Crosby, D. T. Walty, J. T. Lim, J. Babin & P. A. Chambers, 1990. Reduction of Phosphorus and chlorophyl-*a* concentrations following $CaCO_3$ and $Ca(OH)_2$ additions to hypertrophic Figure Eight Lake, Alberta. Envir. Sci. Technol. 24: 1252–1258.

Province of North Holland, 1990. Waterhuishoudingsplan. Natuurlijk, Water. Ecologische aspecten en normdoelstellingen (in Dutch). Haarlem.

Province of South Holland, 1991. Waterhuishoudingsplan. Leven door Water. (in Dutch). 's-Gravenhage.

Sas, H., 1989. Lake restoration by reduction of nutrient loading: Expectations. Experiences. Extrapolations. St. Augustin 1: Akademia Verl., Richarz. ISBN 3-88345379-X.

Schönfeld, J. C. & C. Kranenburg, 1981. Density currents and internal waves. Department Civil Engineering. University of Technology Delft.

Sünderman, J., 1979. Numerical modelling of circulations in lakes. Hydrodynamics of lakes. In: W. H. Graf & C. H. Mortimer (eds), Developments in Water Science, Vol. 11, Elsevier Sci. Publ. Comp.

Van der Does, J. & S. P. Klapwijk, 1985. Phosphorus removal and effects on water quality in Rijnland. H_2O 18: 381–387 (in Dutch with an English summary).

Van der Does, J. & S. P. Klapwijk, 1987. Effects of phosphorus removal on the maximal algal growth in bio-assay experiments with water from four Dutch lakes. Int. Revue ges. Hydrobiol. 72: 27–39.

Van der Does, J. & P. De Jong, 1992. Integrale eutrofiëringsbestrijding in the Nieuwkoopse Plassen verloopt succesvol (in Dutch with and English summary) H_2O 25(1): 2–10.

Van der Vlugt, J. C., P. A. Walker, J. Van der Does & A. J. P. Raat, 1992. Fisheries management as an additional lake restoration measure: biomanipulation scaling-up problems. Hydrobiologia 233: 213–224.

Van Liere, L., R. D. Gulati, F. G. Wortelboer & E. H. R. R. Lammers, 1990. Phosphorus dynamics following restoration measures in the Loosdrecht Lakes (The Netherlands). Hydrobiologia 191: 85–87.

Van Rijn, L. C., 1989. Handbook Sediment Transport by Currents and Waves. Delft Hydraulics, Delft.

Verstraelen, P. J. T., J. Wisserhof, Lj. Rodić & R. Eijsink, 1992. Eutrophication control strategies for three shallow Vecht lakes in the province of North Holland. Hydrobiologia 233: 235–245.

Yousef, A. Y., W. M. Mc Lellon & H. H. Zebutch, 1980. Changes in phosphorus concentrations due to mixing by motorboats in shallow lakes. Wat. Res. 22: 841–852.

Hydrobiologia **233**: 211–212, 1992.
L. Van Liere & R.D. Gulati (eds), Restoration and Recovery of Shallow Eutrophic Lake Ecosystems in The Netherlands.

Fixation of phosphorus in lake sediments using iron(III)chloride: experiences, expectations

Paul Boers,[1] Joop Van der Does,[2] Marinus Quaak,[3] Jo Van der Vlugt[4] & Paddy Walker[5]
[1]*Institute for Inland Water Management and Wastewater Treatment, P.O. Box 17, 8200 AA Lelystad, The Netherlands;* [2]*Waterboard of Rijnland, P.O. Box 156, 2300 AD Leiden, The Netherlands;* [3]*Volker Stevin Dredging, P.O. Box 2695, 3000 CR Rotterdam, The Netherlands;* [4]*National Institute of Public Health and Environmental Protection, P.O. Box 1, 3720 BA Bilthoven, The Netherlands;* [5]*Organization for Improvement of the Inland Fisheries, P.O. Box 433, 3430 AK Nieuwegein, The Netherlands*

Key words: eutrophication, lake restoration, phosphorus, sediments, internal loading, chemomanipulation

Abstract

After a reduction of the external phosphorus loading to a lake, an internal loading from the sediments may delay the improvement of the water quality. The accepted method to combat internal loading is careful dredging of the upper sediment layers (Cooke *et al.*, 1986), but this method is costly and time consuming. Addition of phosphorus binding agents to the sediments might offer an alternative. In the Netherlands the use of aluminum compounds, the most common phosphorus binding agent, for water quality improvement purposes is not favoured. Therefore a sediment treatment with a solution of iron(III)chloride was tested. Iron was chosen because it is considered to be a natural binder of phosphate. 100 g m^{-2} of Fe^{3+} were added to the sediments of the shallow (1.75 m average depth) and eutrophic Lake Groot Vogelenzang (The Netherlands) in October and November 1989. The iron(III)chloride solution was diluted 100 times with lake water and mixed with the surface sediments with a water-jet.

Following the addition the concentrations of total phosphorus (Fig. 1), chlorophyll-*a* and suspended solids decreased. This improvement of the water quality lasted for three months. After this time the total phosphorus concentration increased again, but remained at a lower level than in spring and summer of 1989. The phosphorus release rate from the sediments as measured from intact sediment cores decreased from 4 to 1.2 mg P m^{-2} d^{-1} ($n = 5$), and the bioavailability of the sediment phosphorus, as measured with bioassays, decreased from 34 to 23% ($n = 5$) shortly after the treatment. One year after the treatment the release rate was increased to 3 mg P m^{-2} d^{-1} ($n = 5$). Before treatment, the lake was thought to have a residence time of over one year. However, the chloride added to the lake disappeared according to a dilution rate of 0.03 d^{-1} or a retention time of about 35 days. A high external loading due to rapid flushing with phosphorus-rich water from surrounding lakes possibly prevented a more durable improvement in water quality. Another possibility is that the iron addition has lost its phosphate binding capacity due to reduction or binding with other anions like carbonate or sulphide. Therefore the suitability of the method to reduce internal loading and especially the long term availability of added iron to bind phosphorus needs additional proof.

The treatment of the 18 ha area of Lake Groot Vogelenzang took three weeks. The operational costs were about US$ 125000. This is fast and cheap compared to dredging. Application of the technique is limited to those cases where the sediments are not polluted with micro-pollutants and the water depth need not be increased.

We could not detect an influence on the macrophytes, because they remained absent due to the poor light climate.

Phosphorus inactivation should be applied only after the external phosphorus loading is reduced. It can be used as stand-alone technique or in combination with other lake restoration techniques such as dredging or fish stock manipulation.

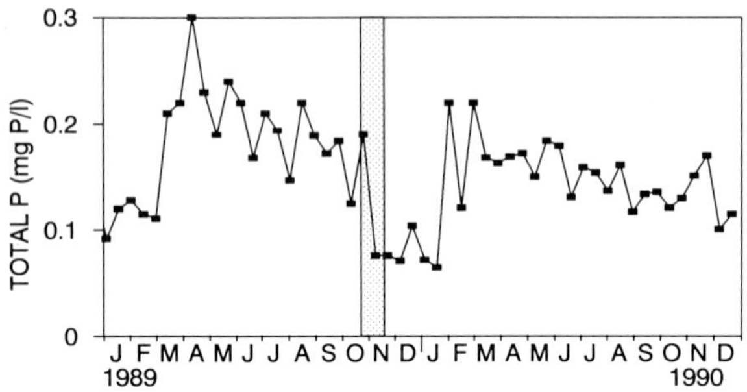

Fig. 1. Concentrations in mg l^{-1} of total phosphorus in Lake Groot Vogelenzang during 1989 and 1990. The shaded column represents the treatment period.

References

Cooke, G. D., E. B. Welch, S. A. Peterson & P. R. Newroth, 1986. Lake and reservoir restoration. Butterworths publishers, Boston. 392 pp.

Hydrobiologia **233**: 213–224, 1992.
L. Van Liere & R.D. Gulati (eds), Restoration and Recovery of Shallow Eutrophic Lake Ecosystems in The Netherlands.

Fisheries management as an additional lake restoration measure: biomanipulation scaling-up problems

J. C. Van der Vlugt,[1] P. A. Walker,[2] J. Van der Does[3] & A. J. P. Raat[2]
[1] *National Institute of Public Health and Environmental Protection, P.O. Box 1, 3720 BA Bilthoven, The Netherlands;* [2] *Organization for the Improvement of Inland Fisheries, P.O. Box 433, 3430 AK Nieuwegein, The Netherlands;* [3] *Waterboard of Rijnland, P.O. Box 156, 2300 AD Leiden, The Netherlands*

Key words: lake-restoration, turbidity, biomanipulation, zooplankton, fish removal, internal loading, eutrophic lakes, bream

Abstract

Restoration of the highly eutrophic Reeuwijk lakes (*ca.* 700 ha) started in 1986 by reducing the external phosphorus loading. As an additional measure to improve the quality of the lake water, the structure of the fish population in Lake Klein Vogelenzang (18 ha) was altered in 1989 by the removal of *ca.* 100 kg ha^{-1} bream from the lake in April and December. This constituted about 50% of the total bream biomass in the lake.

The fish-stock reduction in April, 1989, was initially followed by high phosphorus concentrations, probably the result of considerable phosphorus release from the sediments. The resulting heavy algal blooms that occurred reduced the transparency to very low values. During the summer the zooplankton population increased markedly in numbers coinciding with reductions in total suspended matter including (blue-green) algae. A great improvement in Secchi-disc transparency was observed and by the end of December, 1989, the bottom of the lake (1.5–2.0 m) was visible. After heavy storms in January and February 1990, transparency dropped to < 1 m as a result of resuspension of high concentrations of suspended matter from the bottom sediments. Although transparency over the rest of 1990 was higher than in 1988, i.e. the year preceding the removal of fish (biomanipulation), it was lower than expected, based on the results of 1989. The study shows that technical and biological factors can cause serious management problems for the implementation of biomanipulation in larger water bodies.

Introduction

The eutrophication of the Reeuwijk lakes (The Netherlands) is a typical example of the problems brought about by increased phosphorus and nitrogen emissions in the lower Rhine catchment area. After decades of increasing emissions of these nutrients, remedial actions were taken and the effects of reduced phosphorus loading studied extensively, starting in 1973 with pilot studies (Van der Vlugt, 1976; Bannink & Van der Vlugt, 1978a, b; Bannink *et al.*, 1980; Van der Vlugt & Aldenberg, 1982).

The restoration of the Reeuwijk lakes, which started in April 1986, was planned in two phases. The first phase involved the reduction of the external P-loading. This was achieved by the removal of phosphorus from waste water discharges at the sewage treatment plant of the village of Reeuwijk. These reduction measures did not lead to significant improvements in water quality (Van der Vlugt & Klapwijk, 1988, 1990). In the second phase (1989–1992) the reduction of the phosphorus load is supplemented by in-lake measures, both biomanipulation and chemomanipulation. The latter involves in-lake addition of iron compounds and is discussed recently in detail by Boers *et al.* (1992).

The biomanipulation measure, as carried out in this study, involves the removal of large numbers of planktivorous and benthivorous fish. This measure has been used as a tool in the management of water quality in eutrophic waters and has been practised in several countries throughout the world with varying degrees of success (Shapiro *et al.*, 1975; Andersson, 1988; Benndorf, 1990; Van Donk & Gulati, 1989; Gulati *et al.* 1990). In The Netherlands the history of biomanipulation dates back to 1985 to the feasibility study of Richter (1985, 1986). This study was followed by various experiments in ponds as well as whole lake experiments (Meijer *et al.*, 1989; Van Donk *et al.*, 1990a, b). In small (0.1 ha) ponds it was shown that 0^+ cyprinids could have an effect on the composition of the zooplankton and the transparency of the water (Meijer *et al.*, 1990b; Raat, 1989). In somewhat larger ponds (0.4 ha) it was demonstrated that 0^+ predatory fish (northern pike, European catfish and pike-perch) realised a considerable production on a diet composed of zooplankton and young cyprinids (Raat, 1990). This study showed that these 0^+ predators have a potential as a management tool in the regulation of abundant 0^+ cyprinids, but with the present knowledge of their dynamics the predators could not be used as the only tool in biomanipulation.

Large scale removal of planktivorous and benthivorous fish has resulted in a considerable improvement of the water transparency in small lakes (Meijer *et al.*, 1989; Van Donk *et al.*, 1989b, 1990a). Experiments in deep lakes in Sweden and North America have shown that changes in the fish population in larger water bodies have been successful, at least for a period of up to four years (Shapiro, 1990; Gophen, 1990; Walker, 1989). Experiments on a larger scale in shallow Dutch lakes, however, have so far not been successful, for example in lake Breukeleveen (180 ha) where the water has remained turbid, despite the large scale removal of bream (Van Donk *et al.*, 1990b). The reasons for the failure of the biomanipulation measure are complex and depend on the interaction of abiotic and biotic factors as well as management problems.

Table 1. Water bodies in The Netherlands in which biomanipulation was implemented.

Pond/Lake	Area (ha)	Authors
Drainable ponds;	0.1	(Raat, 1989; Meijer *et al.*, 1990b)
Drainable ponds;	0.4	(Raat, 1990; Meijer *et al.*, 1990b)
Zwemlust;	1.5	(Van Donk *et al.*, 1989b, 1990a)
Bleiswijkse Zoom;	3.1	(Meijer *et al.*, 1989, 1990a)
Noorddiep;	4.5	(Meijer *et al.*, 1990a)
Klein Vogelenzang;	18	(Van der Vlugt *et al.*, 1990)
Breukeleveen;	180	(Van Donk *et al.*, 1990b)
Wolderwijd;	2700	(Meijer, 1989; Hosper & Jagtman, 1990)

Because of the promising results in drainable ponds and small lakes, biomanipulation was carried out in one of the smaller lakes in the Reeuwijk Lakes area. This study in Lake Klein Vogelenzang forms part of a series in which biomanipulation was implemented in increasingly larger lakes in The Netherlands (Table 1).

In the winter 1990–1991 the restoration of Lake Wolderwijd (2 700 ha) has been started as planned (Hosper & Jagtman, 1990). The first results in spring 1991 are promising. Transparency has increased from 40 cm to 100 cm in the course of the winter, and to 1.7 m at the start of summer.

This paper deals with a study in Lake Klein Vogelenzang, one of the Reeuwijk lakes of 18 ha, in which the fish-stock was manipulated. This study shows that technical and biological factors can cause serious management problems for the implementation of biomanipulation in larger water bodies.

Description of sites studied

The Reeuwijk lakes

The Reeuwijk lakes area consists of nine shallow lakes situated at 52° 2′ N and 4° 45′ E in the

Reeuwijk polder near the city of Gouda in the western part of The Netherlands (Van der Vlugt & Klapwijk, 1990). The lakes originated from peat mining carried out in the 16th and 17th centuries; they have a total surface area of 7 km^2 and an average depth of about 2 m. They are all interconnected except for Lake Broekvelden/Vettenbroek, which is completely isolated from the other lakes.

The shallow and peaty Reeuwijk lakes have become highly eutrophic as a result of decades of discharge of household waste water and the inlet of nutrient-rich water from the adjacent Breevaart canal. The water in the canal originally comes from the river Rhine, via the Hollandse IJssel and the canals of the city of Gouda. Moreover, the sewage treatment plant of the village of Reeuwijk (with 6800 inhabitants) discharges its nutrient-rich effluent into the Breevaart canal, through which the level of the lakes is regulated and kept constant at 2.2 m below mean sea level. Excess water in the polder is pumped out near Gouda and Bodegraven. Since April 1986, the effluent from the sewage treatment plant has been dephosphorized and consequently the average yearly total phosphorus concentration in the Breevaart canal has decreased from about 0.8 mg l^{-1} to about 0.4 mg l^{-1} (Van der Vlugt & Klapwijk, 1988, 1990).

Lake Klein Vogelenzang

Lake Klein Vogelenzang is a highly eutrophic small lake in the centre of the Reeuwijk lakes. It is interconnected with the lakes Groot Vogelenzang and Ravensberg. Its morphological and hydrological characteristics, as well as the phosphorus loading, are summarized in Table 2.

Table 2. Hydrological information on Lake Klein Vogelenzang.

Water surface area:	0.2 km^2
Mean water depth:	1.5 m
Maximum water depth:	2.0 m
Residence time:	2.0 years
Gross P-load:	0.5 gP $\cdot m^{-2} \cdot y^{-1}$
General sediment composition:	peat

Materials and methods

Sampling and water analyses

The lake water was sampled fortnightly for three years (1988–1990) at three depths (0–0.5 m, 0.5–1.0 m, 1.0–1.5 m) with a 0.5 m plexiglass tube sampler of 3 l. These samples were mixed to make a composite sample which was used for chemical and biological analyses. Samples were transported to the laboratory in dark polyethylene jerrycans, within three hours of sampling. Nitrate + nitrite nitrogen ($NO_2 + NO_3$-N), ammonium nitrogen (NH_4-N) and orthophosphate (PO_4-P) were measured with a Technicon AA II autoanalyser. Kjeldahl-nitrogen (Kj-N) and total phosphorus (total-P) were measured after digestion with, respectively, sulphuric acid + Wieninger selenium mixture and sulphuric acid + persulphate followed by analysis for NH_4-N and PO_4-P as stated above. Total nitrogen was calculated as the sum of Kj-N and $NO_2 + NO_3$-N. Particulate phosphorus (part-P) was calculated as the difference between total-P before and after filtration. Particulate nitrogen (part-N) was calculated by substraction of NH_4-N from Kj-N.

Chlorophyll *a* (Chl *a*) was determined spectrophotometrically, after extraction of glass-fibre filtered algae with 80% ethanol (75 °C), according to a standard method of the Netherlands Normalisation Institute (NEN 6250). C-phycocyanine was determined spectrophotometrically at 615 nm after extracting samples in distilled water by freezing and thawing it ten times. Water samples used for the determination of C-phycocyanine were concentrated by continuous flow centrifugation at 15000 rpm (Van Liere, 1979). Samples for seston weight and particulate organic matter were collected on 1.2 μm membrane filters; these parameters were determined gravimetrically as dry weight (DW) after drying at 75 °C and ash free dry weight (AFDW), as loss of weight, after ignition at 600 °C. Transparency was measured with a Secchi-disc. For the estimation of the zooplankton fresh weight (FW) a modified Utermöhl microscopical method (Van Heusden, 1972) was used.

Fisheries management in Lake Klein Vogelenzang

Three aspects of the fisheries management programme in Lake Klein Vogelenzang are: (1) the estimation of the fish-stock prior to the biomanipulation measure; (2) the actual management measure, i.e. the implementation of fish removal; (3) the evaluation of the development of the fish-stock after biomanipulation. These aspects will be discussed below.

Fish-stock estimation

In 1976 and 1980 the fish-stock was assessed by the Operational Group from the Ministry for Agriculture and Fisheries in the lakes Elfhoeven and 's Gravenbroek. In these years the fish-stock was composed primarily of pike-perch (*Stizostedion lucioperca*) and bream (*Abramis brama*) and the growth was poor. Information from professional fishermen bore this out and the fish-stock was described as one-sided in favour of planktivorous species (Van Gemeren, 1987).

Little is known about the fish-stock in Klein Vogelenzang before January 1989, but it is assumed to be similar to stocks found in the lakes Elfhoeven and 's Gravenbroek above. The Cooperation of Professional Fishermen from the Reeuwijk lakes has stocked elvers (*Anguilla anguilla*) yearly in Klein Vogelenzang for the past 15 years. Several hundred kilogrammes of bream were removed from the lake each December in 1985, 1986 and 1987. The professional fishery is based on eel (*Anguilla anguilla*) and pike-perch. The first part of the fisheries management programme, the fish-stock assessment, was carried out in January 1989. Fifteen different species of fish were caught, and the size of the fish population was estimated using the mark-and-recapture technique (Ricker, 1975). Estimations of the numbers and biomass of the most dominant fish species are shown in Table 3. The biomass of the total fish-stock was *ca.* 260 kg ha^{-1}.

The pike-perch population was composed of two relatively strong year-classes, from 1985 and 1987. The young pike-perch from 1988 were small, < 10 cm, suggesting that having been unable to switch to piscivory, they had remained planktivorous. Few pike-perch individuals > 50 cm were caught, which indicates an intensive and efficient fishery. The pike (*Esox lucius*) population in Klein Vogelenzang was small and limited to the inshore area. Most of the pike were caught in the vegetation on the west side of the lake.

The most dominant cyprinid was the bream which, with a biomass of 230 kg ha^{-1}, comprised nearly 90% of the biomass of the entire fish-stock (Fig. 1). A major part of the biomass (*ca.*

Table 3. Estimations of the fish-stock composition and size in Lake Klein Vogelenzang, January, 1989 (before biomanipulation). (YOY = young of the year)

Fish size		Numbers (SD)	Biomass (kg)
Pike-perch:			
Stizostedion	YOY	1547 (909)	6
lucioperca	> 14 cm	217 (104)	196
Pike:			
Esox lucius	< 30 cm	53 (25)	4
	> 30 cm	105 (69)	91
Perch:			
Perca fluviatilus	YOY	few	
	> 10 cm	592 (339)	57
Bream:			
Abramis brama	YOY	few	
	10–26 cm	3726 (668)	603
	> 26 cm	4646 (358)	3416
White bream:			
Blicca bjoerkna	YOY	few	
	7–26 cm	81 –	5
Roach:			
Rutilus rutilus	YOY	few	
	8–24 cm	5430 (1467)	153
Rudd:			
Scardinius	YOY	few	
erythrophthalmus	7–24 cm	1387 (511)	75
		369 –	48
Eel:			
Anguilla anguilla			

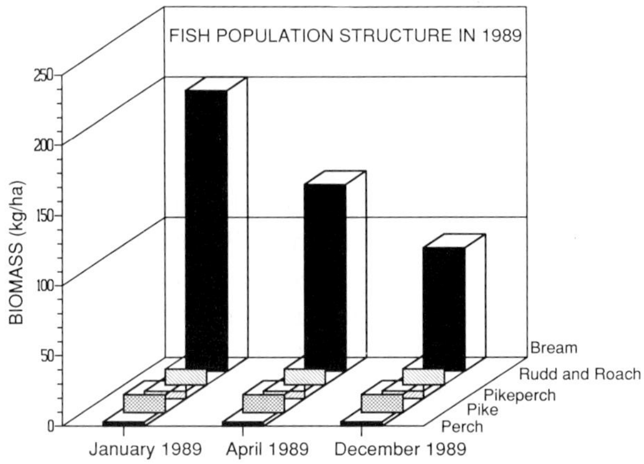

Fig. 1. Fish population structure in Lake Klein Vogelenzang in 1989.

145 kg ha^{-1}) was contributed by bream > 32 cm. The growth of the bream was poor, as was that of the bream caught in the lakes 's-Gravenbroek and Elfhoeven of the Reeuwijk lakes area.

Fisheries management

It was decided to approach the management in three ways. Firstly, to remove a significant portion of the bream population, especially the large bream. This should reduce the amount of sediment material being suspended by foraging activities and relieve predation pressure on the zooplankton. Secondly, to prevent the recruitment of large numbers of young bream from the adjoining lakes, the lake was shut off by two moveable sluice doors. Thirdly, young pike were stocked in the inshore area of the lake, in the hope that the pike population would reproduce naturally as the amount of vegetation increased following biomanipulation. Thus the recruitment of young bream would be regulated by young pike.

The Cooperation of Professional Fishermen removed 1 200 kg of bream from the lake in April 1989, and another 800 kg in December 1989 (De Laak *et al.*, 1989). At the beginning of May, 1989, 1 500 young of the year (YOY) pike were released into the lake. The sluice doors were put in place in June 1989, prior to which the lake had been shut off from the other lakes using nets.

Results

Development of the fish-stock

The lake was sampled for young fish (YOY) during the summer of 1989. The size of the population of young fish, as well as their growth, was followed (Kemper & Raat, 1989). The most dominant species of young fish were pike-perch, perch, bream and roach (*Rutilus rutilus*). The main peak in 0^+ fish density (*ca.* 60 000 ind. ha^{-1}) and biomass (*ca.* 50 kg ha^{-1}) was found at the beginning of June (Fig. 2). By the end of August the numbers had dropped to *ca.* 10 000 ind. ha^{-1} and the biomass to *ca.* 25 kg ha^{-1}. Bream and roach 0^+ fish constituted 15 kg ha^{-1} of this biomass. Pike-perch was numerically the most dominant fish species and the 0^+ fish remained planktivorous, which would suggest a lack of young fish as prey. This is because numbers of 0^+ bream and roach were low and the fish grew fast, making them unavailable as prey fish.

If the relatively low recruitment of the bream and roach was caused by the biomanipulation measure is difficult to assess. The pike-perch born in 1989 will probably not form a strong year-

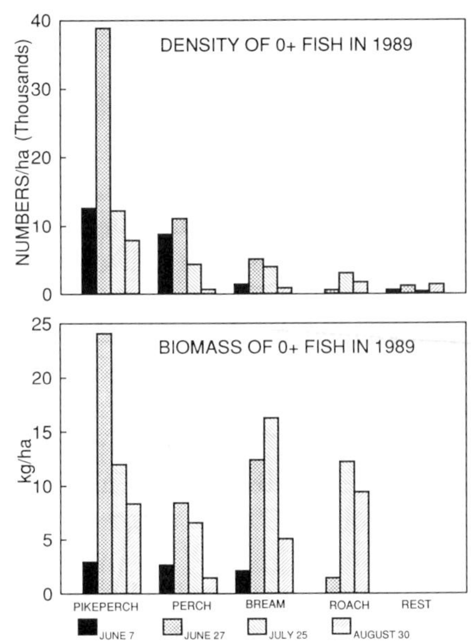

Fig. 2. Density and biomass of 0^+ fish in Lake Klein Vogelenzang in 1989.

February that they remained unstable throughout the summer of 1990. Wind action and foraging by fish would lead to resuspension of the sediments. Although Lake Klein Vogelenzang was shut off from the adjoining lakes during and after the manipulation measure it is probable that fish have migrated into the lake. The sluice doors, separating the lake from the surrounding lakes, were repeatedly and intentionally left open. The lakes are used by sailors, surfers, fishermen and others for recreational purposes and the sluice doors were a hindrance to their freedom of movement. Observations by personnel from the OVB (Organization for the Improvement of Inland Fisheries) whilst fishing suggest that well-fed and under-fed bream can be distinguished in the lake. The underfed fish have been observed in Lake Groot Vogelenzang (De Laak, personal observation), where the availability of benthic organisms is very low following the treatment of the sediments with iron-chloride in order to bind phosphorus (Boers *et al.*, 1992). It is possible that these fish migrated into the adjacent Lake Klein Vogelenzang and are responsible for the resuspension of bottom sediments during their search for food. This illustrates the need for effective measures for isolating the fish-stock, which take into account the needs of other users of the lake(s) in question.

The removal of large numbers of planktivorous and benthivorous fish in the spring has resulted in a number of lakes in an immediate improvement in water clarity (e.g. Bleiswijkse Zoom (Meijer *et al.*, 1989; Noorddiep (Kemper *et al.*, 1990). These were shallow lakes with a clay/sand bottom in which aquatic macrophytes developed within several weeks of fish removal (Meijer *et al.*, 1990a). This colonisation with macrophytes has been defined as one of the stabilising factors in maintaining clear water, either by forming a refuge for zooplankton or by acting as a competitor for nutrients with blue-green algae (Ozimek *et al.*, 1990; Scheffer, 1990). In Lake Klein Vogelenzang the situation is quite different. The peat layer is overlain with a (moveable) layer in fine mud, forming a lake bottom which is easily resuspended and in which it is difficult for macrophytes to take root. In order for macrophytes to grow there has to be sufficient light at the lake bottom or in the inshore area. In Lake Klein Vogelenzang the bottom of the lake was only visible during the winter following the second removal of fish. This is an unsuitable time for plant growth. This lack of aquatic macrophytes could be one of the factors contributing to the slow speed of improvement.

The results in Lake Wolderwijd (2700 ha) two months following fish removal show that it is possible to increase water transparency in large lakes within weeks of a biomanipulation measure. Colonisation by *Potamogeton pectinatus* has occurred and it is hoped that the lake will remain clear.

The specific conditions in peat lakes with respect to bottom substrate and wind action make them 'risky' subjects for biomanipulation measures. The problems described in this paper are not only scaling-up problems, especially considering Lake Wolderwijd, but rather the application of a management measure without taking the particular requirements of the lake into consideration. The extrapolation of results from clay/sand ponds to peat lakes has led to incorrect expectations as to magnitude of results and the time required for acquiring them.

Summarizing, the problems described in this study are two-fold:

- the isolation of interconnected lakes which are used for recreation purposes requires careful management;
- the use of biomanipulation as a management measure in peat lakes requires more detailed research.

Conclusions

The transparency of the water in Lake Klein Vogelenzang did not improve immediately following the removal of *ca.* 50% of the bream population in April and December 1989. A bloom of blue-green algae was responsible for the lack of an immediate improvement in water transparency in 1989. The warm weather conditions in the summer of 1989 caused anoxia at the sediment-water interface and the consequent release of phosphorus from the sediments. This phosphorus was immediately fixed by the blue-green algae. In the late summer of 1989 the water transparency did improve and coincided with a peak in *Bosmina* and *Daphnia* and a decrease in phytoplankton. The heavy storms in January and February, 1990, caused the bottom sediments of the lake to be resuspended. The sediments remained loose and easily resuspendable after the storms. Despite this, an improvement in transparency was observed in 1990. The migration of fish into the lake, as a consequence of others leaving the sluice doors open intentionally, may mean that the bream population has reached its former level.

Acknowledgements

The authors acknowledge the valuable assistence of the fishermen from the Lake Reeuwijk area. Frans Jacques and Gerard de Laak of the Organization for the Improvement of Inland Fisheries were responsible for the fish monitoring in Lake Klein Vogelenzang.

References

Andersson, G., 1988. Restoration of Lake Trummen Sweden. Effects of sediment removal and fish manipulation. In G. Balvay (ed.), Eutrophication and Lake Restoration. Water Quality and Biological Impacts. French-Swedish Symposium, Thonon-les-Bains: 205–214.

Bannink, B. A. & J. C. Van der Vlugt, 1978a. Erfahrungen mit dem Zusatz von Fällmitteln im Grote Rug Speicherbecken und vergleichende Untersuchungen in 3 Modell-Reservoiren nach Lund. DVGW-Schriftenreihe Wasser 16: 216–243.

Bannink, B. A. & J. C. Van der Vlugt, 1978b. Hydrobiological and chemical response to the addition of iron and aluminium salts, studied in three Lund-type butylrubber reservoirs. Verh. int. Ver. Limnol. 20: 1816–1821.

Bannink, B. A., J. H. M. van der Meulen, J. C. H. Peeters & J. C. Van der Vlugt, 1980. Hydrobiological consequences of the addition of phosphate precipitants to inlet water of lakes. Hydrobiol. Bull. 14: 73–89.

Benndorf, J., 1990. Conditions for effective biomanipulation; conclusions derived from whole-lake experiments in Europe. Hydrobiologia 200/201: 187–203.

Boers, P. C. M., J. Van der Does, M. P. Quaak, J. C. Van der Vlugt & P. A. Walker, 1992. Fixation of phosphorus in lake sediments using iron(III)chloride: experiences, expectations. Hydrobiologia 233: 211–212.

De Laak, G. A. J., A. J. P. Raat & J. H. Kemper, 1989. Uitdunningsvisserij in Klein Vogelenzang in april 1989. Nieuwegein, Organisatie ter Verbetering van de Binnenvisserij. OVB-Onderzoeksrapport 1989–16, 10 pp.

Gliwicz, Z. M., 1990. Why do cladocerans fail to control algal blooms? Hydrobiologia 200/201: 83–97.

Gophen, M., 1990. Biomanipulation: retrospective and future development. Hydrobiologia 200/201: 1–11.

Gulati, R. D., E. H. R. R. Lammens, M.-L. Meijer & E. Van Donk (eds), 1990. Biomanipulation tool for water management. Developments in Hydrobiology 61, 628 pp.

Hosper, S. H., 1989. Biomanipulation, a new perspective for restoring shallow lakes in The Netherlands. Hydrobiol. Bull. 23: 5–10.

Hosper, S. H. & E. Jagtman, 1990. Biomanipulation additional to nutrient control for restoration of shallow lakes in The Netherlands. Hydrobiologia 200/201: 523–534.

Kemper, J. H. & A. J. P. Raat, 1989. Broedbemonstering in Klein Vogelenzang zomer 1989. Nieuwegein. Organisatie ter Verbetering van de Binnenvisserij. OVB-Onderzoeksrapport 1989–17, 9 pp.

Meijer, M.-L., (ed.), 1989. Toepassing van Actief Biologisch Beheer in het Wolderwijd-Nuldernauw. Rapportage van projectgroep Actief Biologisch Beheer. DBW/RIZA nota 89-057, Lelystad.

Meijer, M.-L., A. J. P. Raat & R. Doef, 1989. Restoration by biomanipulation of Lake Bleiswijkse Zoom (The Netherlands): first results. Hydrobiol. Bull. 23: 49–57.

Meijer, M.-L., M. W. De Haan, A. W. Breukelaar & H. Buiteveld, 1990a. Is reduction of the benthivorous fish an important cause of high transparency following biomanipulation in shallow lakes? Hydrobiologia 200/201: 303–315.

Meijer, M.-L., E. H. R. R. Lammens, A. J. P. Raat, M. P. Grimm & S. H. Hosper, 1990b. Impact of cyprinids on zooplankton and algae in ten drainable ponds: preliminary results. Hydrobiologia 191: 275–284.

Raat, A. J. P., 1989. Growth and production of 0^+ bream (*Abramis brama*), 0^+ roach (*Rutilis rutilis*) and 0^+ carp (*Cyprinus carpio*) in 10 drainable 0.1 ha ponds. Hydrobiol. Bull. 23: 67–72.

Raat, A. J. P., 1990. Production, consumption and prey availability of northern pike (*Esox lucius*), pikeperch (*Stizostedion lucioperca*) and European catfish (*Silurus glanis*): a bioenergetics approach. Hydrobiologia 200/201: 497–509.

Richter, A. F., 1985. Mogelijkheden van biomanipulation ten behoeve van het waterkwaliteitsbeheer in Nederland. Verslag 1985–8. Limnologisch Instituut Oosterzee. 109 pp.

Richter, A. F., 1986. Biomanipulation and its feasibility for water quality management in shallow eutrophic water bodies in The Netherlands. Hydrobiol. Bull. 20: 165–172.

Ricker, W. E., 1975. Computation and interpretation of biological statistics of fish populations. Bull. Fish. Res. Bd Can. 191, 382 pp.

Shapiro, J., 1990. Biomanipulation: the next phase – making it stable. Hydrobiologia 200/201: 13–27.

Shapiro, J., V. Lamarra & M. Lynch, 1975. Biomanipulation: an ecosystem approach to lake restoration. In: P. L. Brezonik & J. L. Fox (eds) Water Quality Management Through Biological Control, Report No. ENV-07-75-1, University of Florida Gainsville, 85–96.

Scheffer, M., 1990. Multiplicity of stable states in freshwater systems. Hydrobiologia 200/201: 475–486.

Van der Vlugt, J. C., 1976. Comparative limnological research in the 'Grote Rug' and model reservoirs. Hydrobiol. Bull. 10: 136–144.

Van der Vlugt, J. C. & T. Aldenberg, 1982. Blooms of algae and their control by phosphorus precipitation/inactivation. (In Dutch with English summary). H_2O, 15: 73–79.

Van der Vlugt, J. C. & S. P. Klapwijk, 1988. Dose-effect relationships between phosphorus concentration and phytoplankton biomass in the Reeuwijk Lakes (The Netherlands). Verh. int. Ver. Limnol. 23: 482–488.

Van der Vlugt, J. C. & S. P. Klapwijk, 1990. the Reeuwijk Lakes: a five years water quality study in an eutrophic ecosystem. Hydrobiol. Bull. 24: 57–67.

Van der Vlugt, J. C., S. P. Klapwijk & A. J. P. Raat, 1990. First results of biomanipulation in the Reeuwijk Lakes. (In Dutch with English summary). H_2O, 23: 118–121.

Van Donk, E. & R. D. Gulati (eds), 1989a. Biomanipulation in the Netherlands: application in fresh-water ecosystems and estuarine waters. Hydrobiol. Bull. 23, 99 pp.

Van Donk, E., R. D. Gulati & M. P. Grimm, 1989b. Food web manipulation in Lake Zwemlust: positive and negative effects during the first two years. Hydrobiol. Bull. 23: 19–34.

Van Donk, E., M. P. Grimm, R. D. Gulati & J. P. G. Klein Breteler, 1990a. Whole-lake food-web manipulation as a means to study community interactions in a small ecosystem. Hydrobiologia 200/201: 275–289.

Van Donk, E., M. P. Grimm, R. D. Gulati, P. G. M. Heuts, W. A. De Kloet & L. Van Liere, 1990b. First attempt to apply whole-lake food-web manipulation on a large scale in The Netherlands. Hydrobiologia 200/201: 291–31.

Van Gemeren, J., 1987. Mens en natuur in het Reeuwijkse Plassengebied: het leven van de mens met veen en water, vogels en vissen in het heden en verleden. Reeuwijk, John van Gemeren, 48 pp.

Van Heusden, G. P. H., 1972. Estimation of biomass of plankton. Hydrobiologia, 39: 165–208.

Van Liere, L., 1979. On *Oscillatoria agardhii* Gomont, experimental ecology and physiology of a nuisance bloom-forming cyanobacterium. Thesis University of Amsterdam.

Walker, P. A., 1989. The feasibility of lake restoration through biomanipulation, a literature study. DBW/RIZA nota nor. 89-033 M & W Aquasense & DBW/RIZA, Lelystad. 60 pp.

Hydrobiologia **233**: 225–233, 1992.
L. Van Liere & R.D. Gulati (eds), Restoration and Recovery of Shallow Eutrophic Lake Ecosystems in The Netherlands.

The effects of dredging and fish stocking on the trophic status of shallow, peaty ditches

I. R. M. Hovenkamp-Obbema & W. Fieggen
Waterboard of Uitwaterende Sluizen in Kennemerland and West-Friesland, P.O. Box 15, 1135 ZH Edam, The Netherlands

Key words: ditches, fish stock, dredging, biomanipulation, eutrophication, phosphorus

Abstract

In the polder 'Wormer, Jisp and Nek', which forms part of a grossly eutrophicated region in the province North Holland, the effects of dredging and biomanipulation were studied in ditches. Four areas, including 5.7 ha surface water in total, were dammed up and dredged. It was feasible to remove the loose top-layer of sediments with a high content of phosphorus by modifications of the dredging machine. In two areas the fish stock was managed.

The measures resulted in an improvement of the water quality: an increase of the transparency and a decrease of suspended solids, phosphorus, nitrogen and algae. The chlorophyll-*a* concentration decreased nearly to 100 $\mu g\, l^{-1}$ (summer average). The relative densities of diatoms, green algae and cyanobacteria did not change. The zooplankton community was dominated by small rotifers and crustaceans both before and after restoration measures. Cladocerans were scarcely present. In the second year after the measures, submerged macrophytes developed.

Introduction

The polder 'Wormer, Jisp and Nek' forms a part of the peat-area of Holland. It derives its name from the names of three villages within the polder. It contains three shallow lakes. The meadows are separated by a labyrinth of ditches. In the polder the objectives for water quality are not met. Due to algal bloom the water is coloured green. A thick layer of sludge covers the sediment. The origin of this sludge is unclear. Up to fifty years ago the water was brackish. The intrusion of fresh water since 1935 probably has intensified the mineralization of the peat-bottom of the ditches. A further factor could be the disappearance of submerged macrophytes. This situation is similar to that found in the Norfolk Broads, a group of English shallow lakes where the submerged plant communities have disappeared completely since the late sixties (Balls *et al.*, 1989). It is unlikely that nutrient addition alone is the cause of a switch from dominance by submerged plant communities to dominance by phytoplankton.

The waterboard of Uitwaterende Sluizen in Kennemerland and West-Friesland has the task to improve the water quality. The main strategies to control eutrophication in fresh waters have been the reduction of external and internal load of nutrients (e.g. Björk, 1985). Its effects can be enhanced by the control of the internal ecological processes by fish stock management (McQueen *et al.*, 1986; Benndorf, 1988) i.e. the reduction of the biomass of both planktivorous and benthivorous fish. Particularly, bream has a negative ef-

Table 2. Estimated numbers, size range and biomass of the fish stock in the polder based on seine and electro fishing autumn 1989 (Raat & Laak 1989).

Species	Number	Size range (cm)		Total weight (kg)
		Min.	Max.	
Bitterling (*Rhodeus sericeus amarus*)	24	3.2	4.8	
Bullhead (*Cottus gobio*)	34	3.8	3.8	
Bream (*Abramis brama*)/ White bream (*Blicca bjoerkna*) O$^+$	1214	3.3	8.2	3.1
Bream (*Abramis brama*)>0$^+$	2233	9.0	46.0	652.0
Carp (*Cyprinus carpio*)	29	7.9	57.0	15.9
Eel (*Anguilla anguilla*)	58	15.0	37.0	2.6
Gudgeon (*Gobio gobio*)	1		10.2	
Perch (*Perca fluviatilis*) O$^+$	1116	6.1	9.0	6.3
Perch (*Perca fluviatilis*)>O$^+$	154	10.0	27.0	8.1
Pike-perch (*Stizostedion lucioperca*) O$^+$	785	5.6	20.0	4.4
Pike-perch (*Stizostedion lucioperca*) >O$^+$	135	21.0	85.0	74.8
Roach (*Rutilis rutilis*) O$^+$	866	4.6	7.1	1.0
Roach (*Rutilis rutilis*) >O$^+$	104	9.0	19.0	4.5
Rudd (*Scardinius erythrophthalmus*) O$^+$	64	3.2	5.8	0.1
Rudd (*Scardinius erythrophtalmus*) >O$^+$	221	7.0	19.0	7.4
Ruff (*Gymnocecephalus cernua*)	86	8.0	14.0	1.0
Smelt (*Osmerus eperlanus*)	42	7.7	10.2	
Stickleback (*Gasterosteus aculeatus*)	408	3.4	5.5	
'Vetje' (*Leuciscus delineatus*)	2	2.5	3.9	
White bream (*Blicca bjoerkna*) >O$^+$	284	7.0	28.0	20.1
Total	7860			801.3

Table 3. Fish set out in the four experimental sites; March 1990 (Walker *et al.*, 1990)
Experimental sites I and II without biomanipulation (= the fish composition in the polder); experimental sites with biomanipulation (= the fish composition in clear water with aquatic plants in peaty areas); (See Table 2 for Latin fish names).

Site	Fish species	Ratio biomasses	Ratio numbers
I	Pike-perch / Bream + White bream + Roach	1/8	1/16
II	Pike-perch / Bream + White bream + roach	1/9	1/18
III	Pike / Roach + White bream + Rudd	1/12	1/78
IV	Pike / Roach + White bream + Rudd	1/12	1/70

gen, and chlorophyll-*a*. The transparency of the water was also determined. The samples were analysed according to the Dutch Standard Methods (NEN), which comply with the standards of the International Organization for Standardization (ISO).

The plankton was sampled in May, July, August and September. For the enumeration of the phytoplankton one litre of surface water was conserved in 4% formalin, sedimented and counted. For the zooplankton counting 10 litre sample was used. The zooplankton was concentrated after filtering the water over a plankton net of 50 μm mesh size. The concentrate was preserved in 5% formalin. The phytoplankton and zooplankton were counted using an inverted microscope. Macrofauna was sampled in May and August with a net, which was pulled alongside the littoral zone of the ditch over a distance of 5 m. The concentrate in the net corresponded to the sampling of about 250 litre of water. This concentrate was preserved in 5% formalin and *Neomysis integer* was counted.

Results

Sediment phosphorus

A sludge layer of *ca.* 35 cm thick was removed out of the experimental sites. Using a modified dredging machine the resuspension of sludge was prevented and the loose top layer could be removed almost completely. The phosphorus content decreased with depth (Fig. 3). As a result of dredging the phosphate content of the top layer was lowered. The phosphate content of the deeper layers differed from that before dredging due to small local differences in phosphorus-content of the bottom.

Waterquality

The waterquality data are summarized in Table 4. This table gives average results for the experimental and reference sites. The effects of biomanipulation in the biomanipulated waters (sites III and IV) are compared with the controls (sites I and II) in Table 5 for transparency, suspendid solids and chlorophyll-*a*.

Transparency and suspended solids

The measures taken in the experimental sites improved the transparency of the water. At the outset the average Sechhi depth was 15.6 cm and after the measures 29.4 cm. These figures compare to 15.9 and 18.4 cm for the reference sites, respectively. The quantity of suspended solids decreased considerably, namely from an average of 61.4 mg l^{-1} (dry weight) to 24.3 mg l^{-1}, compared to 62.2 and 65.0 mg l^{-1} for the reference sites. The data in table 5 indicate that the differences due to biomanipulation between sites I/II and III/IV are not significant.

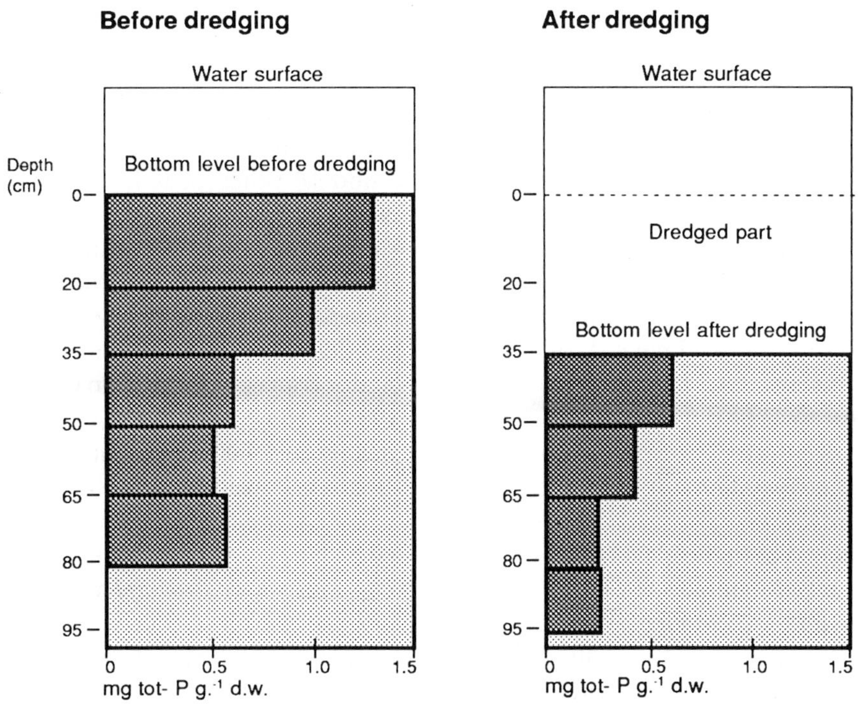

Fig. 3. Phosphorus profiles of the bottom of experimental sites before and after dredging, measured in November 1989 and March 1991.

Table 4. Average water quality data for summer periods 1988–1990 (April–September, for algae densities May–September).

	Secchi-disc depth (cm)	Suspendid solids ($mg\,l^{-1}$)	t-P ($mg\,l^{-1}$)	t-N ($mg\,l^{-1}$)	Chlorophyll-*a* ($\mu g\,l^{-1}$)	Algae density (ind. $ml^{-1} \times 10^3$)
Experimental sites 1988						
mean:	14.9	69.0	0.60	6.0	148	38.3
s.d.:	0.2	7.4	0.07	0.3	20	1.0
Reference sites 1988						
mean:	14.7	70.0	0.67	5.9	147	36.7
s.d.:	0.5	2.7	0.12	0.3	18	1.5
Experimental sites 1989						
mean:	16.3	53.8	0.52	5.6	163	49.3
s.d.:	1.1	2.6	0.12	0.9	27	8.8
References sites 1989						
mean:	17.1	54.3	0.52	5.7	183	48.3
s.d.:	0.6	9.8	0.08	0.6	19	13.4
Experimental sites 1990						
mean:	29.4	24.3	0.3	3.7	106	63.3
s.d.:	4.1	8.7	0.09	0.9	36	20.3
Reference sites 1990						
mean:	18.4	65.0	0.43	5.0	209	90.3
s.d.:	0.7	4.4	0.04	0.5	27	11.2

Phosphorus and nitrogen

In the experimental sites the phosphorus and the nitrogen concentrations decreased. After the measures the summer averages were 0.33 mg total-phosphorus l^{-1} and 3.7 mg total nitrogen l^{-1}. The external phosphorus load also had decreased in 1990, compared with the preceding years due to diminished use of phosphorus in detergents. This is indicated by the summer means of the reference sites of 0.67 $mg\,l^{-1}$ of total-phosphorus in 1988 and 0.43 $mg\,l^{-1}$ in 1990.

Algal biomass and composition

The measures reduced both the algal biomass and densities. In the experimental sites in 1990 the average chlorophyll-*a* was 106 $\mu g\,l^{-1}$ compared to 209 $\mu g\,l^{-1}$ in the reference sites. The data in Table 5 indicate that the differences due to biomanipulation between sites I/II and III/IV are not significant. Algae density in the experimental sites in 1990 the average chlorophyll-*a* was $63.3 \times 10^3\,l^{-1}$ compared to $90.3 \times 10^3\,l^{-1}$ in the reference sites. In the summer the phytoplankton was dominated by green algae (Fig. 4). There was no permanent dominance of cyanobacteria, as found in other peat lakes in Holland, *e.g.* the Nieuwkoopse and Reewijkse plassen (Hovenkamp *et al.*, 1987). For this experiment important species are the filamentous forms like *Lyngbya limnetica*, *Oscillatoria agardhii* and *Oscillatoria redekei*. These species are not easily grazed by cladocerans (Dawidowicz, 1990 & Gliwicz, 1990). They were the dominant cyanobacterial species in 1988, 1989, but also in 1990, although their levels were lower.

Zooplankton and macrofauna

Cladocerans, the most important grazers of algae in many biomanipulated Dutch lakes (Gulati, 1989; 1990), were almost absent. The highest density measured in spring was 24 ind. l^{-1}.

Table 5. Average transparency, suspendid solids and chlorophyll-*a* for sites I/II and III/IV (without resp. with biomanipulation)

	I/II	III/IV	Difference
Secchi-disc depth (cm)			
April	25.0	37.5	12.50
May	25.0	35.0	10.00
June	35.0	32.5	– 2.50
July	25.0	27.5	2.50
August	20.0	25.0	5.00
Sept.	32.5	32.5	0.00
		Mean:	4.58
		St. dev.:	5.79
Suspendid solids (mg l^{-1})			
April	39.0	34.0	– 5.00
May	31.5	18.0	– 13.50
June	23.0	18.0	– 5.00
July	23.00	13.0	– 10.00
August	40.0	23.0	– 17.00
Sept.	11.0	16.0	5.00
		Mean:	– 7.58
		St. dev.:	7.76
Chlorophyll-*a* (μg l^{-1})			
April	192.5	87.5	– 105.0
May	41.5	90.5	49.0
June	58.0	12.0	– 46.0
July	71.0	89.0	18.0
August	222.0	173.0	– 49.0
Sept.	99.5	85.0	– 14.5
		Mean:	– 24.6
		St. dev.:	54.5

Small-sized species of cladocerans were also scarce: only empty exuviae of *Bosmina* spp. were found. A few individuals were met of the following species: *Bosmina longirostris*, *Chydorus globulosus*, *C. sphaericus*, *C.* sp., *Daphnia hyalina*, *D.* sp., *Eurycercus lamellatus*, *Pleuroxus* sp., *P. trigonellus*, *Pseudochydorus globosus*, *Scapholeberis mucronata*, *Sida crystallina* and *Simocephalus expinosus*.

Copepods – less efficient grazers of algae – were found frequently and were far more abundant than cladocerans. Their maximum density amounted to 800 ind. l^{-1} (May 1990). Rotifers – eating mainly detritus and nanoplankton – were also abundant; their maximum number amounted to about 1000 ind. l^{-1} in April and May 1990. The density of the zooplankton (cladocerans, copepods and rotifers) was not influenced by the restoration measures. *Neomysis integer* was found in different numbers (Table 6).

Macrophytes

Submerged plants were absent during the years up to 1990 in all experimental sites. In May/June 1991 submerged macrophytes and filamentous green algae developed. Species which were noticed are *Potamogeton crispus*, *P. pusillus*, *Zannichellia palustris*, *Callitriche* spp., *Chara globularis* and *Elodea nuttallii*.

Discussion

The dredging of the bottom-layer, with a minimum of resuspension of sludge turned out to be

Table 6. Estimated numbers of *Neomysis integer* (individuals per sample of 250 litre).

	1988		1989		1990	
	May	August	May	August	May	August
Site I	3	0	3	10	0	55
Site II	12	166	133	12	43	7
Site III	0	0	2	10	64	12
Site IV	0	2	2	41	88	0
Reference 1	17	19	1	39	1	44
Reference 2	17	48	54	52	54	23

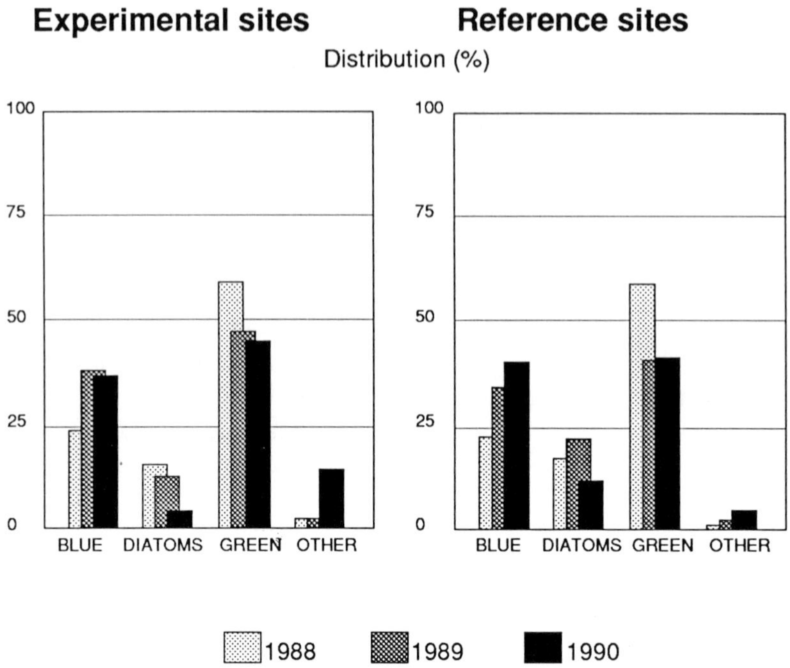

Fig. 4. Percentage average distribution of dominating algae groups in summer (period May–September).

possible after some modifications of the dredging-machine. The method, however, was laborious and expensive. The measures have improved the water quality, especially with respect to transparency and density of algae. The results agree reasonably well with the predictions based on the pilot study (Schöller *et al.*, 1985), especially regarding the phytoplankton biomass measured as chlorophyll-*a*. As the summer of 1990 was relatively dry the inlet of more water than normal was necessary. This inlet water had a high phosphorus content (Table 4) and thus reduced the effect of measures, which have been more effective in a normal year. A further disturbing factor might have been the irregular inlet of water in the year 1990, leading sometimes to the inlet of large quantities of nutrient-rich water in a short period.

In the summer of 1990 no submerged plants developed. An explanation could be insufficient stabilization of the bottom sediment as dredging ended not before February 1990. Secondly in the summer light penetration in the water was not optimal (Secchi-depth 30 cm) because of still too high concentrations of algae and suspended solids. The change-over from an algae-dominated turbid water to a clear water should be followed by the development of aquatic plants (Scheffer, 1990). Already in May/June 1991 six species of submerged macrophytes developed without introducing them.

So far fish stock management has not shown any positive results. This is possible because: 1. the fish stock in the experimental sites I and II did not differ much from that in the control sites III and IV; and 2. contrary to the expectations, large bodied cladocerans were scarce even after dredging and fish stock management. This could have been caused by the composition of phytoplankton namely high concentrations of filamentous cyanobacteria (Dawidowicz, 1990; Gliwicz, 1990; Bernardi & Giussani, 1990). An other causal factor might be predation by *Neomysis integer*. In lake Washington predation by *Neomysis integer* was found to be important in reducing the density of zooplankton (Edmonson & Abella, 1988). *Neomysis integer* is omnivorous, but prefers animal food (Mauchline, 1971; Hanazato, 1990). It feeds on detritus, algae, suspended organic material and hunts zooplankton. In the absence of aquatic plants the cladocerans lack refuges from predators.

On the whole the measures taken have reduced the levels of nutrients, the density of algae and has increased the transparency. The recent return of submerged plants indicates that the prospects for further improvements in the ecosystem are good.

Acknowledgement

We thank Dr R. D. Gulati and Dr E. Van Donk for making helpful comments on an earlier version of this manuscript.

References

Balls, H., B. Moss & K. Irvine, 1989. The loss of submerged plants with eutrophication. I. Experimental design, water chemistry, aquatic plant and phytoplankton biomass in experiments carried out in ponds in the Norfolk Broadland. Freshwat. Biol. 22: 71–87.

Benndorf, J., 1988. Objectives and unsolved problems in ecotechnology and biomanipulation: A preface. Limnologica (Berlin) 19: 15–18.

Björk, S., 1985. Lake restoration techniques. In: Proceedings International Congress 'Lakes Pollution and Recovery', Rome; p. 281–292.

Dawidowicz, P., 1990. Effectiveness of phytoplankton control by large bodied and small bodied zooplankton. Hydrobiologia 200/201: 43–47.

De Bernardi, R. de & G. Giussani, 1990. Are blue-green alga a suitable food for zooplankton? An overview. Hydrobiologia 200/201: 29–41.

Edmondson, W. T. & S. E. B. Abella, 1988. Unplanned Biomanipulation in lake Washington. Limnologica 19: 73–79.

Gliwicz, Z. M., 1990. Why do cladocerans fail to control algal blooms. Hydrobiologia 200/201: 83–97.

Gulati, R. D., 1989. Structure and feeding activities of zooplankton in lake Zwemlust in the two years after biomanipulation. Hydrobiol. Bull. 23: 5–10.

Gulati, R. D., 1990. Structural and grazing responses of zooplankton community to biomanipulation of some Dutch water bodies. Hydrobiologia 200/201: 99–118.

Hanazato, T., 1990. A comparison between predation effects on zooplankton communities by *Neomysis* and *Chaoborus*. Hydrobiologia 198: 33–40.

Hosper, S. H., 1990. Biomanipulation, new perspectives for restoration of shallow eutrophic lakes in the Netherlands. Hydrobiol. Bull. 23: 5–10.

Hovenkamp-Obbema, I. R. M., S. P. Klapwijk & Y. Scheffer-Ligtermoet, 1987. Toepassing van het raamwerk van Caspers en Karbe voor de waterbeoordeling in Noord- en Zuid-Holland. In: Biologische Waterbeoordeling: Instrument voor Waterbeheer; symposium 2 en 3 april 1987, Rijksinstituut voor Natuurbeheer, Leersum.

Irvine, K., B. Moss & H. Balls, 1989. The loss of submerged plans with eutrophication. II. Relationships between fish and zooplankton in a set of experimental ponds, and conclusions. Freshwat. Biol. 22: 89–107.

Lammens, E. H. R. R., R. D. Gulati, M. L. Meijer & E. Van Donk, 1990. The first biomanipulation conference: a synthesis. Hydrobiologia 200/201. R. D. Gulati, E. H. R. R. Lammens, M. L. Meijer & E. Van Donk (eds). Proceedings of an international conference held in Amsterdam, The Netherlands, 8–11 August, 1989; Kluwer 638 pp.

Mauchline, J., 1971. The biology of *Neomysis integer* (Crustacea, Mysidacea). J. mar. biol. Ass. U.K. 51: 347–357.

McQueen, D., J. R. Post & E. L. Mills, 1986. Trophic relationships in pelagic ecosystems. Can. J. Fish. aquat. Sci.: 1571–1581.

Raat, A. J. P. & G. A. J. Laak, 1989. De visstand in de polder Wormer, Jisp en Nek najaar 1989. Hoogheemraadschap/ O.V.B. 32 pp.

Scheffer, M., 1990. Multiplicity of stable states in freshwater systems. Hydrobiologia 200/201: 475–486.

Schöller, M., G. Van Straten & W. Fieggen, 1985. Eutrofieringsmodel ten behoeve van het waterkwaliteitsbeheer van de polder Wormer, Jisp en Neck. H_2O 26: 548–554.

Smeitink, D. J. W., G. Van Straten & A. G. Brinkman, 1988. Het effect van baggeren op de fosfaatafgifte van het sediment in de polder Wormer, Jisp en Nek. Universiteit van Twente, 32 pp.

Walker, P. A., A. J. P. Raat & G. A. J. De Laak, 1990. De afvissing en de bezetting van de proefvakken in de polder Wormer, Jisp en Nek, voorjaar 1990; O.V.B. rapport, Nieuwegein.

Waterboard of Uitwaterende Sluizen in Kennemerland and West Friesland, 1988. Onderzoek van de waterbodem in de polder Wormer, Jisp en Nek. Rapport, Edam. 31 pp.

Hydrobiologia **233**: 235–245, 1992.
L. Van Liere & R.D. Gulati (eds), Restoration and Recovery of Shallow Eutrophic Lake Ecosystems in The Netherlands.

Eutrophication control strategies for three shallow Vecht lakes in the province of North Holland

P. J. T. Verstraelen,[1] J. Wisserhof,[2] Lj. Rodić[3] & R; Eijsink[4]
[1] *Water Pollution Control Authority Amstel & Gooiland P.O. Box 1061, 1200 BB Hilversum, The Netherlands;* [2] *Delft University of Technology, Faculty of Civil Engineering, P.O. Box 5048, 2600 GA Delft, The Netherlands;* [3] *International Institute for Hydraulic and Environmental Engineering, Department of Aquatic Ecology, P.O. Box 3015, 2601 DA Delft, The Netherlands;* [4] *University of Utrecht, Department of Environmental Studies, P.O. Box 80115, 3508 TC Utrecht, The Netherlands*

Key words: lake restoration, eutrophication, water balances, nutrient loading, sediment, alternative supply options

Abstract

Like many shallow surface waters in the Netherlands the North Holland Vecht lakes, formerly known for their rich variety of flora and fauna, now face a serious eutrophication problem. Nutrient enrichment has been mainly in the form of (treated) wastewater discharges, and the continuing ingress of nutrient-loaded water from the river Vecht. Yet, this water has to be supplied in order to compensate for water shortages resulting from (*i*) changes in the groundwater flow pattern due to reclamation of the deep polder Horstermeer, (*ii*) extensive groundwater extraction in the Gooi hills, and (*iii*) extensive drainage for agricultural purposes.

The present policy of eutrophication abatement and restoration of the Ankeveen and Kortenhoef lakes ecosystems is focused on eliminating wastewater discharges and Vecht water supply. It also allows for additional dredging measures. Because of the un-suitable major ion composition of the Vecht, the aim is to compensate for this water supply by (*i*) partial restoration of the original groundwater flow from the Gooi hills and (*ii*) periphere additional supply with fresh seepage water from the skirts of the Horstermeer polder. However, uncertainty exists about the amounts of water needed.

Water balances and phosphorus budgets have been established to ascertain the water demands of the lakes and to gain a detailed insight into the nutrient fluxes through the lakes. A groundwater flow model is used to assess the beneficial effects of the proposed measures.

The results obtained, question the current unilateral restoration objectives. Calculations reveal that, both in the present situation and after (total) reduction of groundwater extractions in the future, the available quantity of fresh seepage water from the skirts of the polder Horstermeer is not sufficient to replace the inlet from the river Vecht into the Kortenhoef lakes. Additional supply options are available but the ones favoured from an ecological viewpoint are either the most expensive or less favoured from a social point of view. Although the sediments of the lakes appear to be a major source of eutrophication, the possibility of dredging the lakes will be considered only after reviewing results of a pilot-dredging project in the Hollands Ankeveen lakes in 1991.

Introduction

Man-induced eutrophication of lakes and reservoirs is considered to be one of the most widely spread water quality problems in the world. Eutrophication can result in (*i*) detrimental changes in water quality and aquatic ecosystems, and (*ii*) it can severely limit human use of water as a resource. The impacts of eutrophication very often call for costly management measures. The most effective strategy for controlling eutrophication is considered to be reducing or eliminating the input of nutrients in the waterbody (Sas & Ahlgren, 1989; Ryding & Rast, 1989). In the majority of cases, phosphorus is the growth limiting nutrient indicating that a control strategy should include phosphorus control (OECD, 1982).

The eutrophication problems of the shallow Vecht lakes Ankeveen and Kortenhoef (Fig. 1) became apparent in the late 50's and the early 60's but find their origin in (*i*) the reclamation of lake Horstermeer in 1883, (*ii*) the groundwater extractions in the Gooi hills that started in 1890 and total ca. 16 million $m^3 y^{-1}$ in the 80's, and (*iii*) ongoing lowering of the polder water levels to keep up with the peat bottom subsidence (Ver-

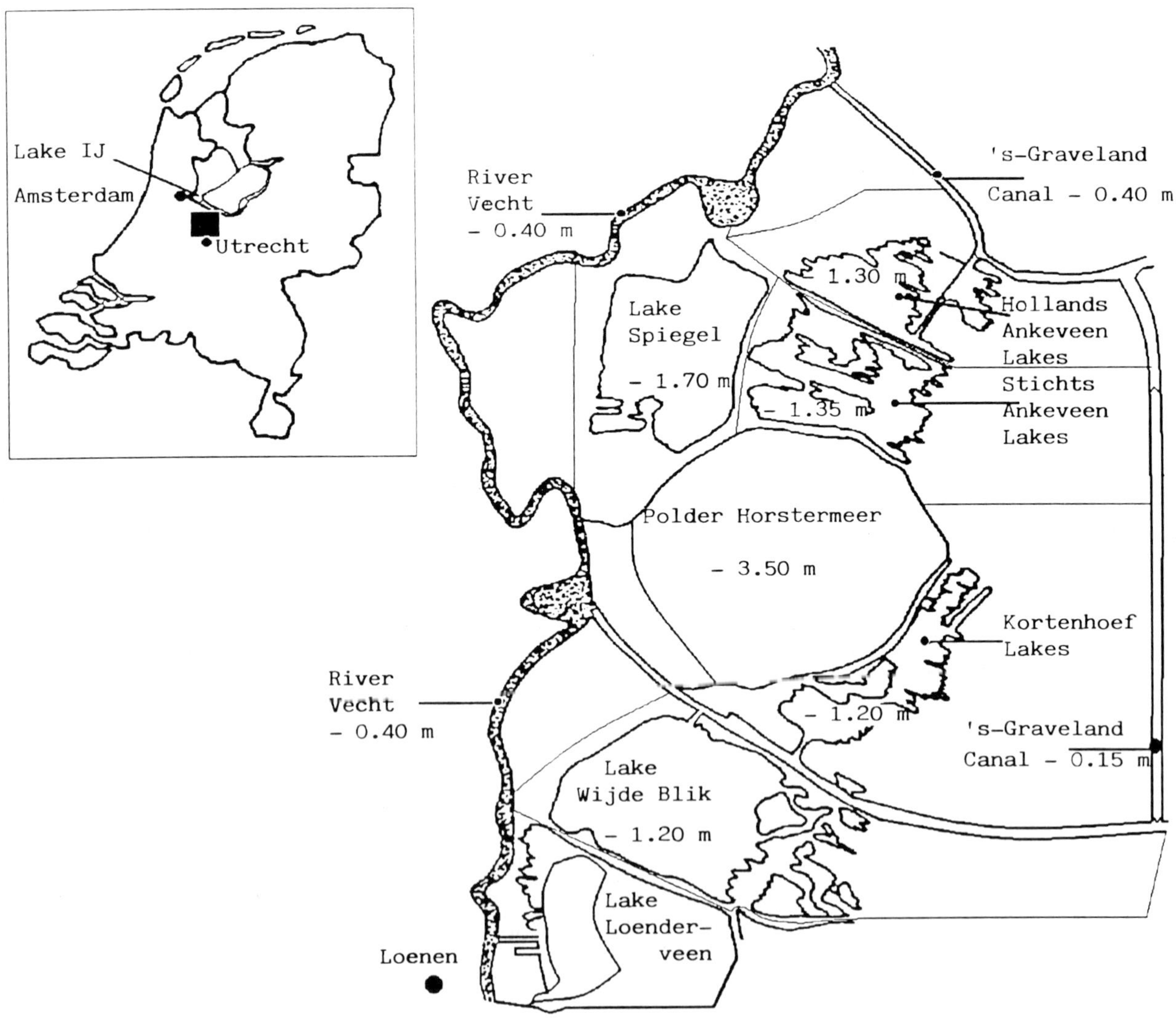

Fig. 1. The research area.

straelen *et al.*, 1989; Gast, 1989). Before these actions the area used to be recharged by groundwater flow originating from the Pleistocene glacial-formed Gooi ridges in the east. This groundwater seepage, with its characteristic chemical composition, has proved to be essential for the very specific ecosystems that developed (Barendregt *et al.*, 1989; Wassen, 1990). However, because of the interferences with the original hydrology, seepage flow to the Vecht lakes decreased substantially and extraneous (surface) water supply became inevitable inflicting major impacts on the nutrient load and chemical characteristics of the lakes.

Thus, the originally oligo-mesotrophic lakes were polluted by nutrient-loaded water from the river Vecht. With time, the accumulation capacity of the lakes was exceeded, and in the last three decades, the Ankeveen and Kortenhoef lakes experienced serious eutrophication. The number of plant species as well as the species characteristic of an oligo-mesotrophic habitat, have declined in their abundance (van den Berg & de Smidt, 1985; Roijackers & Verstraelen, 1988). Submerged water plants have completely vanished from some parts of the area, while blue-green algae and floating leaved macrophytes with rhizomes have become far more common.

Several measures to solve the eutrophication problem have been taken e.g., diverting municipal effluent discharges and halting the extraneous water supply to lake Hollands Ankeveen by preserving local water within the system. Other measures are in preparation e.g., supplying the system with fresh seepage water from the polder Horstermeer, and decreasing the groundwater extractions in the Gooi by 50% (Provincial Board of North-Holland, 1986; Schouten *et al.*, 1988; Roijackers & Verstraelen, 1991).

Since the original conservation value of the Vecht area is linked to good quality water supply, it is essential to determine water quantity and quality demands of the polders and, thus, enable selection of appropriate restoration measures.

In the following sections, water balances and phosphorus budgets are discussed, and the results of the groundwater flow model calculations are presented. Based on the results several management options are considered.

Methods

Water balances

In general, a water balance can be described by the following equation:

$$Q_{in} - Q_{out} = \delta V ,$$

where

Q_{in} is the input;
Q_{out} is the output;
δV is the storage change.

The boundaries of the Ankeveen and Kortenhoef polders (the western part of which form the Ankeveen and Kortenhoef lakes respectively) were chosen to be the boundaries of the balance system, because the polders are reasonably well defined hydrological units. Q_{in} is comprised of precipitation (P), upward seepage (S_u) and inlet of surface water (I_s). Q_{out} is comprised of evapotranspiration (E), infiltration (F_d) and outlet of surface water (O_s). Finally δV consists of surface water (δV_s) and groundwater (δV_g). Each of these components is discussed below.

Two main criteria determined the choice of the period 1985–1989 as a time span to be considered throughout the work. Firstly the period is representative for the situation analyzed; thus all the relevant variables are true for the present state of the system. Secondly the data necessary for the calculations were available. The balance equations have been established for each month during the five-year period.

Precipitation data representative of the area were obtained from the Royal Dutch Meteorological Institute, gauging station Loenen, situated about 5 km from the study area. The seepage term (S_u) in the balance equation was calculated according to the following formula (based on Darcy's law):

$$S_u = \delta h / c , \quad \delta\eta > 0 ,$$

where

δh is the difference in metres between the piezometric head of the groundwater in the first (Pleistocene) aquifer and the surface water level;

c is the vertical flow resistance in days of the covering (Holocene) layer.

Monthly averages of daily records of the surface water levels in the Ankeveen and Kortenhoef polders, have been used for the calculations. A map of contour lines of the groundwater piezometric level in the first aquifer averaged over the period 1974–1980 was obtained from the Department of Environment and Water, Province of North Holland. Use has also been made of the findings of Witmer on the vertical flow resistance of groundwater in the covering layer (Witmer, 1986 and 1989).

The inlet of surface water actually depends on the water demand in the polders, and as such, is crucial for the water economy of the polders. Water enters the polders in several ways:

- deliberate inlet by operation of sluices and pumping stations;
- leakage of sluice gates;
- ship passages through the sluices;
- percolation through the soil under and along the sluices;
- drainage through the dikes along 's-Graveland Canal (Fig. 1);
- deliberate inlet through connections along the 's-Graveland Canal.

Since accurate monitoring of most of the water flows is not feasible precise amounts of water that enter the polders are not known. Therefore, the balance equations are arranged so that the amount of surface water inlet is computed. Simulations have been carried out to assess the values for the different items of the inlet.

Data on evaporation were obtained as decanal values from the monthly climatological bulletins of the Royal Dutch Meteorological Institute at De Bilt. The meteorological station at Schiphol was taken to be the most representative of the evaporation conditions in the polders concerned. Evaporation rates of open water surface, built-up areas, pasture and swamp forrest are distinguished. For the open water surface, Penman's data are used. For surfaces covered by vegetation, the reference-crop evapo-transpiration approach according to Makkink is applied (Hooghart, 1987; CHO-TNO, 1988). The formula used for the calculation of the I_d infiltration term (as well as the input data) is analogous to the one used for seepage, i.e.

$$I_d = \delta h/c\ , \quad \delta h < 0\ ,$$

with the terms δh and c as already defined.

O_s takes place by means of pumps, except for the polder Stichts Ankeveen. Pumping hours are registered on a routine basis by the Water Quantity Control Authority Amstel & Vecht.

δV_s is calculated as the difference in surface water level between sequential months, multiplied by the area of the open water surface. Analogous to this, δV_g is calculated as the difference in groundwater piezometric level between successive months multiplied by the area of land surface and the storage coefficient of peaty soils.

Values for some of the factors incorporated into the water balance calculations are questionable. Therefore, to check the results obtained (mainly concerning groundwater flow), chloride balances for the polders were established. This approach was possible due to the conservative properties of chloride ions in natural waters. The change in storage of chloride consists of two terms

$$\delta Cl = V_i * [Cl]_i - V_{i-i}[Cl]_{i-i}\ ,$$

where δCl is the change in chloride content per month in g, V_i and V_{i-1} are the volumes of water in m^3 contained in the polder in the ith-month and the previous month respectively, and $[Cl]_i$ and $[Cl]_{i-1}$ are the chloride concentrations ($mg\,l^{-1}$) in the polder water in the i-month and the previous month respectively.

The three main chloride components are precipitation, surface water and groundwater. The data on the chemical composition of precipitation were taken from the nearest data collection point to the polders concerned i.e., the meteorological station at De Bilt. Chemical analyses of the surface water samples in the lakes concerned are carried out monthly by the Water Pollution Control Authority Amstel & Gooiland. The chloride balance and the phosphorus budget for the Kortenhoef polder, enable to distinguish between the deep lake Wijde Blik and the shallow lakes in the rest of the polder. For the groundwater chloride content of the seepage water coming from the first aquifer a detailed outline is available from Schot (1989); an average value was taken (P. P. Schot, pers. comm.).

Phosphorus budgets

Although the phosphorus budget equations were established analogous to the chloride balance equations, there are some notable differences. The balance systems are defined so as to include only surface water bodies in the polders, since the nutrient content of surface water in the region is of main concern. Compared with the water balance calculations, there are two additional input terms:

- surface runoff and interflow;
- domestic waste water discharges and combined sewer overflow.

The change in storage term is also different. It consists of:

- a change in storage of surface water;
- an exchange with the bottom sediment.

The data on the phosphorus contents of each particular term of the equations were obtained in a similar way as those of the chloride balance. For nutrient loading by surface runoff and interflow estimates from literature were used (Steenvoorden, 1988). Equations were arranged so that the exchange at the water-sediment interface can be derived from the calculations. It should be noted that no biochemical transformations of phosphorus have been taken into account.

Groundwater flow model

The regional hydrological system was modelled by using the quasi-three-dimensional groundwater flow program Micro-Fem (Hemker & Elburg, 1988). The modelled area comprises parts of the Gooi, the Vecht area and the Eem valley (Fig. 1). The model schematisation relies to a considerable extent on a similar study carried out by Witmer (1989), although input data have been updated. The hydrological system has two aquifers, divided by an aquitard. For the water balances of the Vecht lakes only the first (upper) aquifer and the covering layer are of interest. Groundwater flow in the aquifers is assumed to be horizontal. It is described by the following equation:

$$\frac{\delta}{\delta x}\left(kD * \frac{\delta h}{\delta x}\right) + \frac{\delta}{\delta y}\left(kD * \frac{\delta h}{\delta y}\right) + Q = 0\,,$$

where:
kD is transmissivity ($m^2\,d^{-1}$);
k is permeability ($m\,d^{-1}$);
h is hydraulic head (m);
Q is groundwater recharge/discharge ($m\,d^{-1}$),

In Micro-Fem the groundwater flow equations are resolved using the finite element method. In the covering layer only vertical flow of groundwater is assumed to take place (for the formula see 'Water Balances').

For modelling, the area studied is divided into elements that are considered to be homogeneous. The applied network consists of 2242 nodal points and 4211 triangular elements. The model input comprises the following parameters: vertical flow resistances and transmissivities, fixed hydraulic heads at the outer boundaries of the modelling area, polder water levels, groundwater withdrawals and groundwater recharge by net-precipitation. The major outputs of the model are hydraulic heads (in the first aquifer) and water balances.

Results and discussion

Water balances

The water and chloride balance data (Table 1) relate to the year 1986, an average year with regard to the annual amount of precipitation (data

Table 1. Water balances of Ankeveen and Kortenhoef polders in $m^3\ 10^6\ y^{-1}$ (1986).

	Hollands Ankeveen	Stichts Ankeveen	Kortenhoef
Input			
Precipitation	3.3	4.3	12.6
Seepage	4.0	0.7	1.0
Inlet from			
– Vecht	–	–	29.7
– 's-Graveland Canal	0.2	0.1	2.4
– lake Spiegel	–	2.4	–
Output			
Evapo(transpi)ration	2.4	3.1	10.1
Infiltration	3.2	4.1	22.4
Outlet to			
– River Vecht	–	–	7.3
– 's-Graveland Canal	2.0	–	5.5
– lake Spiegel	–	0.8	–
Annual balance	– 0.1	– 0.5	0.4

series 1910–1989; Royal Dutch Meteorological Institute, 1985 + updates).

The discussion focuses mainly on the inlet and outlet of surface water, the seepage and infiltration being less certain terms. A sensitivity analysis reveals that the outcome is most sensitive to the uncertain variables, namely the amount of infiltration.

Polder Hollands Ankeveen. On an annual basis the water balance between the 's-Graveland Canal and the Hollands Ankeveen polder appears to be negative (Table 1). This is confirmed by the data of the pumping station i.e., only sporadically surface water is let into the Hollands Ankeveen polder, which has a positive net-seepage. Although the calculated amounts of seepage and infiltration bear uncertainties of the input data, the resulting net-seepage is confirmed by the biological data collected the Department of Environment and Water, Province of North Holland. Unlike the Stichts Ankeveen and Kortenhoef polders, in the Hollands Ankeveen polder several plant species characteristic of seepage conditions are found e.g., *Veronica beccabúnga*, *Equisétum fluviátile*, *Spargánium emérsum*, *Carex rostráta*, *Potamogeton acutifólius*, *Potamogeton obtusifólius* and *Hottonia palústris*.

Polder Stichts Ankeveen. The Stichts Ankeveen polder is supplied with surface water both from the 's-Graveland Canal and from the former sand pit Lake Spiegel (Fig. 1). The inlet from the 's-Graveland Canal takes place only in dry periods via connections which are managed by farmers. Water from Lake Spiegel enters the Stichts Ankeveen polder via a pumping station. The data of the pump capacity are not reliable because the pump capacity was measured years ago. The calculated water and chloride balances indicate that the pump capacity is overestimated by 15–30%.

Surplus surface water from the Stichts Ankeveen polder is pumped to Lake Spiegel, but the volume discharged is not registered. The simulations are made to correspondent to fluctuations in the release from the polder to the lake and to the 's-Graveland Canal. The infiltration in the Stichts Ankeveen polder is much larger than the one in the Hollands Ankeveen polder. This is explained by the proximity on the south of the Stichts Ankeveen polder (–1.35 m below MSL) to the lower Horstermeer polder (–3.50 m below MSL; Fig. 1). The acceptable waterlevel fluctuation in the lakes are insufficient to prevent inlet of surface water both into the Hollands and Stichts Ankeveen polders (A. Barendregt, pers. comm.).

Polder Kortenhoef. The calculated surface water inlet from the river Vecht into the Kortenhoef polder ranges between 25.1 and 32.3 $10^6\ m^3\ y^{-1}$, with an average of 29.7 $10^6\ m^3\ y^{-1}$. The resulting inlet is determined almost solely by the value for infiltration (Table 1). But, as mentioned, the input data for the calculation of infiltration are not completely reliable. Vertical flow resistance in the covering layer (c) is derived from modelling (Witmer, 1986). The contour lines of the piezometric head in the first aquifer are derived by linear interpolation between (scarce) measuring points (Department of Environment and Water, Province of North Holland). The pattern of the contour lines based on the groundwater flow model differs significantly from the former. With these piezometric head values, infiltration would amount to not more than 8.2 10^6 m^3 y^{-1}. Since inflows from the 's-Graveland Canal through connections and by drainage through the dike along the canal, are not monitored, an estimated 2.4 10^6 m^3 y^{-1} (Witmer, 1989) is used. This value, however, could easily be twice as high in an unusually dry year like 1989; the calculated inlet from the river Vecht will decrease by the same extent, due to the linear nature of balance equations.

Phosphorus budgets

The phosphorus budget calculations (Table 2) do not include the input term domestic wastewater discharge because the households were connected to the sewerage system in 1985 and 1986. Data for the winter months are lacking.

Hollands Ankeveen Lakes. The phosphorus budget of the Hollands Ankeveen lakes has a net-seepage gain as against the net infiltration assessed for the water balance of the polder as a whole. This is due to the lakes being situated in the west of the polder that is an infiltration area, as opposed to the east of the polder that is a seepage area.

Table 2. Phosphorus budgets of Ankeveen and Kortenhoef lakes in kg P y^{-1} in the period April–September (1986).

	Hollands Ankeveen	Stichts Ankeveen	Kortenhoef
Input			
Precipitation	12.1	16.2	55
Seepage	37.9	–	5
Inlet from			
– Vecht	–	–	20256
– 's--Graveland Canal	24.4	39.0	1346
– lake Spiegel	–	138.1	–
Surface run-off and interflow	10.5	28.8	510
Output			
Infiltration	184.8	279.8	376
Outlet to			
– Vecht	–	–	364
– 's-Graveland Canal	144.0	–	1475
– lake Spiegel	–	2.2	–
Annual balance			
– surface water	3.3	– 731.0	– 1437
– exchange with bottom	247.2	– 790.9	– 21394

There appears to be P-release from the sediment; the average release rate is −1.7 mg m^{-2} d^{-1} in summer time. The exchange with the lakes sediment is the main input into the lakes (Table 2). The external P-load is only 0.06 g m^{-2} y^{-1}. Since the P-concentration in the lakes (Fig. 2) is higher than the calculated limiting level of *ca.* 0.08 mg l^{-1} (Lijklema *et al.*, 1989) the phosphorus release from the sediment needs to be reduced.

Lakes Stichts Ankeveen. Similar to the Hollands Ankeveen lakes, the Stichts Ankeveen lakes are situated in an infiltration area. Although the inlet of surface water from lake Spiegel is *ca.* 17 times larger than from the 's-Graveland Canal (Table 1), the P-inputs from the two are similar. This is caused by the lower P-concentration in Lake Spiegel because of the long residence time of surface water due to a much greater average depth of the lake.

Contrary to the Hollands Ankeveen lakes P-concentration in the surface water of the Stichts Ankeveen lakes (Fig. 2b) mostly complies with the limiting P-concentrations of *ca.* 0.08 mg P l^{-1} estimated by Lijklema *et al.* (1989). This may be because of the large infiltration which may suppress recharge from the sediment. But, the phosphorus concentrations in the southern parts of the lakes are much higher (up to 0.3 mg P l^{-1}) than in the northern parts; they are in open connection with the ditches in the agricultural eastern part of the Stichts Ankeveen polder and receive their P-input mainly from surface run-off, interflow and the 's-Graveland Canal.

Kortenhoef lakes. P concentrations in the Kortenhoef lakes exceed the target concentration of *ca.* 0.08 mg total-P l^{-1} according to Lijklema *et al.* (1989). A reduction is, however, discernible since the sewage treatment plant at Horstermeer started operating and the sewerage system was completed

Table 5. Alternativie Sypply options. Cost-estimates are based on an interest rate of 9%, 75000 guilders per ha. for agricultural land purchase and 15000 guilders per ha. for uncultivated land purchase, a write-off over 30 years for immoveables (covering *ca.* 70% of the building costs) and 15 years for mechanical parts (covering *ca.* 30% of the building costs), 0.05 guilder per cubic m for dephosphorization of supply water, 0.01 guilder per cubic m dephosphorized supply water for processing the simultaneously produced unpolluted sludge and 0.05 guilder per cubic m dephosphorized supply water for processing the simultaneously polluted sludge.
* in accordance with Barendregt *et al.*, in press.
** non-recurrent investment.

Alternative supply options	Available supply or infiltration reduction (in 10^6 m^3 y^{-1})	Estimated costs per potential measure (in guilders per m^3)	Ecological benefit* of measure (expressed in %)	
			Polder Kortenhoef	Polder Stichts Ankeveen
Present River Vecht water supply	>32	0.00	52.7–68.4	60.0–70.9
Dephosphorized Lake IJ water supply	>32	0.10	49.4–53.4	45.8–46.2
Dephosporized Amsterdam-Rhine Canal water supply	>32	0.15	49.5–58.2	42.8–50.6
Dephosphorized River Vecht water supply	>32	0.15	55.0–61.3	–
Dephosphorized + denitriphicated effluent from the waste water treatment plant Horstermeer	10	1.15	55.3–65.0	–
Fresh seepage water from the Horstermeer polder	17	0.15	63.7–72.4	76.5–83.9
75% inundation of the Horstermeer polder	9	0.50**	61.8–69.6	73.0–73.0
100% inundation of the Horstermeer polder	16	1.45**	67.0–74.5	100.0–100.0
Dephosphorized + denitriphicated + desalination discharge water from the Horstermeer polder	>32	1.95	99.7–100.0	–

Canal is not any more the main source of the lakes' eutrophication. Since 1986 when the waste water treatment plant in the Horstermeer polder became operational, the phosphorus dynamics (sedimentation and release) at the sediment is the most important item in the lakes' phosphorus balance. Therefore, dredging of the lakes' nutrient-loaded bottom sediment will be the most effective measures to reduce eutrophication (Van der Does *et al.*, 1992). A pilot-dredging project in this regard is to commence shortly.

In case of the Stichts Ankeveen lakes the surface water supply from lake Spiegel is much larger than the inlet from the 's-Graveland Canal. Considerable amount of water infiltrates from the Stichts Ankeveen polder to the adjacent Horstermeer polder, preventing recharge of the sediment settled phosphorus. To prevent phosphorus release from the sediment and to reduce its eutrophication as well as pollution of groundwater in the Stichts Ankeveen polder, dredging the lakes' sediment appears to be of utmost importance.

For the Kortenhoef lakes the situation is quite similar to that for the Stichts Ankeveen lakes. However, in the Kortenhoef lakes infiltration, and thus settling of phosphorus in the lakes' sediment is found to be much larger. An alternative water supply to these lakes is more necessary than for the Stichts Ankeveen lakes. However, because of the uncertainties of the calculated infiltration rates a range of 8–22 10^6 m^3 y^{-1}, rather than just one value, should be considered.

Groundwater flow-model calculations reveal that reduction of groundwater withdrawals in the Gooi region would not completely eliminate water shortages in the Vecht lakes. Raising the water level in the polder Horstermeer would largely restore seepage conditions that existed in the past and greatly eliminate the necessity of extraneous water supply. However, from a social viewpoint the inundation option does not seem feasible in the near future.

References

Barendregt, A., De Smidt, J. T. & M. J. Wassen, 1985. The impact of groundwater flow on wetland communities. Colloques Phytosociologiques 13, Vegetation et Geomorphologie, Bailleul.: 603–612.

Barendregt, A., Wassen, M. J. & J. T. De Smidt, 1989. Regeneratie en behoud van ecosystemen door integraal waterbeheer. In: Van Liere, L., Roijackers, R. M. M. & P. J. T. Verstraelen (eds.), 1989. Integraal waterbeheer in het Goois/Utrechts stuwwallen- en plassengebied: 45–58.

Barendregt, A., S. M. E. Stam & M. J. Wassen, 1992. Restoration of Fen Ecosystems in the Vecht River Plain: Cost-Benefit Analysis of Hydrological Alternatives. Hydrobiologia (this volume).

Commissie voor hydrologisch onderzoek-TNO, 1988. Van Penman naar Makkink. Een nieuwe berekeningswijze voor de klimatologische verdampingsgetallen. KNMI/CHO, eindrapport van de project- en begeleidingsgroep verdampingsberekeningen. Commissie voor Hydrologisch onderzoek TNO, 's-Gravenhage, 1988. Rapporten en Nota's, 19, 74 pp. (in Dutch).

Gast, M. K. H., 1989. Integraal drinkwaterbeheer in het Gooi en omgeving. In: Van Liere, L., R. M. M. Roijackers, & P. J. T. Verstraelen (eds), 1989. Integraal waterbeheer in het Goois/Utrechts stuwwallen- en plassengebied: 90–102.

Hemker, C. J. & H. Van Elburg, 1988. Micro-Fem. Users manual. MicroComputer MultiLayer Steady Finite Element GroundWater Modelling: 152 pp.

Hooghart, J. C., 1987. Evaporation and weather. CHO-TNO Committee on Hydrological Research. Proceedings and information 39: 106 pp.

Royal Dutch Meteorological Institute, 1985. Climatological Data of Stations in The Netherlands. A climatology of the main observatory at De Bilt for the period 1901–1984, 25, 56 pp.

Lijklema, L., J. H. Janse & R. M. M. Roijackers, 1989. Eutrophication in The Netherlands. Wat. Sci. Tech. 21: 1899–1902.

OECD, 1982. Eutrophication of waters. Monitoring, Assessment and Control. OECD Paris, 154 pp.

Provincial Board of North-Holland, 1986. Provincial Groundwater Management Plan North-Holland: 100 pp. (in Dutch).

Roijackers, R. M. M. & P. J. T. Verstraelen, 1988. Ecological investigations in three shallow lakes of different trophic level in The Netherlands. Verh. int. Ver. Limnol. 23: 489–495.

Roijackers, R. M. M. & P. J. T. Verstraelen, 1991. Eutrophication abatement programme for three North Holland Vecht lakes. Verh. int. Ver. Limnol. 24: 768–772.

Ryding, S-O. & W. Rast, 1989. The control of eutrophication of lakes and reservoirs. Man and the Biosphere series. Published by the UNESCO and the Parthenon Publishing Group, 314 pp.

Sas, H., 1989. Lake restoration by reduction of nutrient loading: expectations, experiences, extrapolations. Academia Verlag Richarz GmbH. 497 pp.

Schot, P. P., 1989. Grondwatersystemen en grondwaterkwaliteit in het Gooi en randgebieden. Rijksuniversiteit Utrecht. Interfacultaire Vakgroep Milieukunde: 44 pp. (in Dutch).

Schouten, P. H., Verstraelen, P. J. T. & J. J. Buntsma, 1988. Restoration of the Vecht lakes ecosystems surrounding the Horstermeer polder: a feasibility study. – Agricul. Wat. Managmt. 14: 471–479.

Steenvoorden, J. H. A. M., 1988. How to reduce nitrogen losses in intensive grasland management? Ecol. Bull. 39: 126–130.

Van den Berg, W. J. & J. T. de Smidt, 1985. De vegetatie van het Oostelijk Vechtplassengebied 1935–1980. Stichting Commissie voor de Vecht en het Oostelijk en Westelijk Plassengebied. 155 pp. + annex.

Van der Does, J., P. J. T. Verstraelen, P. Boers, J. Van Roestel, R. M. M. Rooijackers and G. Moser, 1992. Lake restoration with and without dredging of phosphorus-enriched upper sediment layers. Hydrobiologia 233: 197–210.

Verstraelen, P. J. T., R. M. M. Roijackers & L. Van Liere, 1989. Geohydrologie van de Goois/Utrechtse stuwwal en het Vechtplassengebied. In: Van Liere, L., Roijackers, R. M. M. & P. J. T. Verstraelen (eds), 1989. Integraal waterbeheer in het Goois/Utrechts stuwwallen- en plassengebied: 151–165 (in Dutch).

Wassen, M., 1990. Water flow as a major landscape ecological factor in fen development. Ph.D. Thesis, State University of Utrecht, 199 pp.

Witmer, M. C. H., 1986. Management of groundwater and surface water resources in the infiltration area 'Het Gooi' and adjacent wetlands. Conjunctive Water Use. Proceedings of the Budapest Symposium. IAHS Publ. 156: 534–547.

Witmer, M. C. H., 1989. Integral water management at regional level. An environmental study of the Gooi and Vechtstreek. Ph.D. Thesis, State University of Utrecht, 172 pp.

Hydrobiologia **233**: 247–258, 1992.
L. Van Liere & R.D. Gulati (eds), *Restoration and Recovery of Shallow Eutrophic Lake Ecosystems in The Netherlands.*

Restoration of fen ecosystems in the Vecht River plain: cost-benefit analysis of hydrological alternatives

A. Barendregt, S. M. E. Stam & M. J. Wassen
Department of Environmental Studies, University of Utrecht P.O. Box 80.115, 3508 TC Utrecht, The Netherlands

Key words: fen ecosystems, hydrology, deterioration, restoration, cost-benefit analysis

Abstract

In the Vecht river plain, the vegetation diversity has been diminished during recent decades, due to hydrological changes and resulting eutrophication. Different options are discussed for hydrological management aimed at abating the eutrophication and at stimulating the characteristic mesotraphent succession phases of the fen vegetation. The ecological impact of the management plans are estimated with the hydro-ecological model ICHORS; this model takes into account the relations between the non-biotic environment and the presence of hydrophytes and phreatophytes. The impact of the hydrological options is assessed, after rearrangement of the predicted values of the species to values of the required fen vegetation. These results are incorporated in the management options with the water balances of the polders. Information on the costs of the options is used as a basis for a cost-benefit analysis. Technical hydrological solutions such as purification of the supplied water might contribute to the eutrophication abatement, but they are on their own unlikely to restore the ecosystems. Contrastingly, options involving the restoration of the original hydrology would probably lead to the characteristic mesotraphent fen vegetation once more, and the financial implications of this are likely to be the more favourable in the long-term.

Introduction

Fen ecosystems are endangered, for their location mostly near a river or an estuary can be eutrophicated easily. Moreover, after draining or reclaiming, fen ecosystems become arable lands, desired in agriculture (Williams, 1990). This hydrological change influences the chemistry of water in the ecosystem too, so that the ecological equilibrium is affected both in nutrients and in major ions. The restoration of polluted aquatic ecosystems might be achieved in the hydrological management. The question remains what type of management is profitable.

This problem is investigated in the Eastern Vecht River plain in The Netherlands, a wetland area situated north of the city of Utrecht and southeast of Amsterdam (Fig. 1). Shallow fen lakes (turfponds), marshes and ditched pasture lands are characteristic of the Vecht area. The original fen vegetation, so rich in species, is deteriorating as present. Species characteristic of mesotrophic conditions have disappeared or have become very rare, due to eutrophication of the surface water and to structural changes in the groundwater flow (Barendregt *et al.*, 1989; Wassen, 1990; Roijackers & Verstraelen, 1991). Various water management options have been suggested for restoring the vegetation.

This paper assesses the ecological implications

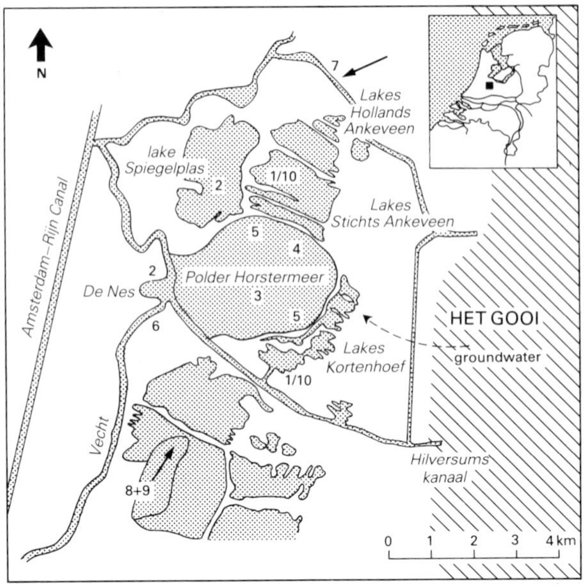

Fig. 1. Part of the Vecht River plain with Lakes Kortenhoef and Lakes Stichts Ankeveen. Numbers correspond to the locations of hydrological options described in Table 1.

as well as the financial consequences of possible changes in water management. Past and present geographical and hydrological conditions are outlined first, while the deterioration of fen vegetation due primarily to nutrient enrichment is discussed secondly. Next, various measures aimed at reversing eutrophication trends – involving management of the hydrological regime – are discussed. Ten options are considered, and here, not only the likely ecological impacts (fen species responses) are taken into account, but also the cost benefits.

Description of the lake areas

The Eastern Vecht area is bordered by the Pleistocene ice-pushed hill ridge Het Gooi to the east and by the Vecht River to the west. It is a polder area below mean sea level, where the level of surface water is artificially controlled. The natural groundwater flow originates from the hill ridge and is directed to the plain, where seepage water discharges at the edge of the ridge. Before cultivation of the area, the high groundwater levels induced the formation of peat as in the flood zone next to the river, where eutrophic woodland peat existed. Also, in the east of the river plain, under the influence of nutrient-poor and calcium-rich seepage water from Het Gooi, mesotrophic peat of *Carex* species was formed. In the intermediate area, the peat rose above the groundwater level and was only influenced by infiltrating rain water. Under these nutrient-poor conditions, *Sphagnum* peat developed.

Over the last few centuries, the natural hydrological regime has changed significantly due to human influence (Witmer, 1989). In the eleventh century, ditches were dug in the wetland area to drain the soil for agricultural purposes. The drainage caused mineralisation and setting of the peat. The natural water flow to the River Vecht diminished. From the sixteenth century onwards, the water was discharged artificially through ditches and canals; peat excavation and dredging could then be intensified. Peat was dredged up to the Pleistocene sand deposits. The excavation resulted in a landscape of elongated turfponds alternating with small residual strips of land, that were washed away. This resulted in the formation of large, shallow peat lakes in the eighteenth century.

The reclamation of Lake Horstermeer in 1882 caused a structural change in the groundwater flow system, such that the surface water level of the new polder was significantly (*ca.* 2.5 m) lower than that of surrounding areas. In order to maintain this low level, 30×10^6 m^3 water is discharged annually from the Polder Horstermeer into the Vecht River. This low level also causes drainage of water from the adjacent polders, which results in water shortages in these polders in summer. At present there is an annual deficit of 20×10^6 m^3 in the polder Kortenhoef and of 3×10^6 m^3 in the polder Stichts Ankeveen (Verstraelen *et al.*, 1992). A secondary effect is the discharge of brackish water from the deeper aquifer to the surface in the centre of the Polder Horstermeer (Witmer, 1989; Schot, 1991). A second cause of the water shortages in the plain is the groundwater abstraction by water supply companies since 1888. The population of Het Gooi increased and, subsequently, the demand for drinking water rose (in 1988 15.6×10^6 m^3; Schot, 1991). This significantly diminished the natural groundwater flow into the wetlands.

To counteract water shortages, the inlet of surface water, mainly from the Vecht River, was inevitable. Because of this partial substitution of ridge water by surface water, the chemical composition of the water changed. Indeed, the quality of wetland water deteriorated because of the poor quality of water entering from the Vecht River. Since 1930, the amounts of nutrients and major ions in the Vecht River have increased due to pollution of the Rhine River and the discharge of wastewater from the city of Utrecht (Barendregt *et al.*, 1989). At present the main source of eutrophication in the Kortenhoef Polder is the water from the Vecht River, which may supply up to 20 tons phosphorus a year (Verstraelen *et al.*, 1992). The supply of nutrient-rich surface water to the lakes over many years has increased the nutrient content of the sediments (Verstraelen *et al.*, 1989). For example the concentration of phosphorus in the mud of the Vecht lakes has doubled since the beginning of this century. Notably, a third of the total amount of stored phosphorus is available to algal growth (Visser *et al.*, 1989).

The ecosystems: their deterioration, restoration and water management options

After the creation of turfponds and larger lakes in the eighteenth century, the successional phases led to the development of a great variety of plant communities. A diverse vegetation with *Chara* species occurred in open water, and in the zone between water and land a mixture of succession phases was observed. When this vegetation growed tightly together forming a floating intertangled mat of plant roots, a species-rich quaking fen developed. At other localities, there were mashes with *Phragmites australis* and *Carex* species. Shrubs and trees appear in succession when plant growth extends above groundwater level. This is how the last successional phase of *Alnus* and *Betula* woodlands starts.

Since the first documented relevés of the Vecht wetland area in 1935, the vegetation has changed (Best *et al.*, 1984; Van den Berg & De Smidt, 1985; Roijackers & Verstraelen, 1988; Barendregt *et al.*, 1990). This is primarily due to changes in hydrology. The decreasing abundance of species that grow under nutrient-poor conditions is observed and the characteristic mesotrophic wetland ecosystems are now restricted to seepage areas that are unaffected by extraneous supply water (Barendregt *et al.*, 1989). Trends observed over the last decades include those of succession, eutrophication and acidification. Thus, there is a general tendency for aquatic vegetation and early succession phases to disappear, and this is followed by an increase of later phases with acid reedland and woodlands. Aquatic plants disappear due to an explosive growth of algae and increasing water turbidity. At the same time species characteristic of mesotrophic fens disappear. Eventually, this diminishes the biological diversity in the wetlands. Nowadays the aquatic vegetation of the Lakes Stichts Ankeveen includes no more than 20 vascular species and 4 *Charas*, in Lakes Kortenhoef only 9 common aquatic plant species occur.

The ecological value of the Vecht River plain and the key role of groundwater supply are recognised by national and provincial authorities (Ministry of Agriculture, Nature conservation and Fisheries, 1990; Province of North-Holland, 1990; Province of Utrecht, 1990). Until now, however, this has not prevented further deterioration of the wetlands. Restoration of the characteristic mesotraphent fen ecosystem requires decisive measures regarding the supply of groundwater and the prevention of eutrophication. The problem can be dealt with in several ways.

(1) The best route to a structural restoration is to revert to the hydrological situation that existed before the reclamation of the Polder Horstermeer. A large amount of water characteristic of the system is drained from the surrounding polders and the hill ridge into the Polder Horstermeer from which it is pumped into the Vecht River and is lost to the wetland

Table 2. Division of species from ICHORS 3.1 and ICHORS 3.2 in ecological groups. Number of species per group in brackets. Group 2, 3 & 4: fresh water species; group 5 & 6: eutraphent species; group 7 & 8: brackish (tolerant) species.

Groups of species: ICHORS 3.1 (turfponds)
1 *Common fen species* (22)
2 *Succession species* (10): more or less endangered, mostly floating species of clear water
3 *Mesotraphent species* (10): endangered species of unpolluted and undisturbed locations in the river plain
4 *Gradient species* (10): species characteristic for the zone between constant seepage of groundwater and infiltration
5 *Eutraphent turfpound species* (10): hydrophytes and phreatophytes indicating eutrophication and disturbance in fresh turfpounds
6 *Bank species* (12): bank species mostly living under eutrophic conditions
Groups of species: ICHORS 3.2 (ditches)
1 *Common fen species* (28)
2 *Hydrophytes of clean fresh water* (13): mesotraphent species of nutrient-poor water
3 *Species of fresh calcium-rich water* (13): endargered mesotraphent species occuring in ditches with water of the $CaHCO_3$-type
4 *Mesotrophic bank species* (28): bank species often prefering nutrient-poor and fresh water
5 *Eutraphent hydrophytes* (10): mostly floating water plant species indicating eutrophic conditions
6 *Eutraphent phreatophytes* (16): phreatophytes tolerant to eutrophic conditions
7 *Species of clear non-eutrophic water* (14): species with some tolerance to brackisch water
8 *Species of low brackisch dynamic conditions* (13): species tolerating or prefering slightly brackisch and dynamic conditions

score was calculated: rarity score = (10 − rarity code) × response. The rarer a species, the higher the score. For each hydrological option the mean rarity score for the fresh water species was calculated, using absolute responses and responses as percentage.

As a result for the fresh water species the means were known of: (1) absolute responses, (2) responses as a percentage, (3) rarity scores based on absolute responses and (4) rarity scores based on responses as a percentage. To obtain one value for fresh water species per hydrological option instead of four, it is not possible to simply calculate a combined average of these means. They were obtained by four ways of transformation and have different dimensions. This was overcome by converting, by way of transformation, the means into percentage, taking the highest mean (optimal hydrological option) as 100 percent. Finally one average value for fresh water species per hydrological option was obtained.

In the calculation of the cost-benefits the final results of freshwater species in the turfponds and the ditches are integrated in one value of potential ecological improvement; as especially the turfponds need ecological stimulation, they have a double weight in this calculation. The potential ecological improvement of a mixture of two different types of chemistry (for instance: option with less water supply) is a linear interpolation of the ecological impact of these two types of chemistry. It should be noted that the final value of potential ecological improvement is a weighted mean which is based on ecological principles. This value is expressed, however, only on a relative scale.

Ecological impact of the hydrological options

The results of the calculations are summarized in Table 4 (the means of responses per species group) and in Fig. 2 (second method, with the fresh water species).

Lakes Kortenhoef

The supply of unpurified water from the Vecht River is a very unfavourable option. The responses of the fresh water species (24–33%) are the lowest and the responses of the eutraphent species (67–69%) are high (Table 4). The present supply, therefore, needs to be discontinued.

The most favourable option is supply of desalinated surface water from the Polder Horster-

Table 3. An example of ICHORS input (water chemistry) and ICHORS output (response of species). Responses of species in reference to four different water supply options of the Lakes Kortenhoef: Vecht = Vecht River water; A/B = dephosphorized water from Amsterdam-Rijn Canal and Polder Bethune. Seep = fresh seepage water from Polder Horstermeer; Grw = (sub-) regional groundwater.

ICHORS 3.2 input: Chemical composition of the supplied water in mg l^{-1}

	Vecht	A/B	Seep	Grw
pH	7.8	7.2	6.7	6.9
Cl	222	139	100	25
HCO_3	172	188	250	170
SO_4	86.5	58.1	45.0	15.0
Na	127	86	60	15
K	9.3	6.0	2.0	1.9
Ca	85.6	80.9	80.0	45.0
Mg	13.3	10.8	5.0	5.5
Fe	0.08	0.50	10.0	2.5
Si	1.2	2.5	7.0	7.0
NH_4	0.5	0.9	2.5	0.8
NO_3	14.6	13.8	0.3	0.4
PO_4	0.98	<0.01	0.40	0.36

ICHORS 3.2 output: Absolute responses of species (range 0–1).

	Vecht	A/B	Seep	Grw
Mesotraphent species				
Achillea ptarmica	0.008	0.008	0.066	0.116
Cardamine flexuosa	0.018	0.021	0.051	0.044
Elodea canadensis	0.030	0.022	0.022	0.031
Equisetum fluviatile	0.191	0.271	0.451	0.457
DD-Equisetum fluviatile	0.032	0.052	0.106	0.107
Hottonia palustris	0.000	0.001	0.003	0.066
Potamogeton obtusifolius	0.020	0.040	0.230	0.123
Potamogeton trichoides	0.034	0.033	0.043	0.071
Potentilla palustris	0.013	0.009	0.204	0.121
Ranunculus flammula	0.347	0.339	0.217	0.408
Veronica beccabunga	0.008	0.011	0.017	0.025
Chara globularis	0.000	0.001	0.000	0.003
Mean value ($n = 12$)	0.059	0.067	0.118	0.131
Eutraphent species				
Hydrophytes (mean, $n = 10$)	0.345	0.158	0.245	0.202
Phreatophytes (mean, $n = 16$)	0.248	0.269	0.266	0.250
Mean value ($n = 26$)	0.297	0.214	0.256	0.226
Brackish species				
Mean value ($n = 13$)	0.156	0.109	0.058	0.047

meer. Compared with the supply from the Vecht River, this would considerably increase the chances for both common species and fresh water species. Another favourable option is groundwater supply. That would substantially stimulate fresh water species besides common and bank species (Fig. 2). Supply of fresh seepage water derived from the Polder Horstermeer would have

Table 4. Summarized results of calculations with ICHORS for all hydrological options (numbers 1–10: see Table 1), for both Lakes Kortenhoef and Lakes Stichts Ankeveen. Mean results (responses converted in % of maximum value) of fresh, common, eutrophic, bank and brackish species for turfponds (t) and ditches (d). In brackets the number of species per group.

		1	2	3	4	5	6	7	8	9	10
Lakes Kortenhoef		Pres	Vecht	HM	RWZI	Seep	V	IJ	A	A/B	Grw
Fresh-water	t (30)	31	24	53	31	41	30	26	29	29	53
species	d (40)	45	33	64	33	49	36	33	34	34	59
Common	t (22)	56	55	65	48	53	63	60	57	56	58
species	d (28)	63	46	71	59	58	51	53	50	51	65
Eutraphent	t (10)	38	70	46	53	43	72	60	60	58	59
species	d (26)	50	67	45	54	54	54	41	50	48	51
Bank species	t (12)	75	66	50	56	47	66	64	63	60	59
Brackish species	d (27)	52	49	36	39	23	56	82	49	46	22
Lakes Stichts Ankeveen		Pres	Spie			Seep		IJ		A/B	Grw
Fresh-water	t (30)	48	25			61		28		32	73
	d (40)	57	43			62		36		36	79
Common	t (22)	75	64			56		66		64	71
species	d (28)	66	62			72		56		56	71
Eutraphent	t (10)	43	50			36		64		64	58
species	d (26)	49	66			65		44		55	63
Bank species	t (12)	79	82			52		68		66	79
Brackish species	d (27)	50	59			25		84		49	25

a positive impact on fen vegetation. In addition, those options would have a positive effect by substantially reducing the changes of brackish species.

The supply from the Vecht River near the sluice Het Hemeltje (Fig. 1) could be replaced by supply with dephosphorized water from the Vecht River south of De Nes (without effluent of the treatment plant). This would have positive effects on both common species and fresh water species. A negative effect of dephosphorization using $FeCl_3$ is a small increase in chloride concentration, increasing the chances for brackish species. The effects on vegetation of supply with dephosphorized water from the Amsterdam-Rijn Canal, with or without water from the Polder Bethune, are similar. Compared with the present supply, the chances for eutraphent species would diminish; for fresh water species of turfponds and common species the situation would improve somewhat. In general, inlet of this water would result in a slight improvement of chances for fen vegetation.

Supply of dephosphorized water from Lake IJmeer would affect the fresh water species similar to that of Vecht River water now. Responses of common species would increase somewhat. Since the most important impact of such supply would be a substantial increase of brackish species, this alternative must be discouraged.

Lakes Stichts Ankeveen

For fresh water species, the supply water from Lake Spiegelplas is the most unfavourable option

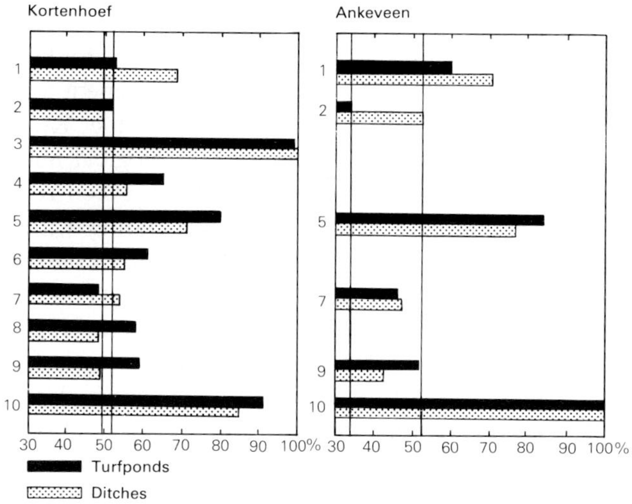

Fig. 2. Ecological impact (expressed in % of the maximum option; x-axis) in turfponds and ditches of ten hydrological options (see Table 1; y-axis) in Lakes Kortenhoef and Lakes Stichts Ankeveen on fresh water species. Vertical lines indicate the impact of present supply (option 2).

(25–42% in Table 4). Besides, the brackish species score high with this option. For these reasons the quality of this water does not permit the development of a rich fen vegetation. Improvement of conditions for fen vegetation can be achieved by supply of fresh seepage water from the Polder Horstermeer or by groundwater supply (option 5 & 10, Fig. 2). The best scoring option for fresh water species in both turfponds and ditches is groundwater supply. In addition to this positive effect on fen vegetation, these options show the lowest scores for brackish species.

Supply of dephosphorized water from the Amsterdam-Rijn Canal and the Polder Bethune will improve the responses of fresh water species of turfponds to a minor extent. However, for fresh water species of ditches an opposite effect is predicted. The responses of brackish species will be somewhat reduced. Summarizing, no marked change in conditions for fen vegetation is expected. The effects on fresh water species of the supply of dephosphorized water from the Lake IJmeer are comparable to those for dephosphorized water from Amsterdam-Rijn Canal and Polder Bethune. However, this option would vigorously stimulate brackish species and should therefore be firmly discouraged.

Cost-benefit analysis

Based on the likely impact of the options, we can set priorities on the restoration measures. The decision to implement an option will, however, depend also on its financial consequences. In this respect a cost-benefit analysis is necessary. This analysis requires data on the hydrology (Verstraelen *et al.*, 1992), the likely ecological impact of the suggested options (this paper) and the estimate costs (Verstraelen *et al.*, 1992). The available amount of the different types of supply water was taken into account. In practice some options will have to be combined with other options, to obtain the required amount of supply water for the polders and lakes. The net infiltration in the Polder Kortenhoef is 20×10^6 m^3 and in the Polder Stichts Ankeveen it is 3×10^6 m^3. Less infiltration in a polder decreases water shortages which consequently leads to a lower amount of supply water needed and a relatively larger contribution of discharged groundwater in the water balance of the polder concerned. An inundation of the Polder Horstermeer, either totally, or for 75%, would induce such changes in the water balance of the surrounding polders. Thus groundwater seepage becomes more important, with unpolluted fresh water extending into the river plain. The inundation of 75% of the Polder Horstermeer, excluding the built-up area, requires an annual investment of 5.9 million guilders for 30 years. Total inundation of Polder Horstermeer requires 20.35 million guilders a year.

By plotting the potential ecological improvement (stimulation of the characteristic ecosystems) against the costs (Fig. 3), the ecological impact of various management options can be evaluated in terms of required investment. Dephosphorized water from the Vecht River south of the treatment plant (option c) is preferable to the other supply options with dephosphorization (a & b), for the costs are almost equal and the ecological improvement is higher. However, with

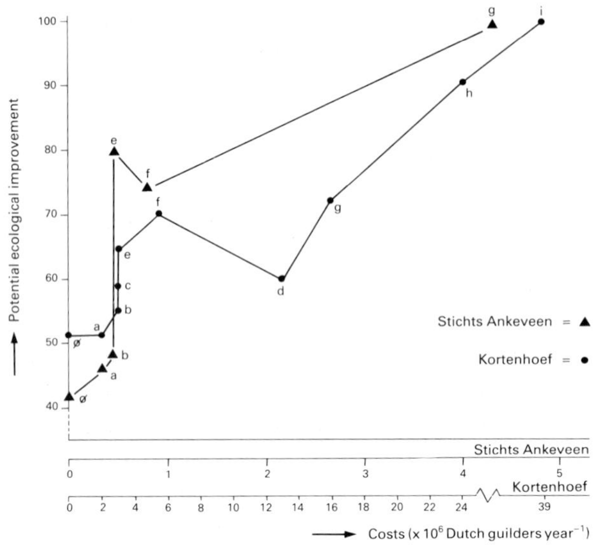

Fig. 3. Cost-benefit analysis of management options and ecological development in fresh fen vegetation. Costs according to Verstraelen *et al.* (1992). Ecological benefit expressed in percentage of maximum option (see text). Codes: ARK = Amsterdam-Rijn Canal + Polder Bethune; HM = Polder Horstermeer).
0 = no change, status quo in water supply (from Vecht River in Kortenhoef and from Lake Spiegelplas in Ankeveen); a, b and c = 100% dephosphorized water from IJmeer, ARK and Vecht, respectively; d = 10×10^6 m^3 dephosphorized effluent and 10×10^6 m^3 dephosphorized water ARK; e = in Kortenhoef: 13×10^6 m^3 fresh seepage water HM and 10×10^6 m^3 dephosphorized water ARK; in Ankeveen: 3×10^6 m^3 fresh seepage water HM; f = in Kortenhoef: 7×10^6 m^3 less infiltration and 13×10^6 m^3 dephosphorized water ARK; in Ankeveen: 1.5×10^6 m^3 less infiltration and 1.5×10^6 m^3 dephosphorized water ARK; g = in Kortenhoef: 10×10^6 m^3 less infiltration and 10×10^6 m^3 dephosphorized water ARK; in Ankeveen: 4×10^6 m^3 less infiltration; h = in Kortenhoef: 20×10^6 m^3 less infiltration (positive water balance); i = in Kortenhoef 20×10^6 m^3 desalinated water from HM.

the same annual amount of money, a higher ecological improvement would be obtained by pumping fresh seepage water from the outskirts of the Polder Horstermeer (e) to the nature reserves in Polder Kortenhoef and Polder Stichts Ankeveen. The supply of extra purificated effluent from the treatment plant (d), that according to the available quantity has to be combined with other supply water, would result in the same ecological improvement as dephosphorized water from the Vecht River (about 60%), but costs about seven times as much and therefore is not a realistic option. The inundation of 75% of the Polder Horstermeer (f) yields an ecological benefit of some percents higher than with (e) in Polder Kortenhoef. Because of the minor hydrological implications the option (f) in Polder Ankeveen requires a comparatively larger sum of money with respect to the ecological benefits. In Polder Ankeveen the total inundation (g) results in a positive water balance and in the optimal ecological impact; in Polder Kortenhoef option (g) offers ecological improvement, although the hydrological implications indicate only 50% reduction of the water shortages. A positive water balance in Polder Kortenhoef is possible with option (h): option (g) with an extra total supplementary investment of 160 million guilders. This creates possibilities for supplementary activities, such as changing of levels of surface water in surrounding polders, so that an ecologically optimal chemistry (resembling groundwater) will be reached. Option (i), that can only be applied in Polder Kortenhoef, obtains the optimal ecological improvement and demands at the same time an extremely high investment.

Discussion

The essential problem of quality management and nature conservation in the Vecht River plain is the structural water shortage and its contaminant water supply with nutrient-loaded water from the river. Solutions to this ecological problem can be formulated in terms of technical options to prevent eutrophication or more sustainable options to prevent a negative water balance.

By the technical options the water is purified, so that the input of nutrients is reduced to meet the demands of eutrophication abatement. Options included in this study are the purification of different allochthonous water sources, further purification of effluent from the treatment plant, and desalinisation of water from the Polder Horstermeer. These supply options can be realised on the short-term. However, the costs of desalinisation are extremely high and the ecological benefit of the other options is relatively small. A supply option that is realistic, both in terms of expenditure and in ecological improvement, is the recycling of fresh seepage water from the outskirts of the Polder Horstermeer. This option utilizes original water from the area. However, all these supply options include purification and the water has to be pumped to the location. These are important disadvantages of the technical solutions, for they involve continuous expenditure and energy, and are not sustainable.

If management objectives include both eutrophication abatement and restoration of the original fen vegetation, restoration of the original hydrology should be given serious consideration. This necessitates structural investment to reduce the negative water balance and the positive nutrient balance to negligible level. To realise these goals, water deficits caused by extreme infiltration have to be avoided. This can only be achieved by inundation (partial or total) of the Polder Horstermeer, together with the stimulation of the discharge of groundwater from the hill ridge. These options (f–h) will lead to a considerable and sustainable improvement of the ecosystems in the long-term. Because restoration of the original hydrology does not completely meet the water demand in summer, an additional supply option has to be implemented, i.e. dephosphorized water from the Vecht River according to this study.

Inundation of the Polder Horstermeer and stimulation of the discharge of groundwater (f–h) will meet the requirements of the ecosystems to a much larger extent than the supply options (a–d) do. The cost-benefit analysis indicates that by inundation the financial investment will be high the first 30 years. However, this problem is a matter of time, for no further financial investment will be needed after 30 years. In contrast, the supply options would require continual funding and therefore deserve lower priority than the sustainable options. The essential choice in policy can be reduced to an investment of little money for hundreds of years to stimulate the ecosystems for a minor part or an investment eight times higher during thirty years to stimulate the ecosystems sustainable to an optimal level.

References

Barendregt, A., J. T. De Smidt & M. J. Wassen, 1986. The impact of groundwater flow on wetland communities. In: Gehu, J. M. (ed.), Colloques Phytosociologiques XIII: 603–912. Kramer, Berlin.

Barendregt, A., M. J. Wassen & J. T. De Smidt, 1989. Regeneration and conservation of ecosystems by integral water management. In: L. Van Liere, R. M. M. Roijackers & P. J. T. Verstraelen (eds), Integral water management in the area with hill ridges and lakes in Het Gooi and Utrecht. CHO-TNO 22: 214–228, 's Gravenhage (in Dutch).

Barendregt, A., M. J. Wassen & A. Van Leerdam, 1990. Changes in hydrology and management adversely affect the vegetation in the area of the river Vecht. Landschap 7: 17–32 (in Dutch).

Barendregt, A. & M. C. Bootsma, 1991. The hydro-ecological model ICHORS (versions 3.1 en 3.2); the relations between aquatic and fen plants and environmental factors in the province of Utrecht. Department of Environmental Studies, University of Utrecht (in Dutch).

Best, E. P. H., D. De Vries & A. Reins, 1984. The macrophytes in the Loosdrecht Lakes: A story of their decline in the course of eutrophication. Verh. int. Ver. Limnol. 22: 868–875.

Ministry of Agriculture, Nature Conservation and Fishery, 1990. Nature Management Plan. Den Haag (in Dutch).

Province of North-Holland, 1986. Provincial Grondwater Management Plan; plan and explanation. Haarlem (in Dutch).

Province of North-Holland, 1987. Wild plants in North-Holland; an research by the Province of North-Holland. Haarlem (in Dutch).

Province of North-Holland, 1990. Provincial Watermanagement. Plan of North-Holland. Haarlem (in Dutch).

Province of Utrecht, 1990. Policy plan for Nature and Landscape in the province of Utrecht. Utrecht (in Dutch).

Roijackers, R. M. M. & P. J. T. Verstraelen, 1988. Ecological investigations in three shallow lakes of different trophic level in the Netherlands. Verh. int. Ver. Limnol. 23: 489–495.

Roijackers, R. M. M. & P. J. T. Verstraelen, 1991. Eutrophication abatement programme for three North Holland Vecht Lakes. Verh. int. Ver. Limnol. 24: 768–772.

Schot, P. P., 1991. Solute transport by groundwater flow to wetland ecosystems. Thesis, University of Utrecht.

Schouten, P. H., P. J. T. Verstraelen, J. J. Buntsma, 1988. Restoration of the Vecht Lakes ecosystems surrounding the Horstermeer polder: a feasibility study. Agric. Wat. Mgmt. 14: 471–479.

Stam, S. M. E. & A. Barendregt, 1990. Assessment of aquatic and fen species by supply of water into the Lakes Stichts Ankeveen and Lakes Kortenhoef. Department of Environmental Studies, University of Utrecht (in Dutch).

Van den Berg, W. J. & J. T. De Smidt, 1985. The vegetation of the Eastern Vecht area 1935–1980. Stichting Commissie voor de Vecht en het Oostelijk en Westelijk Plassengebied. Hilversum (in Dutch).

Verstraelen, P. J. T., R. M. M. Roijackers & A. Barendregt, 1989. Conditions for restoration in the area of the Lakes Ankeveen and Kortenhoef. In: L. Van Liere, R. M. M. Roijackers, P. J. T. Verstraelen (eds), Integral water management in the area with hill ridges and lakes in Het Gooi and Utrecht. CHO-TNO 22: 214–228, 's Gravenhage (in Dutch).

Verstraelen, P. J. T., J. Wisserhof, Lj. Rodić & R. Eijsink, 1992. Eutrophication control strategies for three shallow Vecht lakes in the Province of North Holland. Hydrobiologia 233: 235–245.

Visser, C. M., P. J. T. Verstraelen & R. M. M. Rooijackers, 1989. Report of the lake sediments in the experimental project of the Lakes Ankeveen and Kortenhoef: the dredging considered. Agricultural University, Wageningen (in Dutch).

Wassen, M. J., 1990. Waterflow as a major landscape ecological factor in fen development. Thesis, University of Utrecht.

Williams, M., 1990. Wetlands, A Threatened Landscape. Basil Blackwell, Oxford.

Witmer, M. C. H., 1989. Integral water management at regional level. An environmental study of the Gooi and the Vechtstreek. Thesis, University of Utrecht.

Hydrobiologia **233**: 259–270, 1992.
L. Van Liere & R.D. Gulati (eds), Restoration and Recovery of Shallow Eutrophic Lake Ecosystems in The Netherlands.

Phosphorus eutrophication in the SW Frisian lake district 1. Monitoring and assessment of a dynamic mass balance model

Harry J. W. J. Van Huet*
Limnological Institute, Tjeukemeer Laboratory, De Akkers 47, 8536 VD Oosterzee, The Netherlands;
**Present address: University of Nijmegen, Department of Environmental Sciences, Toernooiveld 1, 6525 ED Nijmegen, The Netherlands*

Key words: eutrophication, lakes, phosphorus, chloride, loading, dynamic mass balance model, sensitivity analysis

Abstract

In 1984 a frequent monitoring programme was started in the hypertrophic S.W. Frisian lake district, with emphasis on total phosphorus (TP) and chloride (Cl^-). The main objectives of the project were: to quantify the phosphorus flows, to gain insight in the process of eutrophication, and to simulate management scenarios. The seasonal variability in the lakes is due mainly to the man-made hydrology: reception of humic-rich polder water in wet periods (winter) and inlet of chloride-rich IJsselmeer water in dry periods (summer). The yearly means of TP concentrations in the lakes (Tjeukemeer, Groote Brekken and Slotermeer) ranged from 0.23 to 0.29 mg l^{-1}. However, much higher concentrations (0.9 mg l^{-1}) were found in periods with high inflow of polder water.

The simulations with a mass balance showed an acceptable similarity between measured and simulated concentrations of TP as well as of Cl^-. Chloride was modelled to verify the accuracy of a hydrodynamic model. A sensitivity analysis of the apparent settling rate in the P model showed that sensitivity was lowest in simulations of Groote Brekken and highest in simulations of Slotermeer, the difference being attributable to the influence of the water residence time. The model was found to be appropriate for simulating management scenarios.

Introduction

In 1984 a study was started to investigate the effects of phosphorus (P)-related eutrophication of S.W. Frisian lakes (Fig. 1). The main objective of the study was to simulate scenarios for policy and management, by means of a P model. In order to support this model, the distribution of P in the lake sediments and the P loads of the lakes from the surrounding polders were investigated. The water transport in the lake area was simulated by use of a hydrodynamic model. Details of these investigations will be published elsewhere (Van Huet & De Haan, in press; Van Huet (1991b), and Van Huet (in press), The preliminary results of the project are reported in Van Huet (1990).

This paper focuses on: a) a brief description of the total phosphorus (TP) concentration and some other water quality parameters of the lakes during 1984–1987; b) the assessment of a simple dynamic mass balance model; and c) the results of this model, with emphasis on TP concentrations.

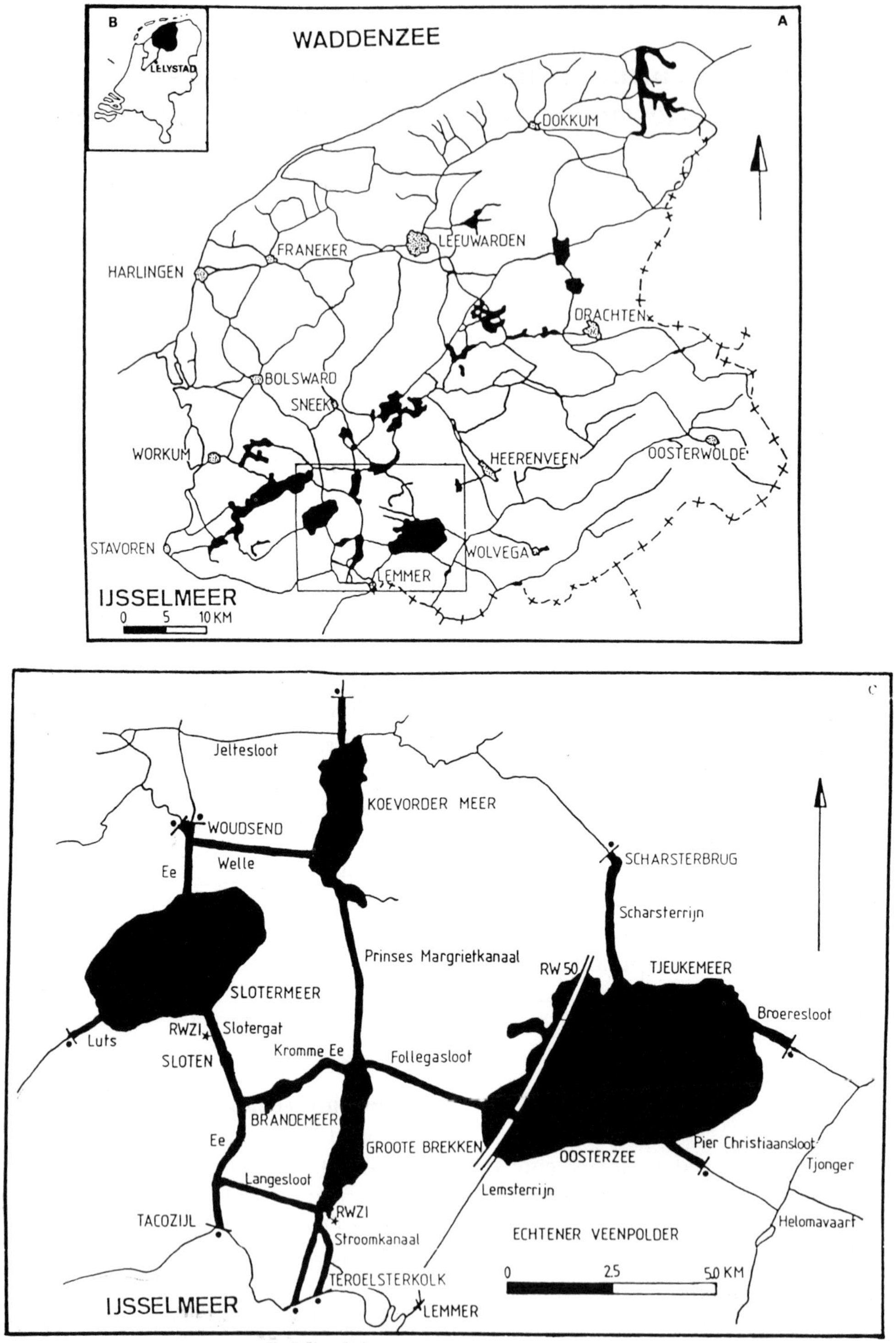

Fig. 1. A. The Frisian 'boezem'. B. Its position in The Netherlands. C. The S.W. lake district (•–:boundaries research area; RWZI: waste water treatment plants; RW50: Motorway 50; from Van Huet, 1990).

Area description and seasonal variation

The S.W. Frisian lakes and the interconnecting canals (water surface area: 4000 ha) are part of the 'boezem', the water network (14 000 ha) in the province of Friesland, used for water table regulation in the surrounding polders and for flushing. Table 1 shows the depth, surface area, catchment area and water residence times of the shallow lakes Tjeukemeer, Groote Brekken and Slotermeer, the lakes modelled here.

In summer (April–September), under typical climatological conditions, a precipitation deficit in the polders occurs and boezem water is let in. To replace the loss, chloride-rich water from IJsselmeer, originating from the river Rhine, is let in at Lemmer, Tacozijl and Stavoren. In winter (October–March), precipitation exceeds evaporation and humic-rich water from the polders is pumped into the boezem system, while, in turn, boezem water is pumped out into IJsselmeer at Lemmer and Stavoren or is released into the Waddenzee (Fig. 1). Thus, man strongly influences the hydrology of the lake system. In addition, the wind has considerable influence on short-term water transport. For detailed information about the boezem system and hydrology, see Leenen (1982a) and Van Huet (1990).

The main differences between the three lakes are given in Table 1. The influence of polder water, reflected in the TP concentrations, in Tjeukemeer in winter is high and relatively low in Slotermeer. The influence of water from IJsselmeer in summer, reflected in the Cl^- concentrations, is great in Groote Brekken and less in Tjeukemeer and in Slotermeer (see also Table 2). The mean water residence time is short in Groote Brekken and long in Slotermeer. Tjeukemeer is a typical polder lake, while Slotermeer is a relatively isolated lake. Water residence time and influence of water from polders and IJsselmeer on Koevorder Meer, the fourth main lake in the area, are similar to those of Groote Brekken. Despite a difference in bottom composition of Tjeukemeer and Slotermeer, sedimentary TP contents do not differ much (Van Huet & De Haan, in press).

Eutrophication and use of models

Frisian lakes are highly eutrophicated. The concentrations of TP and TN (total nitrogen) generally exceed 0.2 and 2.0 mg l^{-1}, respectively, while summer chlorophyll-*a* contents often exceed 150 $\mu g\, l^{-1}$. The lake ecosystems are dominated by the cyanobacteria, mainly *Oscillatoria agardhii* (Moed & Hoogveld, 1982; De Haan & Moed, 1984; Claassen, 1986). The main zooplankton species are: *Bosmina coregoni*, *Chydorus sphaericus* and *Daphnia hyalina*. The main fish species are: bream (*Abramis brama*), and pike perch (*Stizostedion lucioperca*), see Lammens (1986). Eutrophication, chemistry, fish communities and trophic relationships have been described by Beattie *et al.* (1978), De Haan (1982), De Haan & Voerman (1988), Vijverberg & Van Densen (1984), Lammens (1990), Klink & Claassen (1988) and Moed *et al.* (1988).

Table 1. Characteristics of Tjeukemeer, Groote Brekken and Slotermeer (*: after Brinkman *et al.*, 1989; **; Van Huet, in press).

	Tjeukemeer	Groote Brekken	Slotermeer
Surface water area (ha)**	1958	300	1100
Catchment area (ha)	50000	5000	3800
Mean depth (m)**	1.75	2.0	1.65
Bottom structure (> 50%, < 50%)	Peat, sand	Sand, peat	Sand, peat
Mean water residence time (days)**	20	5	54
Relative influence of polder water (winter)	High	Moderate	Low
Relative influence of IJsselmeer water (summer)	Moderate	High	Moderate
Number of lake chloride sampling stations	8	3	6
Number of lake TP sampling stations (mixed sample)	3	3	3

Table 2. Water quality parameters in Tjeukemeer (Tj), Groote Brekken (GB) and Slotermeer (Sl) during 1984–1987 (*: sampling started on May 14, 1984; **: mean value of mixed samples; ***: mean value of all lake stations).

	TP ($mg\,l^{-1}$)**			Cl ($mg\,l^{-1}$)***			Secchi (cm)			pH		
	Tj	GB	Sl	Tj	GB	Sl	Tj	GB	Sl	Tj	GB	Sl
1984*	0.25	0.23	0.24	98	177	130	28	33	28	9.0	8.8	8.8
1985	0.29	0.24	0.24	56	131	101	37	35	27	8.5	8.6	8.6
1986	0.28	0.24	0.26	105	153	120	32	32	29	8.7	8.8	8.6
1987	0.29	0.24	0.23	59	113	96	30	30	25	8.6	8.7	8.7
Apr. 84–Sep. 84*	0.24	0.23	0.25	102	206	122	26	34	27	9.5	9.2	9.0
Apr. 85–Sep. 85	0.30	0.26	0.26	61	146	102	29	33	26	8.8	8.8	8.7
Apr. 86–Sep. 86	0.25	0.22	0.25	99	155	105	34	31	28	9.0	9.1	8.8
Apr. 87–Sep. 87	0.27	0.23	0.22	64	131	96	29	29	25	8.9	8.9	9.0
Oct. 84–Mar. 85	0.27	0.25	0.23	81	125	130	34	32	29	8.1	8.3	8.6
Oct. 85–Mar. 86	0.32	0.24	0.24	45	101	97	47	35	26	7.9	8.2	8.4
Oct. 86–Mar. 87	0.32	0.26	0.27	113	154	145	30	33	30	8.3	8.5	8.3

In Dutch lakes, including the S.W. Frisian lakes, a significant direct correlation between TP concentrations and chlorophyll-*a* contents has been found (Lijklema *et al.*, 1988). There are indications that undesirable cyanobacteria will no longer dominate if TP concentrations can be reduced to $<0.07\ mg\,l^{-1}$. Here, I assume that reducing P levels in the Frisian lakes will eventually lead to its limitation and, finally, to lake restoration.

Golterman *et al.* (1980) used a mass balance model, based on chloride, to quantify the water balance of Tjeukemeer and the P load to this lake during 1969–1975. They concluded that for reliable model results Cl^- concentrations have to be measured frequently (weekly or fortnightly) at several stations on the lakes and in their inflow and outflow canals. Frequent sampling is necessary because of the spatial and temporal variability of Cl^- and TP concentrations, due to inflow of polder water and IJsselmeer water, see also Leenen (1982a), De Haan & Moed (1984).

Materials and methods

From May 1984 to December 1987, 50 stations in the S.W. lake district were sampled weekly with a plexiglass tube (length 1 m) from a boat. Cl^- was measured because time series of Cl^- concentrations can lead to understanding of water transport, and the origin of water in the lakes (due to concentration differences between polder and IJsselmeer water). The Cl^- concentrations were also used to verify the results of a hydrodynamic model (see the section on assessment of the Cl^- model). TP concentrations were measured for the calibration of the mass balance model and to provide an indication of trophic levels. For the latter purpose Secchi disc transparencies and pH values were also measured.

A composite weekly sample of the lake stations was analyzed for TP concentrations, while fortnightly samples of the inflow and outflow sites of the lakes were analyzed for TP. Phosphorus was analyzed colorimetrically with an autoanalyser (molybdate reagent, Murphey & Riley, 1962), following hydrolysis in acedic persulphate. Cl^- was titrated with Ag^+-ions. For a description of the polder sampling programme and results, see Van Huet, 1991b.

Water quality during 1984–1987

Information about the water quality parameters in the three lakes during 1984–1987 is given in Table 2 (see also Van Huet, 1990). It can be concluded that: (a) TP concentrations in polder water were higher than in IJsselmeer water (e.g. the mean values in 1985 were $>0.4 \text{ mg l}^{-1}$ and 0.2 mg l^{-1}, respectively); (b) during dry periods Cl^- concentrations increase, due to inlet of IJsselmeer water, and in wet periods, high TP concentrations occur, due to the pumping of polder water into the system; (c) Tjeukemeer is influenced most by polder water, and Groote Brekken and to a lesser extent Slotermeer by IJsselmeer water; (d) concentrations in Tjeukemeer fluctuated more than in Slotermeer, while concentrations between Tjeukemeer west and Tjeukemeer east differed; and (e) Secchi disc transparencies were low and pH values were high in summer.

Modelling phosphorus dynamics

Model choice

Generally, two types of lake eutrophication models are used: empirical, steady state models and theoretical, dynamic models. The empirical models are derived from statistical analysis of data from large numbers of lakes. In general, these models give good predictions for a group of lakes, but might lead to large errors if applied to individual lakes. Simple empirical models have advantages over complex theoretical models. Few data and simple mathematics are needed and they show generality (Kamp-Nielsen, 1985). However, their limitations are the steady state assumptions and the poor resolution in time and space. The theoretical models are based on (detailed) mathematical descriptions of the nutrient and population dynamics. Hence, these dynamic models can differ in complexity, from simple input-output models using a single net loss term to large ecological n-layer, n-box, n-compartment models, describing spatial variability of several variables; moreover they do not have the limitations of the empirical models, but may have disadvantages in their complexity and use of extensive data sets. Furthermore, they may not be fully adequate, because of lack of knowledge about processes, e.g. sedimentation, resuspension and P-release from sediments. For a review of model types, their steady state solutions, advantages, disadvantages and results see Golterman (1980), Jørgensen (1983), Reckhow & Chapra (1983a, 1983b), Van Straten (1983), Kamp-Nielsen (1985), Ahlgren *et al.* (1988) and Prairie (1989).

The model choice depends on several criteria such as model use, complexity of the system (e.g. occurrence of stratification and the hydrodynamic character of the aquatic ecosystem), availability of data sets, etc. Empirical models are often applied if only mean annually values are of interest, which is not the case in the Frisian situation. Moreover, the steady state approach is not realistic for modelling the P levels in the Frisian lakes because of the seasonal patterns of loading. Therefore a dynamic model was preferred.

An important criterion is the use of the model. Will the model be used for research or for management purposes (or possibly both)? Models for research purposes may have objectives including identification and hypotheses testing. Management models generally focus more on the prediction of one or two variables. Therefore, on average, research models are more complex than the (practical) management models. Modelling the P dynamics in the Frisian lakes is important for management reasons. Hence I used a practical management model approach.

Because of the short residence times and the influence of the wind on the water transport in the lakes (Van Huet, a, in press), it is assumed that the lakes are ideally mixed. Thus, the CSTR-approach (continuously stirred tank reactor) was used. Another reason for choosing a simple dynamic model is that the uncertainty in the data does not support detailed modelling (see Fig. 3, and the section '*Representativeness of phosphorus concentrations*').

The simple dynamic management model is based on the principle of mass conservation:

$$\frac{\mathrm{d}\,(VP)}{\mathrm{d}t} = \text{Input} - \text{Ouput} - \text{Net loss}\,, \qquad (1)$$

,where V = the lake volume; P = the lake TP concentration; t = the time.

Table 1. TP balances of three lakes during three periods;

Tjeukemeer				Groot Brekken				Slotermeer			
Period	1	2	3		1	2	3		1	2	3
		(kg d^{-1})				(kg d^{-1})				(kg d^{-1})	
IN				IN				IN			
Follegasloot	170	129	111	Follegasloot	121	100	198	Ee	54	97	81
Pier Christiaansloot	151	68	41	Langesloot	31	83	46	Slotergat	22	31	36
Broeresloot	281	443	834	Kromme Ee	27	44	35				
Scharsterrijn	104	57	28	Prins. Margrietkan.	222	65	114				
Polders	37	14	33	Polders	9	6	8	Polders	19	7	14
Precipitation	2	3	3	Precipitation	0	4	1	Precipitation	1	2	2
				Waste water plant	9	1	1				
				Inlet IJsselmeer	1	190	36				
				Inlet slucies	1	5	7				
Total	745	714	1050	Total	421	498	446	Total	96	137	133
OUT				OUT				OUT			
Follegasloot	121	100	198	Follegasloot	170	129	111	Ee	21	21	25
Pier Christiaansloot	170	288	484	Langesloot	4	5	12	Slotergat	45	54	56
Broeresloot	286	124	46	Kromme Ee	15	15	24				
Scharsterrijn	32	53	154	Pmrins. Margrietkan.	49	249	228				
Polders	4	22	15	Polders	1	3	2	Polders	2	11	6
Infiltration	2	2	3	Infiltration	0	0	0	Infiltration	1	1	1
				Pumping into IJsselm.	168	75	54				
Net loss	95	115	145	Net loss	13	16	20	Net loss	35	49	48
Total	710	704	1045	Total	420	492	451	Total	104	136	136
Storage	35	10	5	Storage	1	6	−5	Storage	−8	1	−3

polder loads Tjeukemeer can be considered as the typical polder lake in the area, and Slotermeer less so. The contribution of net loss is relatively highest for Slotermeer (Van Huet, 1992).

Phosphorus load reduction scenarios

In scenario studies two approaches are possible. The first approach is to define a scenario and evaluate whether under this scenario a target level is attained. The second starts at a target level and then seeks a combination of scenarios that satisfies this level. For the second method an operational definition on what is meant by 'satisfies a level' is needed, while the set of satisfactory combinations can be very large. Because a limited set of reasonable scenarios can be formulated relatively easily, and the main objective is to illustrate a response that can be achieved by P reduction, the more practical first approach was followed.

The simulated scenarios focus on the reduction of the two main external P-loads: inlet water from IJsselmeer and water pumped from the polders into the system. Peak loads from IJsselmeer occurred in the summer of 1986, and from polders mainly in the winters (especially the indirect polder contribution, see above).

Preliminary simulations showed that small reductions in TP concentration from the main sources and 100% reductions in, for example the waste water effluent, did not result in perceivable changes in P-levels in the system (see also Table 1). Apparently, only large reductions were likely to have effects, therefore the reduction levels in the various scenarios were set arbitrarily at

75% throughout. The model is approximately linear concerning TP concentrations and therefore allows a crude estimate of the effects of larger or smaller reduction levels from the effects. The simulated scenarios are a 75% TP reduction in:

(1) indirectly discharged polder water;
(2) both indirectly discharged and directly pumped polder water;
(3) inlet water from IJsselmeer;
(4) water both from all polders and from IJsselmeer.

The results of the scenario simulations (Figs 2–4) show the following lines: (a) a line that corresponds with the actual basic water quality standard in The Netherlands for the summer period (0.15 mg l^{-1}), (b) a line that corresponds with the target TP concentration of 0.07 mg l^{-1}, (c) a basic line that corresponds with simulation of the actual TP concentrations during the three periods (Van Huet, 1992), and (d) one or more lines that correspond with the 75% reduction scenarios.

Results and discussion

The three simulation periods cover both a normal situation (P-loads from polders in winter and from IJsselmeer in summer), and an unusual situation, such as high polder P-loads in summers of 1985 and 1987, due to precipitation excess.

A 75% reduction of TP concentrations in indirectly discharged polder water, i.e. from Broeresloot and Pier Christiaansloot (scenario 1; Fig. 2) has, as expected, the greatest effect on the P-level in Tjeukemeer. In this lake, TP concentrations below the actual standard of 0.15 mg l^{-1} can be achieved, while the 0.07 mg l^{-1} target level is approximated. In Groote Brekken the 0.15 mg l^{-1} standard is approximated occasionally. In this lake the effect is highest in the third period due to high loads discharged through Follegasloot (198 kg d^{-1}). In Slotermeer the effects of scenario 1 are poor, as expected, because the distance from Broeresloot and Pier Christiaansloot is long. A further reduction of TP concentrations in polder water directly pumped into the system (scenario 2) does not substantially increase the effect of scenario 1 in Tjeukemeer and Groote Brekken, while the effect is relatively highest in Slotermeer (see also the TP-balance, Table 1).

A reduction of TP concentrations in inlet water from IJsselmeer (Fig. 3, scenario 3) has, as expected (see Table 1), the greatest effect in Groote Brekken, as this lake is situated close to IJsselmeer. In Groote Brekken the 0.15 mg l^{-1} standard is achieved, and occasionally the 0.07 mg l^{-1} target level is approximated. In Tjeukemeer only in the summer period of 1986 the 0.15 mg l^{-1} standard is achieved, while the effect in Slotermeer is relatively low.

The effects of all reductions together (scenario 4; Fig. 4) satisfies the 0.15 mg l^{-1} standard in Tjeukemeer and Groote Brekken, while occasionally the 0.07 mg l^{-1} target level can be reached. Slotermeer responds only to a small extent to these measures.

The simulations started in each period with the actual (measured) TP concentrations in the lakes, while the 75% reductions were assumed to come into effect at the beginning of each period. Consequently, the results represent the short-term transient effects for each simulated period separately. In order to get an impression of the measures already in effect before, Fig. 4 also shows the results of the simulations where the initial TP concentrations in each lake are set to 50% of their original value. As expected, the initial condition effect is overruled by the discharged loads within a period of some months.

The dynamic behaviour can be explained by considering the dynamic mass balance equation for each lake, derived from Van Huet (1992):

$$AH\frac{\mathrm{d}P}{\mathrm{d}t} = \Sigma\{Q_{\mathrm{in},j}(P_{\mathrm{in},j} - P)\} + Q_{\mathrm{e}}P - v_{\mathrm{s}}AP\,, \qquad (1)$$

where A = lake surface area (m^2); H = water depth (m); P = lake TP concentration (mg l^{-1}); $P_{\mathrm{in},j}$ = TP concentration in influent j (mg l^{-1}); $Q_{\mathrm{in},j}$ = inflow discharge j (m^3 d^{-1}); Q_{e} = evapo-

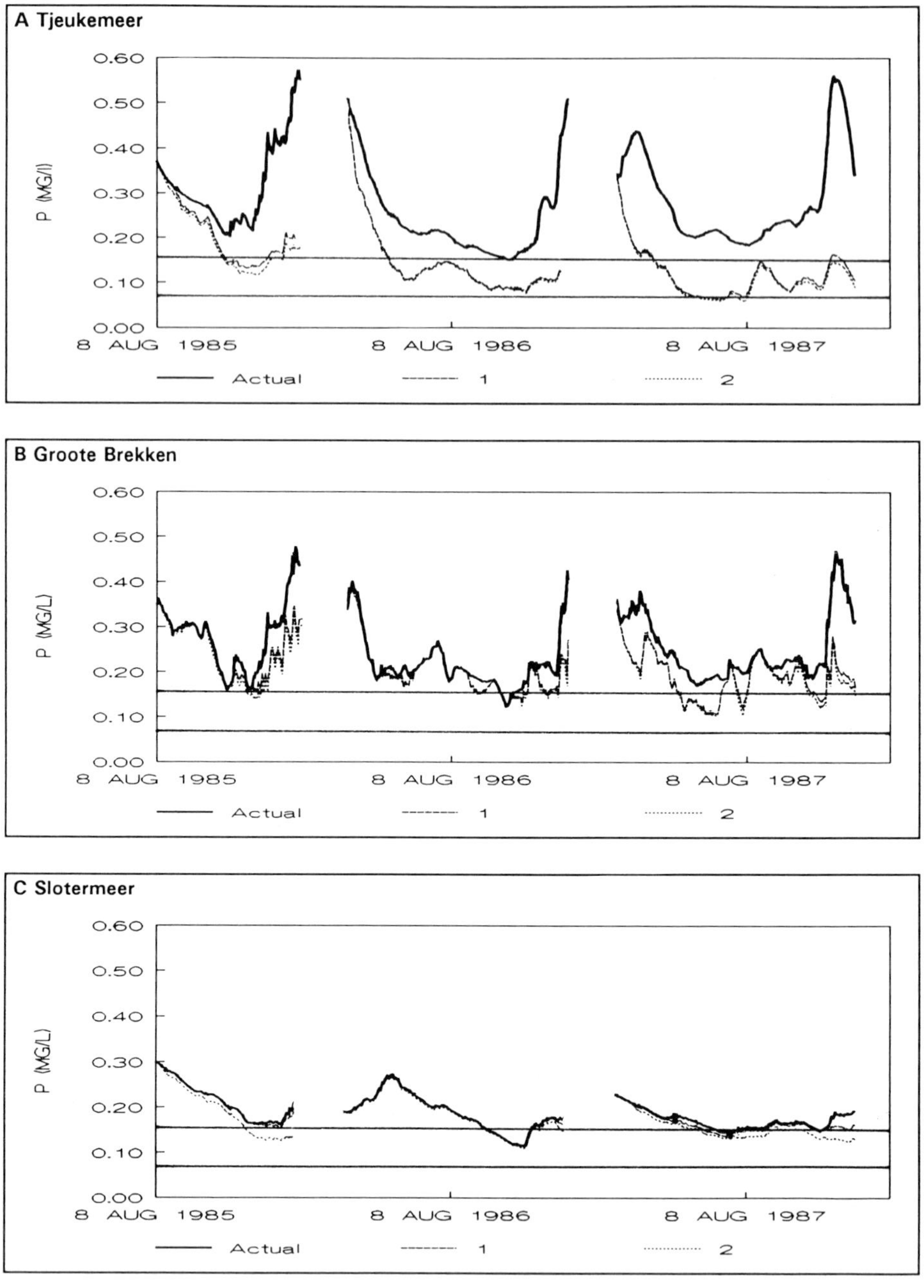

Fig. 2. Effects of 75% TP-reduction in three Frisian lakes in (1) indirectly discharged and (2) both indirectly discharged and directly pumped polder water, during three simulated periods. Actual: simulation results of actual TP concentrations (_).

ration ($m^3 d^{-1}$; v_s = apparent settling rate ($m d^{-1}$).

The transfer function from load discharges and load concentrations to in-lake concentration is generally difficult to evaluate from Eq. (1), because both $Q_{in,j}$, $P_{in,j}$ as well as P and H are functions of time, and consequently Eq. (1) is non-linear. Also, for the three lakes the equations are coupled. A full analysis would require the linearization of the equations around a suitable

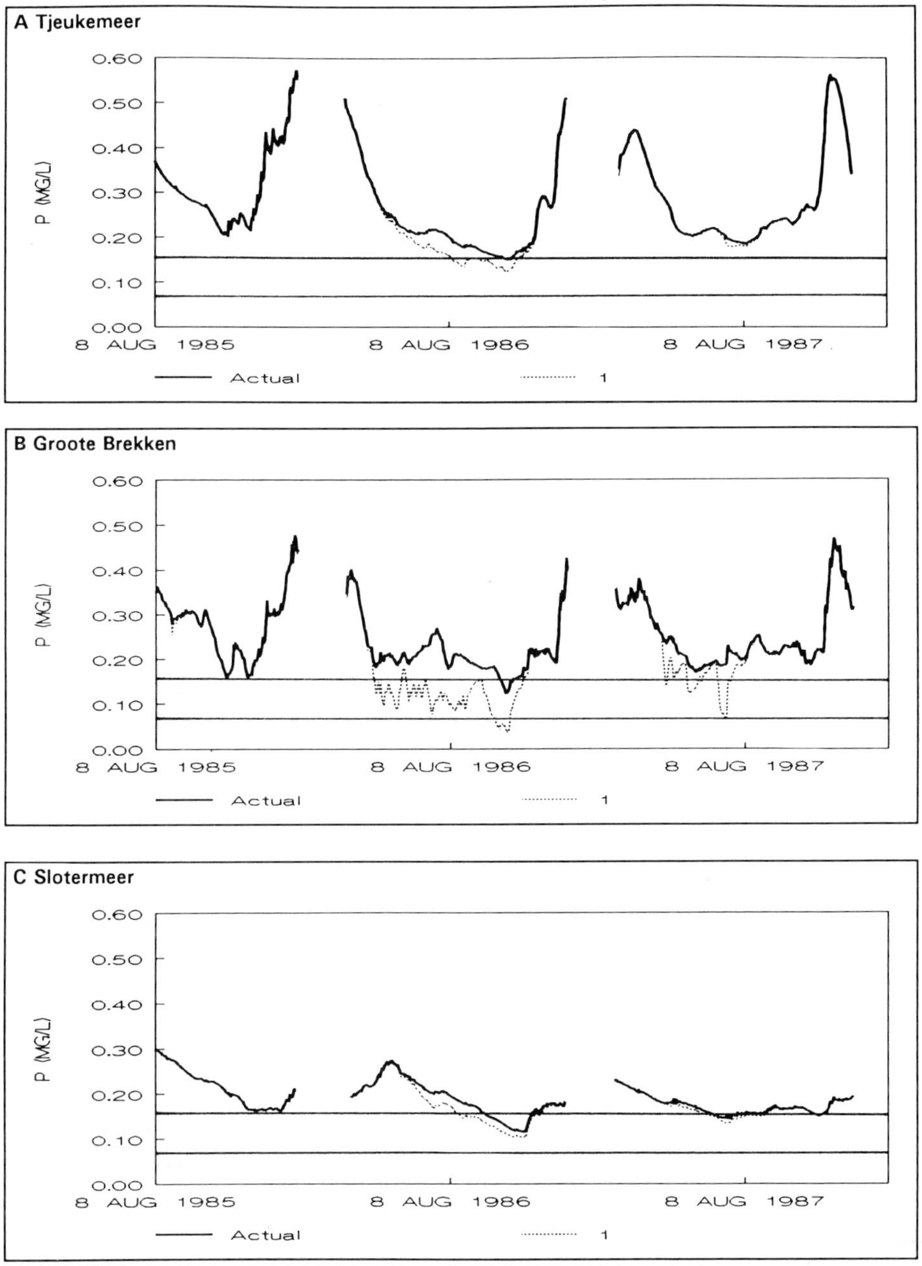

Fig. 3. Effect of 75% TP-reduction in three Frisian lakes in (1) inlet water from IJsselmeer, during three simulated periods. Actual: simulation results of the actual TP concentrations (_).

point, e.g. the period-wise average, followed by calculation of the deviation in output P concentrations vs. deviations in Q_{in} and P_{in} in the Laplace domain. A crude appraisal, however, is possible by observing that $AH = V =$ volume is approximately constant in the long term, and by combining all different loads into one single term:

$$L = \Sigma Q_{\text{in},j} \, P_{\text{in},j} \,, \qquad (2)$$

Treating each lake independently, Eq. (1) can be written in the standard form of a first order dynamic mass system:

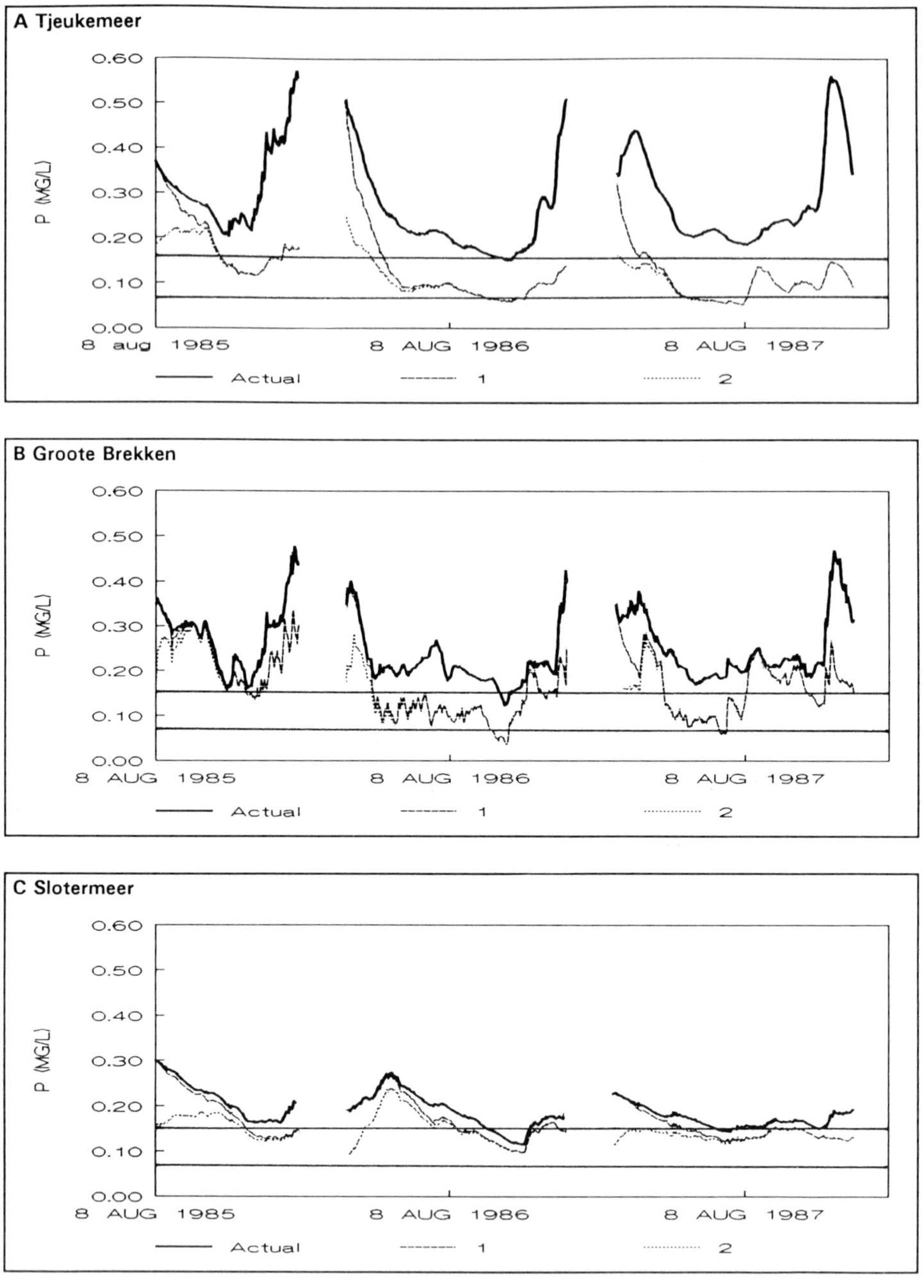

Fig. 4. Effects of 75%-reduction in three Frisian lakes in (1) both polder water and water from IJsselmeer and (2) the same expressed in the case of halving the initial TP concentrations (2), during three simulated periods. Actual: simulation results of the actual TP concentrations (–).

$$\theta \frac{dP}{dt} + P = k_p L \,, \tag{3}$$

where θ = the effective time constant:

$$\theta = \tau \frac{1}{1 + \frac{v_s}{H} \tau} \tag{4}$$

and τ = the adapted hydraulic residence time:

$$\tau = \frac{V}{\Sigma Q_{\text{in},j} - Q_e} \tag{5}$$

and k_p = the static gain:

$$k_p = \frac{\theta}{V}. \tag{6}$$

Equation (3) shows that each lake roughly follows a first order dynamics. The response time of the TP concentration in the lake to a step change in external load is directly related to θ (95% of the ultimate response is reached in about 3θ time units). Equation (4) shows that the effective time constant is at most equal to the adapted hydraulic residence time, but usually smaller because of the net loss to the sediment (if $v_s = 0$ then $\theta = \tau$, in all other cases with positive v_s: $\theta < \tau$). The final concentration effect is k_p units per unit step loading change (i.e. mg l^{-1} concentration effect per kg d^{-1} load reduction).

The values of the effective time constant and static gain were calculated for each of the lakes for each of the three periods (Table 2). These results aid an interpretation of the graphical output. Of all lakes, Groote Brekken has the fastest dynamics with effective time constants <5 days. This explains why P concentrations in this lake show many more high frequency variations than in the other two lakes. Load variations are not as strongly attenuated as in Tjeukemeer and Slotermeer. In particular the P-behaviour of the model for Slotermeer is very smooth, due to the effective time constants in the order of 50 days. It is interesting to note that in contrast to the model the real data show appreciable fluctuations in time. This can occur if (a) there are strong fluctuations in load, or (b) the assumption of complete mixing is occasionally violated. In view of the location of the sampling points, there are, in fact, reasons to believe that the data are influenced by local effects, i.e. intermittent discharges of polder water.

The static gains of the three lakes also differ. The final concentration effect in Slotermeer is about 4 times as large per kg d^{-1} load reduction. That this cannot be observed in Fig. 4 is because a 75% load reduction does not correspond to the similar amount of mass reduction for the three lakes. The mass balance data (Table 1) show that the load to Slotermeer is much less than to Tjeukemeer, and consequently a 75% reduction has much less effect despite the larger static gain. A direct comparison of the load effect calculated from the static gain with that seen in the figures is not easy, because the lakes are interconnected. A true first order response can only be expected if the lake acts as the upstream end of the interconnected system. This is the case with Tjeukemeer, when there is no IJsselmeer inlet, and with Groote Brekken when there is inlet from IJsselmeer. For example, considering the independent loads due to indirectly and directly discharged polder water into Tjeukemeer in period 2, the mass balance data show that these loads amount to 525 kg d^{-1} in the untreated situation. A 75% reduction would thus be about 394 kg d^{-1}, leading to a concentration effect of 0.22 mg l^{-1}, based on the calculated static gain. Comparing this value with the 75% reduction line in Fig. 2 shows that the true effect is smaller, because of the capacity of the connected lakes. Similarly, for Groote Brekken in the second period, when there is con-

Table 2. Calculated values for the adapted hydraulic load, the effective time constant and the static gain.

	V (10^6 m^3)	H (m)	v_s (m d^{-1})	τ (d)			θ (d)			k_p (10^{-3} mg l^{-1})/(kg d^{-1})		
Period				1	2	3	1	2	3	1	2	3
Tjeukemeer	34.4	1.76	0.015	18	23	19	16	19	16	0.45	0.56	0.47
Gr. breken	6.0	2.01	0.015	4	4	6	4	4	6	0.65	0.65	0.96
Slotermeer	18.1	1.64	0.015	54	59	58	36	38	38	2.00	2.12	2.10